Methods in Molecular Biology

Series Editor
John M. Walker
School of Life and Medical Sciences
University of Hertfordshire
Hatfield, Hertfordshire, UK

For further volumes:
http://www.springer.com/series/7651

For over 35 years, biological scientists have come to rely on the research protocols and methodologies in the critically acclaimed *Methods in Molecular Biology* series. The series was the first to introduce the step-by-step protocols approach that has become the standard in all biomedical protocol publishing. Each protocol is provided in readily-reproducible step-by-step fashion, opening with an introductory overview, a list of the materials and reagents needed to complete the experiment, and followed by a detailed procedure that is supported with a helpful notes section offering tips and tricks of the trade as well as troubleshooting advice. These hallmark features were introduced by series editor Dr. John Walker and constitute the key ingredient in each and every volume of the *Methods in Molecular Biology* series. Tested and trusted, comprehensive and reliable, all protocols from the series are indexed in PubMed.

Chaperones

Methods and Protocols

Second Edition

Edited by

Stuart K. Calderwood

Harvard Medical School, Beth Israel Deaconess Medical Center, Boston, MA, USA

Thomas L. Prince

Danville, PA, USA

Humana Press

Editors
Stuart K. Calderwood
Harvard Medical School
Beth Israel Deaconess Medical Center
Boston, MA, USA

Thomas L. Prince
Danville, PA, USA

ISSN 1064-3745 ISSN 1940-6029 (electronic)
Methods in Molecular Biology
ISBN 978-1-0716-3341-0 ISBN 978-1-0716-3342-7 (eBook)
https://doi.org/10.1007/978-1-0716-3342-7

Cover Caption: The predicted protein structure of HSF1, a key component of the heat shock response and guardian of the proteome, is shown on the cover. Model structure generated by AlphaFold (Jumper, J. et al. Nature, 2021; Varadi, M. et al. Nucleic Acid Research, 2021). HSF1 color scheme based on studies by Kijima, T. et al. Scientific Reports, 2018.

This Humana imprint is published by the registered company Springer Science+Business Media, LLC, part of Springer Nature.
The registered company address is: 1 New York Plaza, New York, NY 10004, U.S.A.

Dedication

(SKC) This volume is dedicated to the memory of my sister Carole who passed away in 2021. I would particularly like to thank my wife Laura for her patience and support in the long hours science demands.

Preface

Life is a dynamic and complex interplay of energy and mass within space and time. Biological life depends on proteins to facilitate the processes that allow communities of cells to cooperate and compete to propagate their genetic information. Proteins are linear polymers that fold into intricate three-dimensional structures that provide the essential functions of life. Here nature demonstrates the common notion that structure determines function. Most proteins are able to spontaneously fold into their proper structure given adequate space. However, in the crowded confines of the cell, many proteins require the aid of a special cadre of proteins known as molecular chaperones. These chaperone proteins help other client proteins avoid misfolding, aggregation, and loss of function. Furthermore, molecular chaperones maintain cellular proteostasis by facilitating protein transport and degradation. The cellular stress response is the biological process for enabling the explosive expression of molecular chaperones in response to proteostatic perturbations. And since heat is one of the oldest and most studied forms of proteotoxic stress, many molecular chaperones are referred to as heat shock proteins (HSPs). Expression of HSP genes is initiated by the activation of heat shock factors (HSFs) that bind to the DNA promoters of chaperones and other stress-related genes to initiate or extend transcription. HSPs once expressed may be post-translationally modified as they reestablish proteostasis while also contributing to the modulation of immunity, extracellular communication, and cell growth. Life is shaped by molecular chaperones and the stress response. Almost all aspects of health and disease are affected by the ability to maintain proteostasis. Unchecked chaperone and HSF activity enables malignant cell growth that gives way to cancer. In contrast, reduced chaperone activity allows for accumulated protein aggregation that leads to neurodegeneration. All while, the stress response influences inflammation and immunity.

Our pursuit to better understand molecular chaperones and the stress response along with the desire to pass this knowledge on to future generations of scientist is represented by this book. In this volume, we have complied 23 chapters of methods and topics focused on investigating the role of chaperones. Some chapters are succinct; some chapters are extensive; all address important topics within the field. We appreciate the expertise, time, and effort the authors have provided.

We commence the volume analyzing the initiation and regulation of the stress response. In Chap. 1, Ackerman et al. describe the use of a plasmid-based reporter for monitoring the heat shock response. In Chaps. 2 and 3, Heyoun Bunch analyzes the stress response at the level of transcription and RNA elongation. Holton et al. in Chap. 4 describes their workflow for studying RNA-seq data relative to the stress response. Ritwick Sawarkar in Chap. 5 explores the role of Hsp90 in gene expression through chromosome-immunoprecipitation. Yuka Okusha next details the use of microRNA to regulate HSP levels.

Chapters 7 through 14 revolve around major features of chaperone function and biology: post-translational modifications and protein-protein interactions. Jin et al. explain the effects of key HSF1 phosphorylation sites on transcriptional activity. In Chap. 8, Chakraborty et al. describe a fluorescent complementation assay to analyze Hsp90 and client protein interactions. Edkins and Blatch further elucidate and compare complementation assays for protein-protein interactions studies. Tonui et al. in Chap. 10 explain their protein aggregation assay for measuring chaperone holdase activity. Sager et al. explore the detection of Hsp90 post-translational modifications. In Chap. 12, Ono and Eguchi discuss the challenges of addressing HSP depletion and isoform compensation. Baldan et al. in Chap. 13 describe their proximity ligation assay to analyze HSP protein-protein interaction dynamics. The study of the epichaperone complexes through the use of native polyacrylamide gel electrophoresis is described by Roychowdhury et al in Chap. 14.

Chapter 15 through 21 focus on the emerging role of extracellular HSPs. Borges et al. provide an update on molecular chaperone and extracellular receptor interactions. In Chap. 15, Weng et al. expound the use of Hsp70 in antigen presentation and possible vaccine generation. Votra et al. assess the effect of Hsp90 activity on extracellular proteases. Ono and Eguchi follow up with proteomic profiling of chaperone-rich extracellular vesicles with mass spectrometry. In Chap. 19, Chang et al. provide a protocol for analyzing extracellular Hsp90 levels. Bavisotto et al. in Chap. 20 characterize the role of extracellular Hsp60 in health and disease.

The final Chaps. 22, 23, and 24 explore the use chaperones as biomarkers. Kawai et al develop a method for profiling HSP isoforms in cancer specimen. This is followed by a chapter on chaperone focused cancer database analysis for prognostic biomarkers. Finally, Dezfouli et al. utilize flow cytometry and enzyme-linked immunoassays to profile Hsp70 localization in cancer.

Boston, MA, USA ***Stuart K. Calderwood***
Danville, PA, USA ***Thomas L. Prince***

Acknowledgments

This book is dedicated to the next generation of scientists including my (T.L.P.) boys Jessup and Lincoln along with my wife, Bourdana, and to the memory of my mother Suzanne Prince. I would also like to acknowledge the Thomas Beaver Free Library in Danville, PA, for providing me a quiet place to get work done.

Contents

Contributors

ANDREW ACKERMAN • *Geisinger Clinic, Danville, PA, USA*
DEEMA ALSALIH • *Department of Urology, SUNY Upstate Medical University, Syracuse, NY, USA*
SARAH J. BACKE • *Department of Urology, SUNY Upstate Medical University, Syracuse, NY, USA*
SIMONE BALDAN • *MRC Toxicology Unit, University of Cambridge, Cambridge, United Kingdom*
GREGORY L. BLATCH • *Biomedical Biotechnology Research Unit (BioBRU), Department of Biochemistry and Microbiology, Rhodes University, Makhanda, South Africa; Biomedical Research and Drug Discovery Research Group, Faculty of Health Sciences, Higher Colleges of Technology, Sharjah, United Arab Emirates; The Vice Chancellery, The University of Notre Dame Australia, Fremantle, WA, Australia*
GREGORY L. BLATCH • *Biomedical Biotechnology Research Unit (BioBRU), Department of Biochemistry and Microbiology, Rhodes University, Makhanda, South Africa; Biomedical Research and Drug Discovery Research Group, Faculty of Health Sciences, Higher Colleges of Technology, Sharjah, United Arab Emirates; The University of Notre Dame Australia, Fremantle, WA, Australia*
THIAGO J. BORGES • *Center for Transplantation Sciences, Department of Surgery, Massachusetts General Hospital, Harvard Medical School, Boston, MA, USA*
DIMITRA BOURBOULIA • *Department of Urology, SUNY Upstate Medical University, Syracuse, NY, USA; Upstate Cancer Center, SUNY Upstate Medical University, Syracuse, NY, USA; Department of Biochemistry and Molecular Biology, SUNY Upstate Medical University, Syracuse, NY, USA*
HEEYOUN BUNCH • *Department of Applied Biosciences, Kyungpook National University, Daegu, Republic of Korea*
STUART K. CALDERWOOD • *Molecular and Cellular Radiation Oncology, Beth Israel Deaconess Medical Center, Harvard Medical School, Boston, MA, USA; Department of Radiation Oncology, Beth Israel Deaconess Medical Center, Boston, MA, USA; Beth Israel Deaconess Medical Center, Harvard Medical School, Boston, MA, USA*
FRANCESCO CAPPELLO • *Department of Biomedicine, Neurosciences and Advanced Diagnostic (BiND), Human Anatomy Section, University of Palermo, Palermo, Italy*
CELESTE CARUSO BAVISOTTO • *Department of Biomedicine, Neurosciences and Advanced Diagnostic (BiND), Human Anatomy Section, University of Palermo, Palermo, Italy*
ABIR CHAKRABORTY • *Biomedical Biotechnology Research Unit (BioBRU), Department of Biochemistry and Microbiology, Rhodes University, Makhanda, South Africa*
CHENG CHANG • *Department of Dermatology the Norris Comprehensive Cancer Centre, University of Southern California Keck Medical Center, Los Angeles, CA, USA*
GABRIELA CHIOSIS • *Program in Chemical Biology, Memorial Sloan Kettering Cancer Center, New York, NY, USA*
EVERLY CONWAY DE MACARIO • *Department of Microbiology and Immunology, School of Medicine, University of Maryland at Baltimore-Institute of Marine and Environmental Technology (IMET) – Rita Rossi Colwell Center, Baltimore, MD, USA*

ALI BASHIRI DEZFOULI • *Radiation Immuno-Oncology Group, Center for Translational Cancer Research Technische Universität München (TranslaTUM), Department of Radiation Oncology, Klinikum rechts der Isar, Technische Universität München (TUM), Munich, Germany*

ADRIENNE L. EDKINS • *Biomedical Biotechnology Research Unit (BioBRU), Department of Biochemistry and Microbiology, Rhodes University, Makhanda, South Africa*

TAKANORI EGUCHI • *Department of Dental Pharmacology, Graduate School of Medicine, Dentistry and Pharmaceutical Sciences, Okayama University, Okayama, Japan; Department of Dental Pharmacology, Faculty of Medicine, Dentistry and Pharmaceutical Sciences, Okayama University, Okayama, Japan*

GEMMA A. FOULDS • *John van Geest Cancer Research Centre, School of Science and Technology, Nottingham Trent University, Nottingham, UK*

RICHARD M. GIADONE • *Department of Stem Cell and Regenerative Biology, Harvard University, Cambridge, MA, USA; Harvard Stem Cell Institute, Cambridge, MA, USA; Stanley Center for Psychiatric Research, Broad Institute of MIT and Harvard, Cambridge, MA, USA*

JIANLIN GONG • *Department of Medicine, Boston University School of Medicine, Boston, MA, USA*

KRISTINA M. HOLTON • *Department of Stem Cell and Regenerative Biology, Harvard University, Cambridge, MA, USA; Harvard Stem Cell Institute, Cambridge, MA, USA; Stanley Center for Psychiatric Research, Broad Institute of MIT and Harvard, Cambridge, MA, USA*

XIONGJIE JIN • *Molecular Chaperone Biology, Medical College of Georgia, Augusta, GA, USA; Georgia Cancer Center, Augusta University, Augusta, GA, USA*

RUTH O. JOHN • *Biomedical Biotechnology Research Unit (BioBRU), Department of Biochemistry and Microbiology, Rhodes University, Makhanda, South Africa*

HOTAKA KAWAI • *Department of Oral Pathology and Medicine, Faculty of Medicine, Dentistry and Pharmaceutical Sciences, Okayama University, Okayama, Japan*

TOSHIKI KIJIMA • *Department of Urology, Dokkyo Medical University, Tochigi, Japan*

BENJAMIN J. LANG • *Department of Radiation Oncology, Beth Israel Deaconess Medical Center, Boston, MA, USA*

PHILIPP LENNARTZ • *Radiation Immuno-Oncology Group, Center for Translational Cancer Research Technische Universität München (TranslaTUM), Department of Radiation Oncology, Klinikum rechts der Isar, Technische Universität München (TUM), Munich, Germany*

WEI LI • *Department of Dermatology the Norris Comprehensive Cancer Centre, University of Southern California Keck Medical Center, Los Angeles, CA, USA*

ALBERTO J. L. MACARIO • *Department of Microbiology and Immunology, School of Medicine, University of Maryland at Baltimore-Institute of Marine and Environmental Technology (IMET) – Rita Rossi Colwell Center, Baltimore, MD, USA*

ANATOLI B. MERIIN • *Department of Biochemistry, Boston University, Boston, MA, USA*

NAHID F. MIVECHI • *Molecular Chaperone Biology, Medical College of Georgia, Augusta, GA, USA; Georgia Cancer Center, Augusta University, Augusta, GA, USA; Department of Radiation Oncology, Augusta University, Augusta, GA, USA*

MEHDI MOLLAPOUR • *Department of Urology, SUNY Upstate Medical University, Syracuse, NY, USA; Department of Biochemistry and Molecular Biology, SUNY Upstate Medical University, Syracuse, NY, USA*

DEMETRIUS MOSKOPHIDIS • *Molecular Chaperone Biology, Medical College of Georgia, Augusta, GA, USA; Georgia Cancer Center, Augusta University, Augusta, GA, USA; Department of Medicine, Augusta University, Augusta, GA, USA*

GABRIELE MULTHOFF • *Radiation Immuno-Oncology Group, Center for Translational Cancer Research Technische Universität München (TranslaTUM), Department of Radiation Oncology, Klinikum rechts der Isar, Technische Universität München (TUM), Munich, Germany*

AYESHA MURSHID • *Molecular and Cellular Radiation Oncology, Beth Israel Deaconess Medical Center, Harvard Medical School, Boston, MA, USA*

LEN NECKERS • *Urologic Oncology Branch, Center for Cancer Research, National Cancer Institute, Bethesda, MD, USA*

YUKA OKUSHA • *Beth Israel Deaconess Medical Center, Harvard Medical School, Boston, MA, USA; JSPS Overseas Research Fellow, Tokyo, Japan*

KISHO ONO • *Department of Oral and Maxillofacial Surgery, Okayama University Hospital, Okayama, Japan*

PALAK PANCHAL • *Program in Chemical Biology, Memorial Sloan Kettering Cancer Center, New York, NY, USA*

GEOFFREY J. PILKINGTON • *Brain Tumour Research Centre, School of Pharmacy and Biomedical Sciences, Institute of Biomedical and Biomolecular Sciences, University of Portsmouth, Portsmouth, UK*

A. GRAHAM POCKLEY • *John van Geest Cancer Research Centre, School of Science and Technology, Nottingham Trent University, Nottingham, UK*

THOMAS L. PRINCE • *Ranok Therapeutics, Danville, PA, USA*

FRANCESCA RAPPA • *Department of Biomedicine, Neurosciences and Advanced Diagnostic (BiND), Human Anatomy Section, University of Palermo, Palermo, Italy*

ANNA RODINA • *Program in Chemical Biology, Memorial Sloan Kettering Cancer Center, New York, NY, USA*

TANAYA ROYCHOWDHURY • *Program in Chemical Biology, Memorial Sloan Kettering Cancer Center, New York, NY, USA*

REBECCA A. SAGER • *Department of Urology, SUNY Upstate Medical University, Syracuse, NY, USA*

ANAND R. SANTHASEELA • *Program in Chemical Biology, Memorial Sloan Kettering Cancer Center, New York, NY, USA*

RITWICK SAWARKAR • *Department of Genetics, Medical Research Council (MRC), University of Cambridge, Cambridge, UK*

SAHIL SHARMA • *Program in Chemical Biology, Memorial Sloan Kettering Cancer Center, New York, NY, USA*

MICHAEL Y. SHERMAN • *Department of Molecular Biology, Ariel University, Ariel, Israel*

STEFAN STANGL • *Radiation Immuno-Oncology Group, Center for Translational Cancer Research Technische Universität München (TranslaTUM), Department of Radiation Oncology, Klinikum rechts der Isar, Technische Universität München (TUM), Munich, Germany*

XIN TANG • *Department of Dermatology the Norris Comprehensive Cancer Centre, University of Southern California Keck Medical Center, Los Angeles, CA, USA*

JIMMY THERIAULT • *Molecular and Cellular Radiation Oncology, Beth Israel Deaconess Medical Center, Harvard Medical School, Boston, MA, USA*

RONALD TONUI • *Biomedical Biotechnology Research Unit (BioBRU), Department of Biochemistry and Microbiology, Rhodes University, Makhanda, South Africa*

SarahBeth D. Votra • *Department of Urology, SUNY Upstate Medical University, Syracuse, NY, USA*
Desheng Weng • *Department of Medicine, Boston University School of Medicine, Boston, MA, USA*
Mark R. Woodford • *Department of Urology, SUNY Upstate Medical University, Syracuse, NY, USA; Department of Biochemistry and Molecular Biology, SUNY Upstate Medical University, Syracuse, NY, USA*

Chapter 1

Monitoring of the Heat Shock Response with a Real-Time Luciferase Reporter

Andrew Ackerman, Toshiki Kijima, Takanori Eguchi, and Thomas L. Prince

Abstract

The heat shock response (HSR) is a cellular mechanism for counteracting acute proteotoxic stress. In eukaryotes, transcriptional activation of the HSR is regulated by heat shock factor 1 (HSF1). Activation of HSF1 induces the expression of heat shock proteins (HSPs) that function as molecular chaperones to fold and maintain the three-dimensional structure of misfolded proteins. The regulation of the degree and duration of the HSR is controlled by multiple biochemical mechanisms that include posttranslational modification of HSF1 and numerous protein–protein interactions. In this chapter, we describe a method to evaluate the activation and deactivation of the HSR at the transcriptional level using a short half-life luciferase reporter assay. This assay can be used to further characterize the HSR or as a screen for small molecule inducers, amplifiers, or repressors.

Key words Heat shock response, Heat shock factor 1 (HSF1), Heat shock protein 90 (HSP90), Luciferase assay, Drug screen

1 Introduction

The heat shock response (HSR) is an evolutionarily conserved cytoprotective mechanism induced by proteotoxic stress. The HSR is essential for maintaining proteostasis and plays a role in overall health and longevity [1]. Heat shock factor 1 (HSF1) is the master transcription factor responsible for initiating the HSR and inducing the expression of a variety of genes, most notably heat shock proteins (HSP) [2, 3]. Through a mechanism incompletely understood, proteotoxic stress initiates the release of HSF1 from bound HSP complexes to homotrimerize, translocates from the cytosol into the nucleus, and binds degenerate heat shock elements (HSE, ideally represented as GAAnnTTCnnGAA) across the genome and within promoters of transcriptional target genes [4, 5]. Preloaded and paused RNA polymerase II within HSP

Stuart K. Calderwood and Thomas L. Prince (eds.), *Chaperones: Methods and Protocols*, Methods in Molecular Biology, vol. 2693, https://doi.org/10.1007/978-1-0716-3342-7_1,

genes are activated by HSF1-binding to resume transcription elongation resulting in the rapid production of mRNA transcripts [6, 7]. Once stress has abated and proteostasis restored, HSF1-trimers are dissembled by newly synthesized HSPs such as HSP90 and HSP70 along with other cellular components [8–10]. Throughout this process, HSF1 is heavily posttranslationally modified and interacts with numerous cellular components that may also be modified [11–14]. These modifications and interactions affect the degree and duration of the HSR that is often amplified in cancer and reduced in neurodegenerative diseases [15, 16]. Understanding how to effectively modulate the HSR through small molecule therapies and lifestyle has the potential to positively influence disease outcomes and improve overall health.

Being able to monitor real-time expression of *HSP* gene products is essential for determining the degree and duration of the HSR. Previous versions of luciferase reporter assays are useful for measuring the induction and degree of the HSR but not effective at determining the attenuation of transcription and the actual duration of the response. Here we describe the use of a destabilized luciferase reporter assay driven by the heat-inducible *HSP70B* gene promoter for real-time monitoring of the HSR in mammalian cell culture (Fig. 1). The *HSP70B* gene (*HSPA7* locus) is one of the most strongly expressed human transcripts induced by proteotoxic stress, despite not encoding a known functional protein [17]. The reporter assay is derived from a plasmid developed by Younis et al.

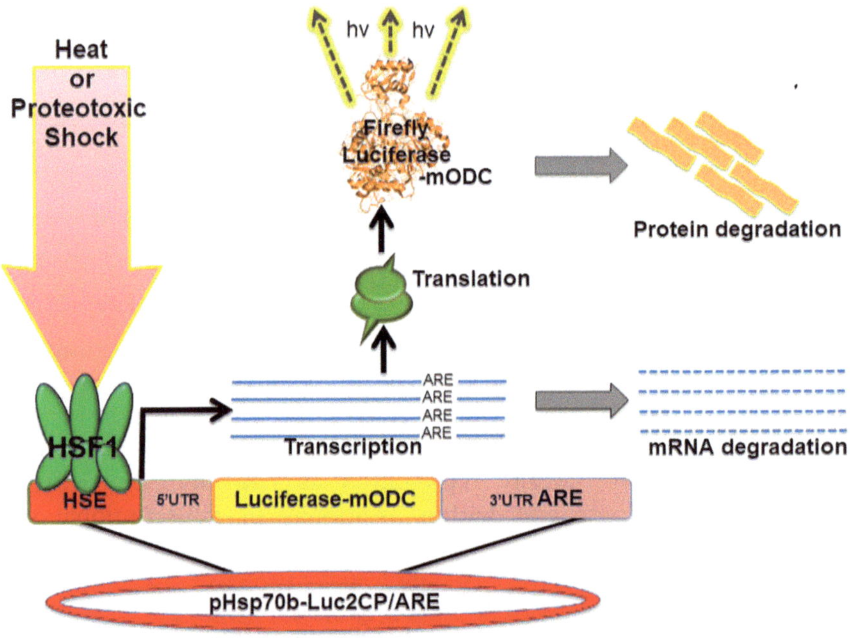

Fig. 1 Cartoon of real-time detection of induction of the HSR using destabilized luciferase reporter

[18]. For this real-time HSR reporter assay, a plasmid encoding an *HSP70B* promoter-driven firefly luciferase linked to a mouse ornithine decarboxylase (mODC) domain open reading frame (ORF) followed by a 3′untranslated region (UTR) AU-rich element (ARE) was constructed. The MODC contains a PEST sequence that is promptly ubiquitinated resulting in degradation of the luciferase, while the 3′-UTR ARE induces decay of the mRNA transcript (Fig. 1). Together these features promote rapid turnover of the luciferase within cells and thereby provide a real-time readout of HSR transcription. With this assay, screens for the effects of small molecule drugs, natural products, or transiently expressed HSPs and HSF1 mutants on the HSR can readily be developed.

Our method essentially involves the transfection of plasmids into cells, treating with drugs (optional), heat shock, and then allowing the cells to recover for a specific period of time before harvesting and assaying for luciferase activity. Differences in luciferase activity at each time point will reflect the intensity and stage of the HSR. In human embryonic kidney (HEK), 293 cell transcription of the *HSP70B* promoter typically attenuates after 2 h when heat shocked for 30 min at 42 °C; however, in other cell lines, these times may vary. Treating the cells with drugs before, during, or after the heat shock may alter the HSR by shortening, extending, and/or amplifying it. Some drugs such as HSP90 and proteasome inhibitors may also induce the HSR, while other small molecules may inhibit it [19–22]. Furthermore, transiently over-expressing other protein components such as HSF1 mutants or HSPs can alter the HSR and provide a mechanism for evaluating the effects of genetic alterations or posttranslational modifications.

2 Materials

All solutions should be prepared with double-deionized water. Chemicals should be molecular biology grade or above. Follow your institution's waste disposal guidelines when discarding used reagents.

2.1 Cell Culture and Transfection

1. HEK293 cells
2. Dulbecco's modified Eagle medium (DMEM), serum-free for transfecting and supplemented with 10% fetal calf serum (FCS), and 100 U/mL penicillin–streptomycin (optional) for growing cells
3. X-tremeGENE 9 DNA Transfection Reagent (Roche)

2.2 Plastic Ware

1. 96-well clear cell culture plates
2. 96-well white assay plates
3. Parafilm

2.3 Plasmids

1. Standard positive control luciferase, pGL3-CMV-Luc (Promega).
2. Transfection normalization β-galactosidase control, pGL3-CMV-βGal (Promega).
3. *HSP70B* promoter luciferase reporter plasmid, pHsp70b-Luc, was made for an earlier project [23].
4. Short half-life luciferase expression plasmid, pCMV-Luc2CP/ARE, with two protein-destabilizing sequences (2CP) at the C-terminus of a luciferase protein along with a 3′UTR AU-rich element (ARE) for rapid mRNA turnover, was made and deposited into Addgene (#62857) by Dr. Gideon Dreyfuss [18].
5. The heat shock-inducible reporter plasmid, pHsp70b-Luc2CP/ARE, was constructed by substituting the CMV promoter sequence in pCMV-Luc2CP/ARE with the *HSP70B* gene promoter sequence by Gibson Assembly [14, 24].

2.4 Buffers and Reagents

1. Reporter lysis buffer (RLB): 25 mM bicine (pH 7.6), 0.05% Tween 20, 0.05% Tween 80. Store at 4 °C.
2. Luciferase assay reagent (LAR): 25 mM glycine, 15 mM KPO_4, 15 mM $MgSO_4$, 4 mM EGTA, 2 mM ATP, 1 mM DTT, 1.7 mM K-luciferin. Aliquot and store at −20 °C.
3. β-Galactosidase assay reagent (BAR): 200 mM $NaPO_4$ (pH 7.3), 2 mM $MgCl_2$, 100 mM β-mercaptoethanol, 1.33 mg/mL o-nitrophenyl-β-D-galactopyranoside (ONPG). Aliquot and store at −20 °C.
4. β-Galactosidase stop (BGS): 1 M Na_2CO_3. Store at 4 °C.
5. 17-AAG (N-terminal HSP90 inhibitor): 1 μM in DMSO.

2.5 Equipment

1. Set of single and multichannel pipettes
2. Cell culture incubator (37 °C, 5% CO_2)
3. Laboratory water bath (42 °C)
4. 96-well plate reader with a luminometer and UV-Vis absorbance
5. −20 °C and/or −80 °C freezer

3 Methods

Care should be taken when working with any biological material. The plasmids and cell lines used in this method are not known to be infectious; however, proper personal protective equipment should be worn. All cell lysate, media, and chemicals should be treated and disposed of according to the institution's guidelines. Prepare all reagents in a sterile environment such as a cell culture hood.

3.1 Cell Plating

A separate 96-well plate is needed for each time point in the experiment. For studying the HSR with this assay, 5 time points suffice: control no HS, 1 h post HS, 2 h post HS, 4 h post HS, and 6 h post HS. This allows the experiment to be done in single a workday; however, as for most experiments, more time points are better. The density at which the cells are plated may influence transfection levels and overall HSR reporter signal. Moreover, each cell line divides at a different rate. Varying the cell plating density to optimize the HSR signal may be a worthwhile preliminary experiment. As a general rule, plating HEK293 cells at 30–40% confluency the night before transfecting works well.

1. Aliquot 12,000 cells in 100 μL of supplemented DMEM per well into each 96-well plate.
2. Incubated overnight this gives a confluency of 50–80% the next day at the time of transfection (*see* **Note 1**).

3.2 Transfection

Transfect cells with a mixture of the following: 0.05 μg of DNA + 0.15 μL of X-tremeGENE +20 μL of serum-free DMEM for each well (*see* **Note 2**). The 0.05 μg of DNA includes all plasmids to be transfected. Adjusting all plasmids to the same concentration is rather helpful. The ratio of each plasmid can be varied to optimize the assay. Typically, less pCMV-βGal is required compared to other plasmids in the experiment. The ratio of phsp70b-Luc2CP/ARE and effector HSF1 or HSP plasmid can be varied to test for mutation or regulatory effects. It is advised to repeat each sample 4 or 8 times per plate when assaying 24 or 12 respective conditions (*see* **Note 3**). Setting up a spreadsheet to calculate plasmid and reagent amounts is advised (Fig. 2a). Plasmids to be used: pCMV-βGal (transfection normalization control), pCMV-LUC (positive luciferase control), phsp70B-Luc (stable HSR control), phsp70b-Luc2CP/ARE (experimental, real-time reporter), and optional pcDNA3-HSF1 or –HSP (effector constructs).

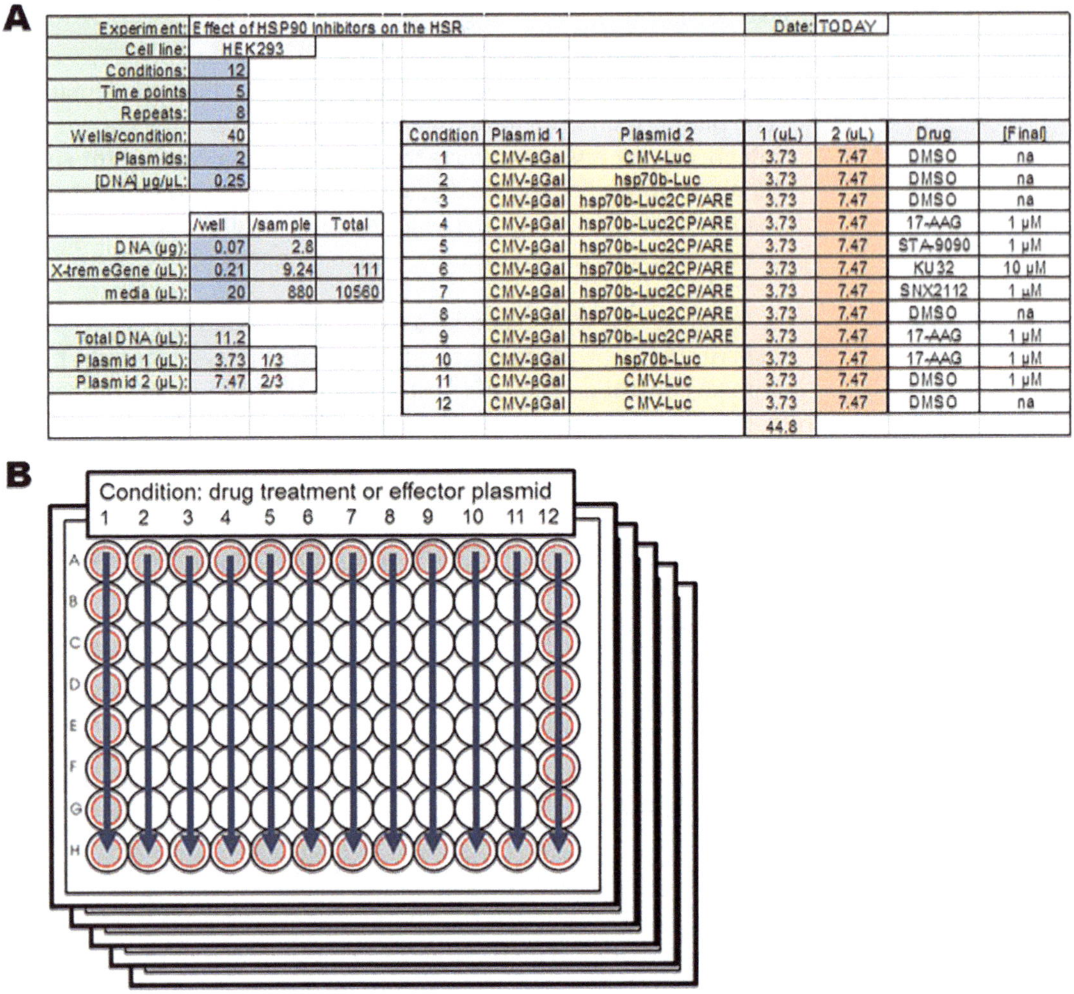

Fig. 2 Preparation of 96-well plates for transfection and drug treatment. (**a**) Sample spreadsheet for aliquoting reagents and arranging conditions. (**b**) Schematic for 12 drug treatment conditions across 5 time points. Gray circles indicate perimeter wells that should not typically be used for experimental testing

1. Aliquot and mix X-tremeGENE, serum-free DMEM, and plasmids in a separate sterile 96-well plate with each well corresponding to a different sample condition.
2. Using a multichannel pipette, distribute the ~20 μL of transfection mix into each well on the 96-well plates containing the cultured cells.
3. Incubate the cells and transfection mix overnight for 16–24 h.

3.3 Drug Treatment (Optional)

To test the effects of drugs on the HSR, transfected cells can be treated with small molecule inhibitors for 1 h before heat shocking. Treatment time can be varied but at least 45 min is needed to allow the drugs to accumulate in the cells. The number of plates and wells per plate to be treated will determine the number of drugs to be pre-diluted in DMEM and aliquoted.

1. For 17-AAG at a final concentration of ~1 μM in 16 wells (2 columns of 8) across 5 plates, a total of 12 μL of 1 mM stock 17-AAG dissolved in DMSO should be added to 1700 μL of DMEM.
2. Once mixed, 20 μL of pre-diluted 17-AAG is added to each well.
3. DMSO is used as the vehicle control (Fig. 2a, b).

3.4 Heat Shock

There are a few different ways to heat shock cells depending on available equipment (*see* **Note 4**):

1. Wrap each plate tightly with Parafilm.
2. Float each plate in a water bath heated to 42 °C for 30 min.
3. If using a common lab water bath, before unwrapping and placing back into the 37 °C incubator, each plate must be thoroughly sprayed and wiped with 70% ethanol and then dried to prevent contaminating the cell culture incubator.

Or

1. Preheat a separate incubator to 43 °C.
2. Place the plates in the incubator for 45–60 min to induce a robust heat shock.
3. Place the cells back into the 37 °C incubator to recover and be harvested.

3.5 Cell Lysis and Sample Storage

To harvest the cells at each time point:

1. Remove the plate from the incubator and aspirate off all media, but be careful not to dislodge the cells from the plate.
2. Add 110 μL of RLB to each well and place in a freezer for at least 45 min. Samples can also be stored until the time course is complete. For longer-term storage, samples should be placed in an −80 °C freezer.
3. A single freeze-thaw cycle should efficiently lyse the cells.

3.6 Luciferase Assay

To determine the level of induced gene expression:

1. Allow all plates to completely thaw on ice.
2. Once thawed, plates can be spun down on a large benchtop centrifuge at 4 °C to aggregate cell debris and ensure uniform lysis (optional).
3. Thaw both LAR and BAR making sure both reagents are warm and completely dissolved.

4. Turn on the plate reader luminometer and load the proper program for analysis.
5. When ready use a multichannel pipette to transfer 50 μL of lysate from each well to the corresponding well in a clean white 96-well plate.
6. Add 50 μL of LAR to each well. Let the plate set for 10 min at room temperature.
7. Read the plate on the luminometer and record the results.
8. Repeat the process of aliquoting the lysate, adding the LAR, allowing it to set for 10 min, and reading the luciferase activity for each plate in the time course.
9. Due to changes in luciferase activity over time, make sure the timing between adding the LAR and reading luciferase activity is consistent for each plate. This will ensure consistent and accurate readings of HSR induction at each time point (*see* **Note 5**).

3.7 β-Galactosidase Assay

To determine the transfection efficiency for each well:

1. For each plate pipette 50 μL of lysate from each well into a corresponding well in a clear 96-well plate. Do this for all the plates in the time course.
2. Add 35 μL of BAR to each well, and incubate all the plates at 37 °C for 20 min or until the lysate mix turns yellow.
3. Add 50 μL of BGS to all wells to terminate the enzyme reaction. This will ensure that the β-galactosidase reaction time is the same across all plates.
4. Read each plate for absorbance at 420 nM and record the results (*see* **Note 6**).

3.8 Normalization and Statistics

To determine the luciferase activity relative to the amount of plasmid transfected into the cells:

1. Divide the luciferase readout value by the β-galactosidase value using a spreadsheet program.
2. Identify the baseline reporter signal, in this case, the control no HS plate and no drugs sample set, and then calculate the average.
3. Divide all the other wells in all other plates by this control value. This makes all reporter readout signals relative to the average of the baseline signal.
4. For each sample condition set of repeats, determine the average and standard deviation.

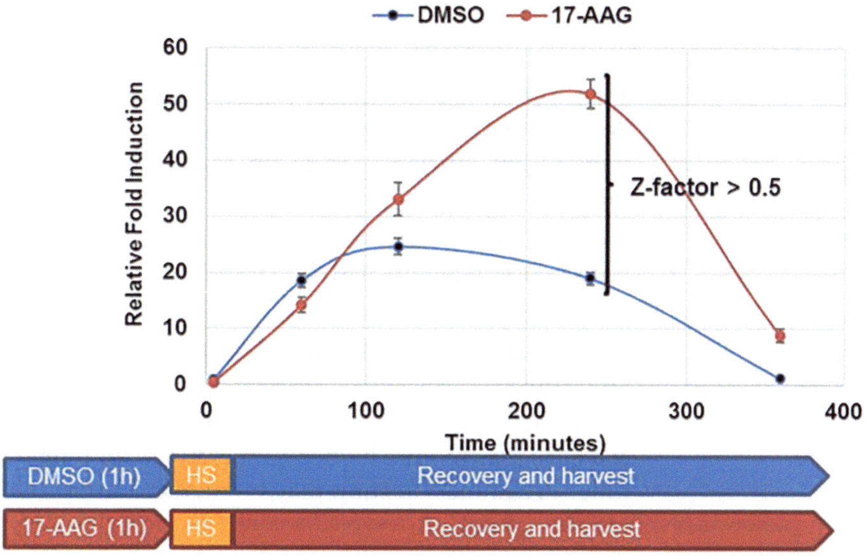

Fig. 3 Effect of 17-AAG pretreatment on HSR degree and duration. Cells were incubated with DMSO or 1 μM 17-AAG for 1 h, then heat shocked for 30 min at 42 °C, allowed to recover, and then harvested at designated time points

5. Sample values can finally be graphed +/− standard deviation at each time point and *p*-values calculated using Student's *t*-test between each time point and/or condition (Fig. 3).
6. If utilizing the assay as a high-throughput screen, a Z-factor score can be calculated for a specific time point where a value greater than 0.5 indicates significance [25].

4 Results

Differences in HSR transcription rates were observed between DMSO and 17-AAG pretreatments in HEK293 cells. For DMSO control cells, the HSR increased over time with a maximum at 2 h then decreased to baseline at 6 h. For 17-AAG pretreated cells, the HSR increased over time with a maximum of around 4 h and remained above baseline at 6 h, indicating that HSP90 N-terminal inhibition increased the degree and duration of the HSR. At 4 h the difference between DMSO and 17-AAG treated cells gave a Z-factor of 0.65 suggesting that 17-AAG warrants further attention as an amplifier of the HSR (Fig. 3). Only the results for DMSO and 17-AAG treatment are shown.

5 Notes

1. Transfection and luciferase/β-galactosidase assays work well in 96-well plates with HEK293 cells. For cell lines with low transfection efficiency, scale up to 12- or 24-well plates to increase cell numbers.
2. We have used X-tremeGENE 9 and lipofectamine 3000 as transfection reagents with equal success.
3. Due to exposure and evaporation issues, use the wells on the perimeter of the plate as control repeats and not for experimental testing.
4. Each cell line is different in regards to inducing the HSR; some cell lines may require higher temperatures for longer periods of time. Preliminary testing may be required using the 2 h post-HS time point.
5. It is essential that all plates are assayed and read in the same manner and timing after the addition of LAR. This is to ensure that the substrate concentration and luciferase activity are the same for each time point. The pre-read incubation time can be adjusted to 5 min and staggered depending on setup.
6. In HEK293 cells, 10–20 min incubation is enough for the β-galactosidase assay. Do not incubate too long because the reaction can be saturated. If you have multiple plates for time-course assay, it is very important to measure the β-galactosidase activity of each plate with the same incubation time.

References

1. Murshid A, Eguchi T, Calderwood SK (2013) Stress proteins in aging and life span. Int J Hyperth 29(5):442–447
2. Prince TL, Lang BJ, Guerrero-Gimenez ME, Fernandez-Muñoz JM, Ackerman A, Calderwood SK (2020) HSF1: primary factor in molecular chaperone expression and a major contributor to cancer morbidity. Cell 9(4): 1046
3. Lang BJ, Guerrero ME, Prince TL, Okusha Y, Bonorino C, Calderwood SK (2021) The functions and regulation of heat shock proteins; key orchestrators of proteostasis and the heat shock response. Arch Toxicol 95(6):1943–1970
4. Mendillo ML, Santagata S, Koeva M, Bell GW, Hu R, Tamimi RM, Fraenkel E, Ince TA, Whitesell L, Lindquist S (2012) HSF1 drives a transcriptional program distinct from heat shock to support highly malignant human cancers. Cell 150(3):549–562
5. Schmauder L, Sima S, Hadj AB, Cesar R, Richter K (2022) Binding of the HSF-1 DNA-binding domain to multimeric *C. elegans* consensus HSEs is guided by cooperative interactions. Sci Rep 12(1):1–19
6. Bunch H, Zheng X, Burkholder A, Dillon ST, Motola S, Birrane G, Ebmeier CC, Levine S, Fargo D, Hu G, Taatjes DJ (2014) TRIM28 regulates RNA polymerase II promoter-proximal pausing and pause release. Nat Struct Mol Biol 21(10):876–883
7. Vihervaara A, Mahat DB, Guertin MJ, Chu T, Danko CG, Lis JT, Sistonen L (2017) Transcriptional response to stress is pre-wired by promoter and enhancer architecture. Nat Commun 8(1):1–6
8. Kijima T, Prince TL, Tigue ML, Yim KH, Schwartz H, Beebe K, Lee S, Budzynski MA, Williams H, Trepel JB, Sistonen L (2018) HSP90 inhibitors disrupt a transient HSP90-

HSF1 interaction and identify a noncanonical model of HSP90-mediated HSF1 regulation. Sci Rep 8(1):1–3

9. Kmiecik SW, Le Breton L, Mayer MP (2020) Feedback regulation of heat shock factor 1 (Hsf1) activity by Hsp70-mediated trimer unzipping and dissociation from DNA. EMBO J 39(14):e104096
10. Pernet L, Faure V, Gilquin B, Dufour-Guérin S, Khochbin S, Vourc'h C (2014) HDAC6–ubiquitin interaction controls the duration of HSF1 activation after heat shock. Mol Biol Cell 25(25):4187–4194
11. Guettouche T, Boellmann F, Lane WS, Voellmy R (2005) Analysis of phosphorylation of human heat shock factor 1 in cells experiencing a stress. BMC Biochem 6(1):1–4
12. Neef DW, Jaeger AM, Gomez-Pastor R, Willmund F, Frydman J, Thiele DJ (2014) A direct regulatory interaction between chaperonin TRiC and stress-responsive transcription factor HSF1. Cell Rep 9(3):955–966
13. Kmiecik SW, Mayer MP (2021) Molecular mechanisms of heat shock factor 1 regulation. Trends Biochem Sci 47(3):218–234
14. Kmiecik SW, Drzewicka K, Melchior F, Mayer MP (2021) Heat shock transcription factor 1 is SUMOylated in the activated trimeric state. J Biol Chem 296:100324
15. Gomez-Pastor R, Burchfiel ET, Thiele DJ (2018) Regulation of heat shock transcription factors and their roles in physiology and disease. Nat Rev Mol Cell Biol 19(1):4–19
16. Cyran AM, Zhitkovich A (2022) Heat shock proteins and HSF1 in cancer. Front Oncol 12: 860320
17. Parsian AJ, Sheren JE, Tao TY, Goswami PC, Malyapa R, Van Rheeden R, Watson MS, Hunt CR (2000) The human Hsp70B gene at the HSPA7 locus of chromosome 1 is transcribed but non-functional. Biochim Biophys Acta 1494(1–2):201–205
18. Younis I, Berg M, Kaida D, Dittmar K, Wang C, Dreyfuss G (2010) Rapid-response splicing reporter screens identify differential regulators of constitutive and alternative splicing. Mol Cell Biol 30(7):1718–1728
19. West JD, Wang Y, Morano KA (2012) Small molecule activators of the heat shock response: chemical properties, molecular targets, and therapeutic promise. Chem Res Toxicol 25(10):2036–2053
20. Kurop MK, Huyen CM, Kelly JH, Blagg BS (2021) The heat shock response and small molecule regulators. Eur J Med Chem 226: 113846
21. Kim D, Kim SH, Li GC (1999) Proteasome inhibitors MG132 and lactacystin hyperphosphorylate HSF1 and induce hsp70 and hsp27 expression. Biochem Biophys Res Commun 254(1):264–268
22. Kijima T, Prince T, Neckers L, Koga F, Fujii Y (2019) Heat shock factor 1 (HSF1)-targeted anticancer therapeutics: overview of current preclinical progress. Expert Opin Ther Targets 23(5):369–377
23. Murshid A, Chou SD, Prince T, Zhang Y, Bharti A, Calderwood SK (2010) Protein kinase A binds and activates heat shock factor 1. PLoS One 5(11):e13830
24. Gibson DG, Young L, Chuang RY, Venter JC, Hutchison CA, Smith HO (2009) Enzymatic assembly of DNA molecules up to several hundred kilobases. Nat Methods 6(5):343–345
25. Zhang JH, Chung TD, Oldenburg KR (1999) A simple statistical parameter for use in evaluation and validation of high throughput screening assays. J Biomol Screen 4(2):67–73

Chapter 2

Studying RNA Polymerase II Promoter-Proximal Pausing by In Vitro Immobilized Template and Transcription Assays

Heeyoun Bunch

Abstract

The immobilized template assay is a versatile biochemical method for studying protein–nucleic acid interactions. Using this method, immobilized nucleic acid-associated or specific proteins can be identified and quantified by techniques such as mass spectrometry and immunoblotting. Here, a modified immobilized template assay combined with in vitro transcription assay to study the function of transcription factors and transcriptional activities at the human heat shock protein 70 (*HSP70*) gene is described. Notably, this method can be applied to study other important genes and transcription factors in vitro.

Key words Transcriptional regulation, In vitro transcription, Immobilized template assay, Transcription factors, Gene regulation

1 Introduction

Immobilized template assays followed by in vitro transcription assays have been used for over a couple of decades [1–3]. The development of one such method for isolating HeLa nuclei and producing the nuclear extracts (HeLa NE) enabled us to study transcription in vitro. This is mainly attributed to the ability of the immobilized template, including the promoter of a gene to assemble the pre-initiation complex (PIC) from HeLa NE. In the case of bacterial in vitro transcription, recombinant 5-subunit-RNA polymerase (2α, β, β', and ω subunits typically associated with σ factor) [4], a DNA template, and dNTPs under proper buffer conditions are sufficient for de novo synthesis of RNA molecules [5]. However, eukaryotic transcription is much more complex and involves an array of general and gene-specific transcription factors and at least 12 subunit-RNA polymerase II (Pol II) [6, 7]. This complexity imposes practical difficulties in studying eukaryotic transcription in vitro. The immobilized template assay to assemble the PIC on a DNA template of interest from HeLa NE bypasses the

Stuart K. Calderwood and Thomas L. Prince (eds.), *Chaperones: Methods and Protocols*, Methods in Molecular Biology, vol. 2693, https://doi.org/10.1007/978-1-0716-3342-7_2,

need to purify all the protein components. To be immobilized, the DNA template requires to be conjugated on the collectable beads. With elaborative modifications, this powerful method can be used to understand the function and interaction of transcription factors in gene regulation in vitro.

HSP70 is one of the most representative and well-studied stimulus-inducible genes [8–11]. Gene regulation of *HSP70* is instantaneous and dramatical, provoking a burst of mRNA synthesis. This rapid response to the increased internal energy is ensured due to a single transcription factor, HSF1, serving as a primary signal transduction regulator that transfers the cytosolic signal to the target gene activation in the nucleus [11–13]. However, upon HSF1 binding to the *HSP70* promoter, the activated gene undergoes several changes in terms of its interacting transcription factors, Pol II modifications, DNA structure, and nucleosome architecture [2, 9, 10, 14–25] (Fig. 1). Notably, many features of transcriptional regulation at *HSP70* are conserved and common in diverse classes of stimulus-inducible genes, thereby establishing *HSP70* as a model gene to study gene regulation mechanisms. Pol II promoter-proximal pausing is one of the critical regulatory features of

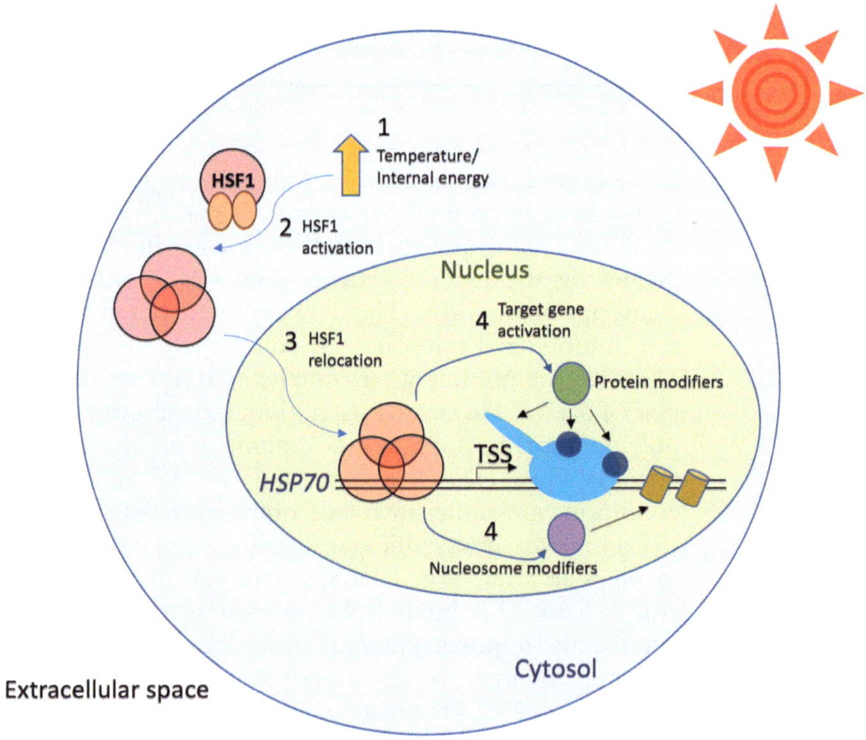

Fig. 1 HSF1-mediated transcriptional activation at *HSP70*. The numbers indicate the order of *HSP70* transcriptional activation. Proteins are shown in circular shapes. Nucleosomes are shown in cylinders. The sun is shown in a circle surrounded by triangles, a source of heat

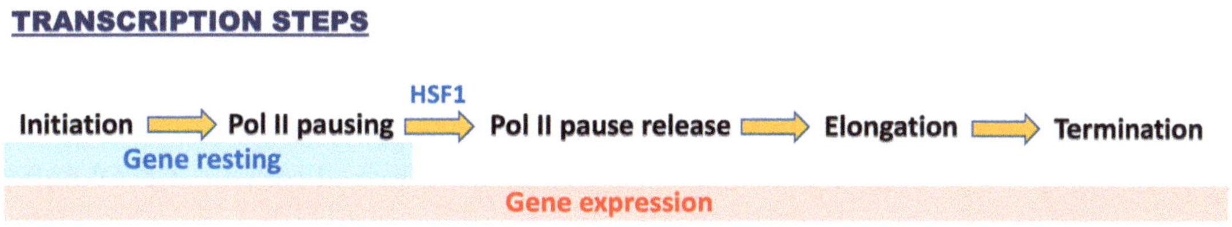

Fig. 2 Transcriptional regulation at *HSP70*. Pol II pausing is an additional critical regulatory step in transcriptional regulation in a large number of genes in metazoan cells

HSP70 and other stimulus-inducible genes, including serum-, neurotransmitter-, and hypoxia-inducible genes [8, 14, 26–28]. In this phenomenon, Pol II is paused at +25–+100 from the transcription start site (TSS) before receiving gene activating signal, which is recognized as a rate-limiting step in transcriptional activation [29, 30] (Fig. 2). In *Drosophila*, it was estimated that approximately 30% of protein-coding genes are estimated to harbor Pol II pausing [29]. It reportedly occurs in both protein-coding and nonprotein-coding genes at relatively higher percentages in higher organisms like mice and humans [28, 31, 32].

In our previous studies, we demonstrated Pol II pausing at the human *HSP70* gene by combining an immobilized template and in vitro transcription assay [10, 33]. Furthermore, we identified a novel Pol II pausing regulator, TRIM28, and characterized its function [33]. These results led us to reveal the noteworthy and critical cross talk between DNA break and damage response signaling and transcriptional activation [14, 34, 35]. Other research groups have employed similar methods to investigate transcriptional regulation at human *HSP70* [2, 36, 37]. Here, I describe the methods used to perform the immobilized template assay and in vitro transcription assay to study Pol II pausing and transcriptional regulation at *HSP70*.

2 Materials

2.1 Preparation of the HSP70 Template DNA

1. PCR with a set of primers, including a 5′-biotinylated forward primer (*see* **Note 1**) that can amplify the *HSP70* segment of interest: For the PCR template in the reaction, the promoter and TSS of *HSP70* (−467 – +216) were amplified from the HeLa cells and inserted into a pCR4-TOPO vector to generate pTOPO-HSP70, as described in our previous study [33].
2. Agarose gel electrophoresis: The PCR reaction is run on a 0.8% agarose gel to run in 1× Tris–HCl–acetate–EDTA (TAE) buffer.
3. DNA extraction from the gel using a Qiagen gel extraction kit. Elute the DNA using ultrapure deionized water.

2.2 Preparation of the HSP70 Immobilized Template

1. Dynabeads M-280 Streptavidin (Invitrogen) are washed and equilibrated with 2× B & W buffer (10 mM Tris-HCl, pH 7.5, 1 mM EDTA, 2 M NaCl, filtered using 0.2 μm filter) (*see* **Note 2**) and incubated with the biotin-conjugated *HSP70* template.
2. The template–bead complex is washed with 1× B & W buffer (5 mM Tris-HCl, pH 7.5, 0.5 mM EDTA, 1 M NaCl, filtered using 0.2 μm filter) and resuspended in 0.1 M Buffer D (20 mM HEPES, 20% glycerol, pH 7.6, pH 7.9, 0.1 mM EDTA, 100 mM KCl, filtered using 0.2 μm filter) to a final DNA concentration of 10 ng/μL.

2.3 Performing the Immobilized Template Assay

1. HeLa NE (>80 mg/mL) can be prepared using the traditional method described in [33, 38, 39]. Per a reaction, 100–120 ng of the *HSP70* template DNA is used and mixed with transcription factor (TF) buffer (12.5 ng/μL dI-dC, 0.075% NP40, 5 mM $MgCl_2$, 250 ng/μL BSA, 12.5% glycerol, 100 mM KCl, 12.5 mM HEPES, pH 7.6, 62.5 μM EDTA, 10 μM $ZnCl_2$; the mixture is filtered through a 0.2 μm filter before supplementing with BSA and dI-dC) to incubate the DNA–bead complex with a DNA binding transcription factor(s) (e.g., HSF1). Recombinant HSF1 can be purified as previously described [33].
2. The protein-bound DNA–bead complex is pulled down using a magnetic stand and resuspended in nuclear extract (NE) buffer (17.5 ng/μL dI-dC, 0.1% NP40, 7.5 mM $MgCl_2$, 1.25 μg/μL BSA, 8.7% glycerol, 8.7 mM HEPES, pH 7.6, 44 μM EDTA, 130 mM KCl, 10 μM $ZnCl_2$; the mixture is filtered through a 0.2 μm filter before supplementing with BSA and dI-dC) (*see* **Note 3**).
3. After incubating the complex with approximately 100 μg of HeLa NE, the bead–DNA–protein complex is washed with transcription wash 1 (TW1) buffer (12.5 mM HEPES, pH 7.6, 12.5% glycerol, 500 mM KCl, 833 nM $MgCl_2$, 62.5 μM EDTA, 0.0125% NP40; the buffer is filtered through a 0.2 μm filter). NE and TW1 buffers are freshly supplemented with protease inhibitors (1 mM benzamidine, 0.25 mM PMSF, aprotinin [1:1000, Sigma A6279], and 1 mM sodium metabisulfite). The specific transcription factor of investigation can be depleted from HeLa NE by a series of immunoprecipitation assays as described in our previous study [33].
4. The DNA-bound proteins are eluted by 2% sarkosyl dissolved in TW1 buffer and are shown by silver staining and quantified by immunoblotting.

2.4 Performing In Vitro Transcription Assay

1. All steps are identical to those of the immobilized template assay until the washing step after NE incubation. The pellet fraction, including the bead–DNA–DNA-bound protein complex, is washed with TW2 buffer (13 mM HEPES, pH 7.6, 13% glycerol, 60 mM KCl, 7 mM $MgCl_2$, 7 mM DTT, 100 μM EDTA, 0.0125% NP40, 10 μM $ZnCl_2$, filtered through a 0.2 μm filter) and resuspended in transcription (TC) buffer (13 mM HEPES, pH 7.6, 13% glycerol, 60 mM KCl, 7 mM $MgCl_2$, 10 μM $ZnCl_2$, 7 mM DTT, 100 μM EDTA, 15 ng/μL dI-dC, 10 mM creatine phosphate, filtered through a 0.2 μm filter before supplementing with dI-dC and creatine phosphate). The TC buffer is supplemented with the protease inhibitors described in point 3 of Subheading 2.3.
2. The NTP mixture is prepared by mixing nonradioactive 250 μM ATP, GTP, and UTP and 10 μM CTP with radioactive CTP (5–10 μCi). Adding the NTP mixture initiates nascent RNA synthesis by Pol II, and hot CTP enables the probing of nascent RNA molecules.
3. To terminate the reaction, stop buffer (0.6 M Tris-HCl, pH 8.0, 12 mM EDTA, 100 μg/mL tRNA; filtered through a 0.2 μm filter before supplementing with tRNA) is used.
4. The supernatant fraction includes the nascent RNAs and is treated with an equal volume of phenol–chloroform–isoamyl alcohol (25:24:1, PCI) solution to remove proteins. After separating the soluble (upper) phase by centrifugation, the RNA molecules are precipitated with 2.6 volumes of 100% (200 proof) ethanol. Additionally, as an extension of the immobilized template assay, the proteins in the pellet and supernatant fractions after the addition of NTPs can be analyzed by silver staining or immunoblotting.

2.5 Running Denaturing Polyacrylamide Gel Electrophoresis

1. The precipitated RNA transcripts in the pellet are completely dried before resuspending in a formamide-based RNA loading buffer (90% deionized formamide, 0.05% bromophenol blue, 0.05% xylene cyanol FF, 1× Tris–boric acid–EDTA [TBE]).
2. The RNA transcripts and standard size markers are separated on denaturing polyacrylamide gels (SequaGel UreaGel 19:1 Denaturing Gel System, National Diagnostics). The gels are exposed to X-ray films or intensifying screens for 16–72 h.
3. The intensity of the bands representing nascent RNA molecules is quantified using ImageJ software (National Institute of Health, USA).

3 Methods

Unless otherwise indicated, all the reactions must be performed at room temperature (RT). The tube containing the mixture with beads must be held upright during the incubation with rotation to retain the mixture at the bottom of the tube.

3.1 PCR Amplification of the HSP70 Template DNA

1. Mix the below components:

   5 μL of 10× Pfu buffer

   5 μL of 2 mM dNTPs

   3 μL of 25 mM $MgSO_4$

   1 μL of 5 μM forward primer (5′biotin-conjugated, *see* **Note 4**)

   1 μL of 5 μM reverse primer (*see* **Note 5**)

   Approximately 2 ng of cut plasmid including the template DNA (e.g., pTOPO-HSP70)

   0.5 μL of Pfu polymerase

   2.5 μL of 5% DMSO

   Make up the reaction volume to 50 μL with ultrapure H_2O.

2. Run PCR cycles as per the listed conditions:

   Initial denaturation at 94 °C for 5 min

   Thirty cycles of denaturation at 94 °C for 1 min, annealing at 62 °C for 30 s, and extension at 72 °C for 1 min

   Final extension at 72 °C for 10 min

3. Run agarose gel electrophoresis, and purify the biotinylated *HSP70* template from the gel band.
4. Measure the DNA concentration by NanoDrop.

3.2 Conjugating Biotinylated DNA with Streptavidin-Coated Beads

1. Take avidin-coated magnetic beads.
2. Remove the storage buffer.
3. Wash four times with 10× volume of 2× B & W buffer (*see* **Note 6**).
4. Resuspend the beads in 2× B & W buffer.
5. Incubate the template DNA with the beads [2 μg DNA/100 μL (1 mg) beads] for 30 min at RT (the duration is critical for consistent results).
6. Remove the supernatant.
7. Wash the bead–DNA complex twice with a 10× volume of 1× B & W buffer.
8. Wash the bead–DNA complex twice with a 10× volume of 0.1 M Buffer D.
9. Resuspend the bead–DNA complex in a 2× volume of 0.1 M Buffer D.

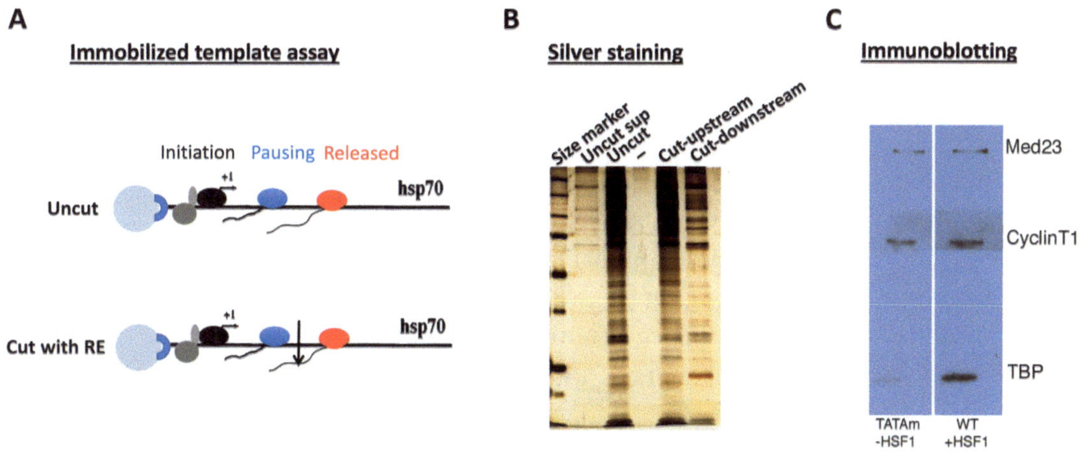

Fig. 3 Immobilized template assay using the *HSP70* template. (**a**) Schematic representation showing Pol II in different stages of transcription on the immobilized template. The protein factors associated with Pol II in different stages can be identified by restriction enzyme (RE) digestion of the template. (**b**) An example of silver staining after the immobilized template assay. Sup represents the supernatant fraction. Uncut and cut-upstream represent the pellet fractions of the RE-uncut and cut template, respectively. Cut-downstream represents the bead-released downstream DNA after RE treatment. (**c**) Immunoblotting identifies and quantifies proteins associated with the *HSP70* template. Med23, CyclinT1 (a subunit of active P-TEFb), and TBP (TATA-binding protein) on TATAm (TATA box mutant) and WT *HSP70*

3.3 Immobilized Template Assay

1. Mix 40 μL of TF buffer with 10 μL (100 ng DNA) of the immobilized DNA template.
2. Add a transcriptional activator(s) (e.g., HSF1) (*see* **Note 7**).
3. Incubate for 30 min at RT.
4. Remove the supernatant.
5. Add 60 μL of NE buffer to the pellet.
6. Add 10–20 μL of HeLa NE or factor-depleted NE, and incubate for 30 min at RT.
7. Wash the pellet four times with 500 μL TW1 buffer.
8. Remove the supernatant.
9. Elute the proteins from the immobilized template–bead complex by incubating (with rocking) in the sarkosyl buffer for 30 min at 4 °C.
10. Collect the supernatant to analyze proteins by silver staining or immunoblotting (Fig. 3).

3.4 In Vitro Transcription Assay

1. Mix 40 μL of TF buffer with 12 μL of the immobilized *HSP70* template DNA.
2. Add HSF1 or control buffer to the mixture and incubate for 30 min at RT.
3. Remove the supernatant with the aid of a magnetic stand.

4. Resuspend the beads with 60 μL of NE buffer.
5. Add 10 μL of HeLa NE (or a factor-depleted one) (*see* **Note 8**).
6. Incubate the mixture for 30 min at RT.
7. Wash once with 50 μL TW2 buffer.
8. Resuspend the beads with 24 μL of TC buffer.
9. HSF1 could be included (after PIC formation and before transcriptional initiation).
10. Add 0.5 μL of G/C/U/A mixture plus 1 μL of radioactive C (10 μCi) to initiate transcription.
11. Allow the synthesis of nascent RNA molecules for 5–15 min at 30 °C.
12. Add HSF1 to assess its function for Pol II pausing and elongation.
13. Incubate for an additional 30 min at 30 °C.
14. Add 125 μL of ice-cold stop buffer.
15. Add 150 μL of PCI to remove proteins, vortex, and centrifuge briefly (8000 rpm for 2 min).
16. Transfer carefully all the aqueous phases into a new tube (*see* **Note 9**).
17. Add 400 μL of ice-cold pure ethanol for precipitation, and vortex the contents thoroughly.
18. Store the mixture at −20 °C overnight.
19. Centrifuge at 14,000 rpm for 30 min and remove the supernatant.
20. Air-dry the pellet and dissolve it by thoroughly vortexing it with 6.5 μL of RNA loading buffer and then briefly spinning down the contents.
21. Boil the mixture for 5 min and centrifuge at 10,000 rpm for 2 min before loading it onto a sequencing gel (Fig. 4).

3.5 Running Denaturing Polyacrylamide Gel Electrophoresis

1. Prepare a 12% gel by mixing the component below:

   33.6 mL SequaGel concentrate

   29.7 mL SequaGel diluent

   7 mL SequaGel buffer

   Add 400 μL APS and 15 μL TEMED immediately before casting the gel

2. Alternatively, the gel could be homemade as below:

   21 g urea

   15 mL 40% acrylamide solution (29:1)

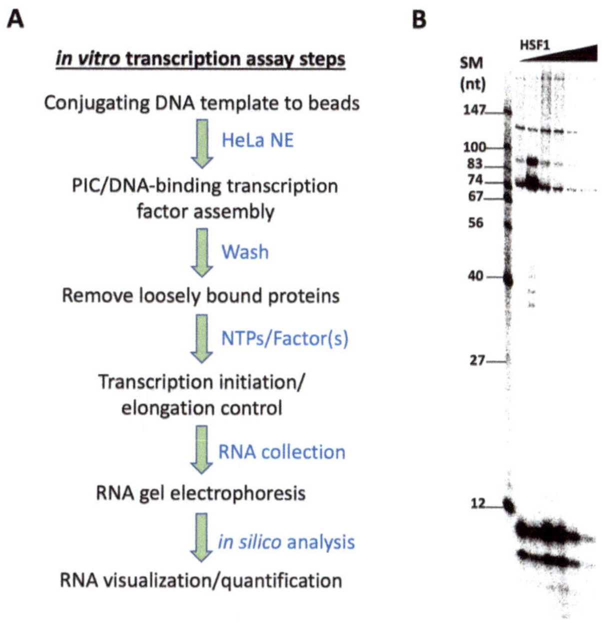

Fig. 4 In vitro transcription assay using the *HSP70* template. (**a**) Steps to perform the in vitro transcription assay. (**b**) An example of in vitro transcription assay result. HSF1 was titrated to identify the ideal amount to stimulate *HSP70* transcription in vitro. SM, RNA size marker in nucleotides (nt)

5 mL 10× TBE

Add pure, deionized water to make up the volume to 50 mL

Syringe-filter the solution through a 0.2 μm filter

Add 400 μL APS and 15 μL TEMED immediately before casting the gel

3. Clean the gel plates (0.6 mm gel is preferred) thoroughly. One of the gel plates should be siliconized for easy detachment.
4. Pour the gel solution using a pipette, pipette aid without introducing air bubbles, and place the comb.
5. Apply pressure (aluminum/metal ice blocks work well) on the comb part, clip the remaining three sides, and solidify the gel overnight.
6. Set up the gel in the sequencing gel apparatus, equilibrate the wells with the running buffer, and pre-run at 1200–1500 V for 30–40 min before loading the samples.
7. Run the gel at 1200–1500 V until the bromophenol blue dye reaches about a fifth of the gel from the bottom (*see* **Note 10**).
8. Expose on an X-ray intensifying screen or X-ray film for a couple of hours to a few days before acquiring the gel image.

4 Notes

1. *HSP70* cloning forward primer: 5′-CTCCTTCCCATTAAGACGGAAAAAACATCCGGGAGAGCCGGTCCG-3′; *HSP70* cloning reverse primer: 5′-ACCTTGCCGTGTTGGAACACCCCCACGCAGGAGTA GGTGGTGCCCAGGTC-3′.
2. B & W buffers (2× and 1×) and Buffer D should be stored at RT.
3. TF, NE, TW1, TW2, and TC buffers should be aliquoted and stored at −70 °C. PCI and ethanol to precipitate RNA must be stored at 4 °C. Stop buffer and protease inhibitors (except for aprotinin at 4 °C) must be stored below −20 °C.
4. *HSP70* template forward: 5′-biotin-GAAAGGACCCAAGGCTGCTCCGTCCTTCAC-3′.
5. *HSP70* template reverse: 5′-GCCGGTGCCCTGCTCTGTGGGCTCCGC-3′.
6. The beads and solution must remain at the bottom of the tube to avoid splashing during the template DNA conjugation, immobilized template assay, and in vitro transcription assays.
7. The amounts of transcriptional activators added to the reaction should be determined empirically. Titration assays ranging from 0.5 to 100 ng, for example, can be performed to determine the amounts.
8. HeLa NE should be aliquoted in small amounts because repetitive freeze-thawing leads to the precipitation of proteins. Before using the HeLa NE, centrifuge it at 14,000 rpm for 30 min at 4 °C, and use the supernatant (soluble fraction) for the experiment.
9. This step should be performed meticulously to include most of the upper phase. Variability while pipetting the upper phase between the reactions can hinder their quantitative comparisons.
10. Remove any air bubbles from the bottom of the gel before electrophoresis.

Acknowledgments

This work was supported by grants from the National Research Foundation of the Republic of Korea (Grant no. 2022R1A21003569). HB appreciates S.K. Calderwood at Beth Israel Deaconess Medical Center/Harvard Medical School and D.J. Taatjes at the University of Colorado for their support in developing the methods described in this study. HB thanks John and D. Bunch and J. Christ for their loving support and encouragement throughout this work.

References

1. Johnson KM, Wang J, Smallwood A, Carey M (2004) The immobilized template assay for measuring cooperativity in eukaryotic transcription complex assembly. Methods Enzymol 380:207–219
2. Brown SA, Imbalzano AN, Kingston RE (1996) Activator-dependent regulation of transcriptional pausing on nucleosomal templates. Genes Dev 10(12):1479–1490
3. Meyer KD, Lin SC, Bernecky C, Gao Y, Taatjes DJ (2010) p53 activates transcription by directing structural shifts in mediator. Nat Struct Mol Biol 17(6):753–760
4. Vassylyev DG, Sekine S, Laptenko O, Lee J, Vassylyeva MN, Borukhov S, Yokoyama S (2002) Crystal structure of a bacterial RNA polymerase holoenzyme at 2.6 A resolution. Nature 417(6890):712–719
5. Strobel EJ, Roberts JW (2015) Two transcription pause elements underlie a sigma70-dependent pause cycle. Proc Natl Acad Sci U S A 112(32):E4374–E4380
6. Ehara H, Yokoyama T, Shigematsu H, Yokoyama S, Shirouzu M, Sekine SI (2017) Structure of the complete elongation complex of RNA polymerase II with basal factors. Science 357(6354):921–924
7. Gnatt A, Fu J, Kornberg RD (1997) Formation and crystallization of yeast RNA polymerase II elongation complexes. J Biol Chem 272(49):30799–30805
8. Rougvie AE, Lis JT (1988) The RNA polymerase II molecule at the 5′ end of the uninduced hsp70 gene of *D. melanogaster* is transcriptionally engaged. Cell 54(6):795–804
9. Zobeck KL, Buckley MS, Zipfel WR, Lis JT (2010) Recruitment timing and dynamics of transcription factors at the Hsp70 loci in living cells. Mol Cell 40(6):965–975
10. Bunch H, Calderwood SK (2015) TRIM28 as a novel transcriptional elongation factor. BMC Mol Biol 16:14
11. Bunch H (2017) RNA polymerase II pausing and transcriptional regulation of the HSP70 expression. Eur J Cell Biol 96(8):739–745
12. Bunch H (2021) HSF1 in RNA polymerase II promoter-proximal pausing and HSP70 transcription. In: Asea AAA, Kaur P (eds) Heat shock proteins in inflammatory diseases. Springer, Cham, pp 489–508
13. Hentze N, Le Breton L, Wiesner J, Kempf G, Mayer MP (2016) Molecular mechanism of thermosensory function of human heat shock transcription factor Hsf1. elife 5:e11576
14. Bunch H, Lawney BP, Lin YF, Asaithamby A, Murshid A, Wang YE, Chen BP, Calderwood SK (2015) Transcriptional elongation requires DNA break-induced signalling. Nat Commun 6:10191
15. Gilchrist DA, Nechaev S, Lee C, Ghosh SK, Collins JB, Li L, Gilmour DS, Adelman K (2008) NELF-mediated stalling of Pol II can enhance gene expression by blocking promoter-proximal nucleosome assembly. Genes Dev 22(14):1921–1933
16. Fujimoto M, Takii R, Takaki E, Katiyar A, Nakato R, Shirahige K, Nakai A (2017) The HSF1-PARP13-PARP1 complex facilitates DNA repair and promotes mammary tumorigenesis. Nat Commun 8(1):1638
17. Kim H, Heo K, Choi J, Kim K, An W (2011) Histone variant H3.3 stimulates HSP70 transcription through cooperation with HP1gamma. Nucleic Acids Res 39(19): 8329–8341
18. Kusch T, Mei A, Nguyen C (2014) Histone H3 lysine 4 trimethylation regulates cotranscriptional H2A variant exchange by Tip60 complexes to maximize gene expression. Proc Natl Acad Sci U S A 111(13):4850–4855
19. Kwon SH, Florens L, Swanson SK, Washburn MP, Abmayr SM, Workman JL (2010) Heterochromatin protein 1 (HP1) connects the FACT histone chaperone complex to the phosphorylated CTD of RNA polymerase II. Genes Dev 24(19):2133–2145
20. Lis JT, Mason P, Peng J, Price DH, Werner J (2000) P-TEFb kinase recruitment and function at heat shock loci. Genes Dev 14(7): 792–803
21. Mazina MY, Nikolenko YV, Krasnov AN, Vorobyeva NE (2016) SWI/SNF protein complexes participate in the initiation and elongation stages of drosophila hsp70 gene transcription. Genetika 52(2):164–169
22. Missra A, Gilmour DS (2010) Interactions between DSIF (DRB sensitivity inducing factor), NELF (negative elongation factor), and the Drosophila RNA polymerase II transcription elongation complex. Proc Natl Acad Sci U S A 107(25):11301–11306
23. Murawska M, Hassler M, Renkawitz-Pohl R, Ladurner A, Brehm A (2011) Stress-induced PARP activation mediates recruitment of Drosophila Mi-2 to promote heat shock gene expression. PLoS Genet 7(7):e1002206
24. Petesch SJ, Lis JT (2008) Rapid, transcription-independent loss of nucleosomes over a large

chromatin domain at Hsp70 loci. Cell 134(1): 74–84

25. Wu CH, Yamaguchi Y, Benjamin LR, Horvat-Gordon M, Washinsky J, Enerly E, Larsson J, Lambertsson A, Handa H, Gilmour D (2003) NELF and DSIF cause promoter proximal pausing on the hsp70 promoter in Drosophila. Genes Dev 17(11):1402–1414
26. Ju BG, Lunyak VV, Perissi V, Garcia-Bassets I, Rose DW, Glass CK, Rosenfeld MG (2006) A topoisomerase IIbeta-mediated dsDNA break required for regulated transcription. Science 312(5781):1798–1802
27. Galbraith MD, Allen MA, Bensard CL, Wang X, Schwinn MK, Qin B, Long HW, Daniels DL, Hahn WC, Dowell RD et al (2013) HIF1A employs CDK8-mediator to stimulate RNAPII elongation in response to hypoxia. Cell 153(6):1327–1339
28. Scheidegger A, Dunn CJ, Samarakkody A, Koney NK, Perley D, Saha RN, Nechaev S (2019) Genome-wide RNA pol II initiation and pausing in neural progenitors of the rat. BMC Genomics 20(1):477
29. Adelman K, Lis JT (2012) Promoter-proximal pausing of RNA polymerase II: emerging roles in metazoans. Nat Rev Genet 13(10):720–731
30. Jonkers I, Lis JT (2015) Getting up to speed with transcription elongation by RNA polymerase II. Nat Rev Mol Cell Biol 16(3): 167–177
31. Rahl PB, Lin CY, Seila AC, Flynn RA, McCuine S, Burge CB, Sharp PA, Young RA (2010) c-Myc regulates transcriptional pause release. Cell 141(3):432–445
32. Bunch H, Lawney BP, Burkholder A, Ma D, Zheng X, Motola S, Fargo DC, Levine SS, Wang YE, Hu G (2016) RNA polymerase II promoter-proximal pausing in mammalian long non-coding genes. Genomics 108(2): 64–77
33. Bunch H, Zheng X, Burkholder A, Dillon ST, Motola S, Birrane G, Ebmeier CC, Levine S, Fargo D, Hu G et al (2014) TRIM28 regulates RNA polymerase II promoter-proximal pausing and pause release. Nat Struct Mol Biol 21(10):876–883
34. Bunch H, Jeong J, Kang K, Jo DS, Cong ATQ, Kim D, Kim D, Cho DH, Lee YM, Chen BPC et al (2021) BRCA1-BARD1 regulates transcription through modulating topoisomerase IIbeta. Open Biol 11(10):210221
35. Bunch H (2016) Role of genome guardian proteins in transcriptional elongation. FEBS Lett 590(8):1064–1075
36. Becker PB, Rabindran SK, Wu C (1991) Heat shock-regulated transcription in vitro from a reconstituted chromatin template. Proc Natl Acad Sci U S A 88(10):4109–4113
37. Szlachta K, Thys RG, Atkin ND, Pierce LCT, Bekiranov S, Wang YH (2018) Alternative DNA secondary structure formation affects RNA polymerase II promoter-proximal pausing in human. Genome Biol 19(1):89
38. Dignam JD, Lebovitz RM, Roeder RG (1983) Accurate transcription initiation by RNA polymerase II in a soluble extract from isolated mammalian nuclei. Nucleic Acids Res 11(5): 1475–1489
39. Ebmeier CC, Erickson B, Allen BL, Allen MA, Kim H, Fong N, Jacobsen JR, Liang K, Shilatifard A, Dowell RD et al (2017) Human TFIIH kinase CDK7 regulates transcription-associated chromatin modifications. Cell Rep 20(5):1173–1186

Chapter 3

Role of Heat Shock Factors in Stress-Induced Transcription: An Update

Heyoun Bunch and Stuart K. Calderwood

Abstract

Heat shock proteins (HSP) are rapidly induced after proteotoxic stresses such as heat shock and accumulate at high concentrations in cells. HSP induction involves primarily a family of heat shock transcription factors (HSF) that bind the heat shock elements of the *HSP* genes and mediate transcription *in trans*. We discuss methods for the study of HSP binding to *HSP* promoters and the consequent increases in *HSP* gene expression in vitro and in vivo.

Key words Heat, Shock, Factor, Binding, Purification, Transcription, Heat shock protein, Nuclear run-on, Chromatin immunoprecipitation

1 Introduction

Heat shock factor (HSF) was first discovered in yeast as a sequence-specific transcription factor that binds to the promoters of heat shock protein (HSP) genes [1]. HSF was shown to bind as a trimer to three inverted repeats of the sequence nGAAn at high affinity, an activity that was later shown in *Drosophila* HSF and human HSF1 [2–5]. In more complex organisms, there are at least four members of the family in avian and mammalian species and multiple members in higher plants [6–11]. The current consensus in mammalian cells is that HSF1 is the most potent regulator of the heat shock response with the remaining factors playing supplementary roles in stress and perhaps more significant roles in development [6, 12–14]. The mechanisms by which HSF1 is triggered by stress are not entirely clear. HSF1 is thought to be constitutively repressed by the products of its transcriptional activity—HSPs through a feedback inhibition mechanism [14–16]. Activation is thus envisaged as a reversal of such inhibition as denatured proteins sequester HSPs

This work was supported by NIH research grants RO-1CA047407, R01CA119045 and RO-1CA094397.

Stuart K. Calderwood and Thomas L. Prince (eds.), *Chaperones: Methods and Protocols*, Methods in Molecular Biology, vol. 2693, https://doi.org/10.1007/978-1-0716-3342-7_3,

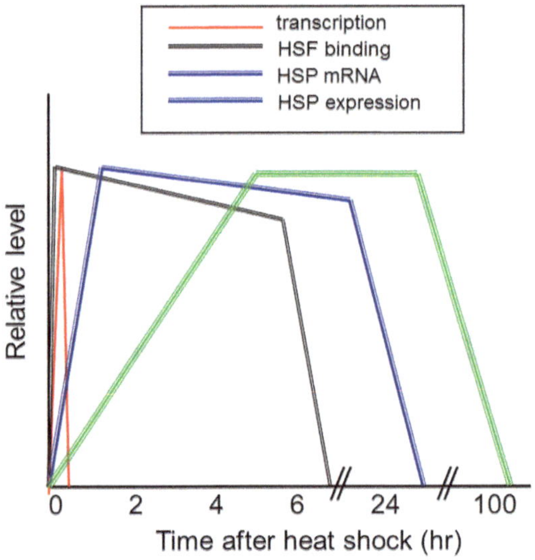

Fig. 1 Heat shock gene expression. Relative kinetics of *HSP* gene transcription, HSF binding, HSP mRNA expression, and heat shock protein expression after heat shock

during heat shock and HSF1 becomes liberated to bind to HSE elements in *HSP* genes [14]. However, alternative/overlapping hypotheses have been proposed involving stress-mediated HSF1 phosphorylation, binding to large non-coding RNA, and regulation at the level of posttranscriptional pausing [17–20]. HSF1 and HSF2 are predicted to encode at least two splicing variants, with HSF2A and HSF2b showing differential expression during erythroid differentiation [6]. The available evidence suggests that HSF2A is active in transcriptional regulation while HSF2B appears to be inactive [6].

Heat shock causes a rapid increase in HSF1 binding to *HSP* gene promoters and an acute elevation in the transcription of *HSP* genes (relative rates are indicated in Fig. 1) [17, 20]. Transcription decays rapidly after initiation, while HSF continues to bind to HSE for several hours [21, 22]. HSP mRNAs are then observed within 1 h of activation and are maintained at these high levels for up to 24 h due to enhanced stabilization after stress [23]. After acute stress, HSP protein expression in mammalian cells is delayed due to initial translational inhibition; then Hsp70, Hsp90, and Hsp110 are observed by 2–6 h after a 43 °C heat shock and can persist in most cells for up to 100 h [24] (Fig. 1). In this report, we are concerned with the early phase of HSP gene expression, involving HSF1 binding to *HSP* genes and activation of transcription.

2 Materials

Purification of HSF2 and Electrophoretic Mobility Shift Assay (EMSA)

Primers

For human HSF2:

Forward primer 5′-3′-GC[GAATCC]ATGAAGCAGAGTTCGA

Reverse primer 5′-3′-AAA[GTCGAC]TTCCTGGGGATTTA GCTA

For murine HSF2:

Forward primer 5′-3′-GG[GAATCC]ATGAAGCAGAGTTCGAA CG

Reverse primer 5′-3′-AGT[GTCGAC]TTGGGAGTTTAACTA TCT

EMSA Oligonucleotides

Hsp70 HSE top strand: 5′-CACCTCGGCTGGAATATTCCCGA CCTGGCAGCCGA-3′

Mutant Oligonucleotides

5′-CACCTCGGCTG**C**A**A**TAA**T**CCCGACCTGGCAGCCGA-3′

Cells: BL21 (DE3) *E. coli*, *Human HeLa*

Columns and Filters

20 mL Glutathione Sepharose (Pierce Chemicals)

Mono-Q HR 5/5 (Pharmacia)

Centricon 10 ultrafilter

Buffers

E. coli lysis buffer: 7 M guanidine–HCl in 0.1 M potassium phosphate buffer at pH 7.4 containing 50 mM DTT and 0.05% NP-40

Dialysis buffer: 50 mM potassium phosphate buffer containing 0.1 M KCl and 2 mM DTT

BSA solution: 0.1 mg/mL bovine serum albumin

EMSA lysis buffer: 10 mM (HEPES), 10 mM NaCl, 0.1 mM EDTA, 1.0 mM dithiothreitol (DTT), 1.0 mM phenylmethylsulfonyl fluoride (PMSF), 2.0 mg/mL aprotinin, leupeptin, 20 mM NaF, and 2.0 mM Na_3VO_4 (pH 7.9)

HSF extraction buffer: Aprotinin, leupeptin, 20 mM NaF, and 2.0 mM Na_3VO_4 (pH 7.9) on ice. Cells are then lysed by the addition of Nonidet P-40 to 0.6%, and lysates are clarified by spinning at 12,000 *g*. Nuclear pellets are solubilized in EMSA incubation buffer.

EMSA incubation buffer: (12 uL) contained 2.0 uL nuclear extract or recombinant protein, 2.0 mg/mL bovine serum albumin, 2.0 mg/mL poly dI-dC, 0.5–1.0 ng ^{32}P-labeled, double-stranded oligonucleotide probe, 12 mM Hepes, 12% glycerol, 0.12 mM EDTA, 0.9 mM $MgCl_2$, 0.6 mM DTT, 0.6 mM PMSF, and 2.0 mg/mL aprotinin and leupeptin (pH 7.9).

ChIP Assay

ChIP dilution buffer (1.1% Triton X-100, 1.2 mM EDTA, 16.7 mM Tris–HCl, pH 8.0, and 167 mM NaCl)

Protein A Agarose Slurry (Sigma Chemicals, St. Louis, Mo)

ChIP washing buffers: *washing buffer 1* (20 mM Tris–HCl, pH 8.0, 150 mM NaCl, 1% Triton X-100, 2 mM EDTA, and 0.1% SDS) *washing buffer 2* (20 mM Tris–HCl, pH 8.0, 500 mM NaCl, 1% Triton X-100, 2 mM EDTA, and 0.1% SDS) *washing buffer 3* (10 mM Tris–HCl, pH 8.0, 250 mM LiCl, 1% NP-40, 1% sodium deoxycholate, and 1 mM EDTA), TE (10 mM Tris–HCl, pH 7.5, 1 mM EDTA).

ChIP elution buffer (1% SDS, 0.1 M NaHCO3)

5 M NaCl

ChIP uncross-linking buffer: 0.5 M EDTA, 10 mL of 1 M Tris–HCl, pH 6.5, and 2 mL proteinase K.

ChIP hsp70.1 primers: Exon region forward primers: {*hsp70.1* exn forward: 5′ ggacatcagccagaacaagc 3′ *hsp70.1* exn reverse: 5′ aagtcgatgccctca aac ag 3′ *hsp70.1* HSE reverse: 5′ cggcttttataagtcgtcgt 3′ *hsp70.1* HSE forward: 5′ aggcgaaacccctggaata 3′}

Run-on Transcription

Run-on lysis buffer: 10 mM Tris–HCl (pH 7.4), 10 mM NaCl, 3 mM $MgCl_2$, and 0.5% Nonidet P-40.

Run-on storage buffer: 50 mM Tris–HCl (pH 8.30), 40% glycerol, 5 mM MgCl2, and 40 units of RNAsin (Roche Molecular Biochemicals)

Run-on reaction buffer: 10 mM Tris–HCl (pH 8.0), 5 mM $MgCl_2$, 0.3 M KCl, 5 mM DTT, 1 mM ATP, 1 mM CTP, 1 mM GTP, and 50 uCi [^{32}P] UTP (3000 Ci/mmol)

Hybridization solution: ULTRAhyb solution (*Ambion*)

Hybridization washing buffers: (1) 2× SSC, 0.1% SDS, (2) high-stringency solution (1× SSC, 0.1% SDS), (3) 2× SSC, 0.1% SDS with 10 ug RNase A

2.1 ChIP Assay Buffers and Reagents

Protease inhibitors: 1 mM DTT, 1 mM sodium metabisulfite, 1 mM benzamidine, 0.25 mM PMSF, and aprotinin [Sigma A6279, 1:1000]

ChIP cell lysis buffer: 5 mM PIPES pH 8.0, 85 mM KCl, 0.5% NP-40, and freshly added protease inhibitors

ChIP nuclei lysis buffer: 50 mM Tris–HCl pH 8.0, 10 mM EDTA, 1% SDS, and freshly added protease inhibitors

ChIP dilution buffer: 50 mM Tris–HCl pH 8.0, 150 mM NaCl, 2 mM EDTA pH 8.0, 1% NP-40, 0.5% sodium deoxycholate, 0.1% SDS, and freshly added protease inhibitors

ChIP-grade protein G magnetic beads: cell signaling

ChIP washing buffers: *washing buffer 1* (0.1% SDS, 1% Triton X-100, 2 mM EDTA, 20 mM Tris–HCl pH 8.0, 150 mM NaCl, and freshly added protease inhibitors), *washing buffer 2* (0.1% SDS, 1% Triton X-100, 2 mM EDTA, 20 mM Tris–HCl pH 8.0, 500 mM NaCl, freshly added protease inhibitors)

ChIP elution buffer: 1% SDS, 0.1 M $NaHCO_3$

RNase A

ChIP-PCR *HSPA1B* primers: Primer sets targeting the promoter, transcription start site, and gene body of *HSPA1B* have been used, and their sequence information for both humans and mice is available [17]. Representative human *HSPA1B* primers for ChIP-PCR are as follows:

Promoter (−167 to +10; forward: GCGGCACCCTGCCCTCT-GATTGGTCCAAGGAAGGC; reverse: GCCGTTTTCCGGACCGCGCCCCTCGGC)

Transcription start site (−19 to +103; forward: GCCGAGGGGCGCGCGGTCCGGAAAACG; reverse: CGGAACCGGGGAAACTCAACACGCCGGTGCC)

Gene body (+1861 to +2010; forward: GGTGTCAGCCAA-GAACGCCCTGGAGTCC; reverse: CTCCTTCCTCTTGT GCTCAAACTCGTCCTTCTC)

3 Methods

(i) *Purification of Heat Shock Factors and* In Vitro *EMSA*

In order to study the properties of HSF family members in vitro, we have prepared purified, recombinant HSF1 and HSF2.

(ii) *Complementary DNA Cloning of Human and Mouse HSF2A and HSF2B.*

RNA was isolated from NIH-3 T3 (mouse) or HeLa (human cells), and messenger RNA was prepared by poly-T affinity chromatography (PolyAtract System, Promega, Madison, WI). cDNA was then prepared from the mRNA using the AMV reverse transcriptase system (Promega), and HSF2 cDNAs were amplified using Taq polymerase and the polymerase chain reaction using the following primer sets mentioned above [6].

Forward primers contain Eco-Rl restriction site consensus sequences [marked in boxes], and the reverse primers contain Sal-1 sequences for subsequent cloning of amplified DNAs into the PGEX5 prokaryotic expression vector (Pharmacia). After transformation and growth of competent bacteria, colonies are screened for either total HSF2 using oligonucleotides (1764–1785; CAGGAGCAAGTTCACATAAATA and 1786–1807; GGCATATCACTATCCAGAGGTG) predicted to detect all forms of HSF2 or for the larger form (HSF2A) using oligonucleotides predicted to hybridize specifically with this species (1420–1440; TTGTATTATTGATGTAATCT and 1392–1412; CATCTGCACAGAACTAG TGA). Oligonucleotides are then end-labeled with ^{32}P ATP and T4 polynucleotide kinase. Plasmids detected using these probes are isolated and screened for the presence of inserts and for the production of HSF2-glutathione transferase fusion proteins from representative cDNAs in bacteria exposed to the inducing agent IPTG (Pharmacia). After induction, bacterial lysates are prepared and screened by immunoblot with anti-GST antibodies (St. Cruz Antibodies), and anti-HSF2 antibody Ab-3158 is prepared in the Calderwood lab. Representative clones from human HSF2A and HSF2B and murine HSF2A and HSF2B are then further analyzed by dideoxynucleotide sequencing.

(iii) *Purification of HSF2 Proteins*

HSF2 variants are cloned into the pGEX-5 expression vector, between the Eco RI and Sal I sites, and the resulting plasmids are used to transform BL21 (DE3) *E. coli* bacteria. HSF2 was thus expressed as a fusion protein with glutathione-S transferase. All purification steps are carried out at 4 °C. Briefly, IPTG-induced bacteria are pelleted and dissolved in *E. coli* lysis buffer and dialyzed against *E. coli* lysis buffer. The samples are centrifuged at 2500 *g* for 5 min, and the supernatant was loaded on a 20 mL volume Glutathione Sepharose column at a flow rate of 0.5 mL/min, washed extensively with dialysis buffer, and eluted with buffer A containing 10 mM reduced glutathione. The eluate was loaded onto a Mono Q HR 5/5 ion exchange chromatography column at a flow rate of 0.8 mL/min and eluted with a 24 mL linear gradient from 0.1 M to 1.0 M KCl final concentration of KCl in buffer A. Absorbance was monitored at 280 nm, and the fractions corresponding to HSF2 were assayed for binding to HSE, pooled, and concentrated with a Centricon 10 ultrafilter in the presence of 0.1 mg/mL bovine serum albumin. Relative concentrations of active HSF2 are estimated by quantitative EMSA (Fig. 1).

GST-HSF1 is purified using a similar protocol [25]. Alternatively, we have described a detailed method for purifying recombinant HSF1 after expression in *E. coli* from the pET7.1 vector [26]. Recombinant HSF1 without a GST tag is prepared by ammonium sulfate precipitation, heparin–agarose affinity, and ion exchange chromatography in pure form as assessed by SDS-PAGE and reverse-phase HPLC [26].

The activity of purified GST-HSF2A, GST-HSF2B, or GST-HSF1 is estimated by EMSA. Proteins are incubated with ^{32}P-labeled HSE at a range of dilutions and then subjected to EMSA analysis as described below. GST-HSF2A and GST-HSF2B are serially diluted 1/2200, 1/660, 1/220, 1/66, 1/22, and 3/22 prior to EMSA.

(iv) *Nuclear Extraction from Tissue Culture Cells and EMSA*

EMSA is carried out using purified recombinant HSF or after the extraction of intracellular HSF complexes from either whole cell or nuclear extracts from heat-shocked cells and incubation of complexes with double-stranded oligonucleotides encoding heat shock elements in HSP genes HSE) (*see* **Notes 1–4**). To prepare HSF from cells growing in vitro, nuclear extracts are prepared according to Schreiber [27]. In our standard assay, cells are incubated for 15 min in 200–800 uL of EMSA lysis buffer on ice. Cells are then lysed by the addition of Nonidet P-40 to 0.6%, and lysates are clarified by spinning at 12,000 *g*. Nuclear pellets are then resuspended in a 25 uL ice-cold EMSA extraction buffer. Extracts containing HSF are then aliquoted and stored at −80 °C.

For incubation with an oligonucleotide probe, each binding mixture (12 uL) contained 2.0 uL nuclear extract or recombinant protein and 2.0 mg/mL bovine serum albumin in EMSA incubation buffer. Samples are incubated at room temperature for 15 min and then fractionated by electrophoresis on 4.0% polyacrylamide, 1× TBE gels. Oligonucleotide hHSE was synthesized, annealed, and labeled by end filling with ^{32}P-dCTP at 6000 Ci/mmol (DuPont, NEN) to an activity of 100,000 cpm/ng. hHSE contains the heat shock element (HSE) from the top strand of the human *HSP70.1* promoter [28]. (We have found that double-stranded oligonucleotide end filling with Klenow fragment or end labeling of single-stranded oligonucleotides with T4 kinase to be equally effective). The oligonucleotide shown in *Materials* (575 ng) and the complementary oligonucleotide (2300 ng) (resulting in 1150 ng double-stranded oligo) are made up to 25 ng/uL in 46 uL of TE buffer, annealed by incubation at 100 °C for 5 min, and cooled overnight. As a control, we carried out the EMSA procedure with a similar oligonucleotide containing mutations in the HSE elements, indicated in bold in the sequence shown in *Materials*. For experiments on cell extracts, a number of controls are used standardly. To determine the specific binding of HSF1 to labeled HSE, we examine the ability to inhibit HSE–HSF association with a tenfold excess of unlabeled wild-type oligonucleotide included in the incubation. In addition, specific binding is further indicated by the failure of a tenfold excess of the mutant HSE shown above to inhibit binding. The protein (HSF) in the HSF–HSE complex can be identified by the addition of specific anti-HSF1 or anti-HSF2 antibodies to the reaction mix. We used a 1: 100 dilution of anti-HSF1 antibody 68-3 prepared in our laboratory to positively identify HSF1 in the complexes [29, 30]. As a control, we use pre-immune antiserum obtained from the same rabbit. For commercially obtained antibodies, a serial dilution approach was used to determine optimal antibody concentrations.

HSF1 from heat-shocked cells is contained in large complexes of at least 600 kD and is fractionated on 4% Tris–borate non-denaturing gels [31, 32]. For most purposes, we found that the mini-gel (Bio-Rad, CA) format was quite adequate for separation, although for supershift assay and higher resolution, a larger format was used [33].

(v) *Measuring the Contribution of HSF1-HSE Binding to Transcription*

(a) *Luciferase Reporter Assays for HSF Activity*

To construct an intracellular reporter of HSF activity (pGL.hsp70B), we used 1.44 kB of the human *HSP70B* gene inserted into the pGL. Basic plasmid (*Promega*). The *HSP70B* gene is almost entirely silent at physiological temperatures but powerfully activated by heat shock [34]. pGL.hsp70B was constructed by digestion with Bgl*II* and Hind*III* and cloning into pGL.Basic. We have also used the human Hsp27 gene by a similar process, inserting the 730 kB Bgl*II* and Hind*III* digest of an HSP27 promoter fragment into pGL.Basic. For overexpression of HSF1, human HSF1 cDNA [35] is inserted into the pcDNA3.1 (-) expression vector (*Invitrogen*) at the *Xho*I and *Eco*RI sites [36]. Human HSF2A was inserted into the pcDNA3.1(+) vector at the *Xho*I and *Eco*RI sites to produce pHSF2A [37].

To assay the HSF1 transcriptional activity in HeLa cells, the cells are maintained in HAM's F-12 (Mediatech) with 10% heat-inactivated fetal bovine serum (FBS). HeLa cells (2.5×10^5 cells/well) in 6-well plates are transfected with the pGL.hsp70B or pGL.hsp27 plasmids [38]. pCMV-*lacZ* plasmid is co-transfected as an internal control for transfection efficiency. pHM6 empty vector is used as a blank plasmid to balance the amount of DNA transfected in transient transfection. Luciferase and beta-galactosidase activity assays are performed after 24 h of transfection according to the Promega protocol. Luciferase activity is normalized to β-galactosidase activity. The results are expressed as the relative luciferase (relative light units) activity of the appropriate control.

(b) *Nuclear Run-on Assay of the Rate of Hsp70 Gene Transcription*

To determine the HSP70 gene transcriptional rate, cells are treated according to the experiment and then quenched in ice-cold phosphate-buffered saline, pH 7.4 (PBS) on ice. Cells are next washed in PBS and lysed in run-on lysis buffer nuclei are collected by centrifuge (500× *g*, 5 min) at 4 °C and resuspended in storage buffer.

To assay the rate of transcription, 100 uL of nuclei and 100 uL of run-on reaction buffer are added, and samples are incubated for 30 min at 30 °C with shaking. RNA is then extracted from the reaction mix using TRIzol (*Invitrogen*) according to the manufacturer's protocol.

The hsp70 DNA-containing cDNA probe [39] or control beta-actin probe is linearized and purified by phenol/chloroform extraction and ethanol precipitation. Probes are then denatured and slot-blotted onto the Hybond N+ membrane. (Membranes are first pre-hybridized with ULTRAhyb solution (*Ambion*) for 2 h at 42 °C, before equivalent counts of newly transcribed RNA (10^6 cpm) are added to the solution). Hybridization is then carried out for 24 h at 42 °C. Membranes are then washed twice for 20 min at 42 °C in low-stringency solution (2× SSC, 0.1% SDS), twice for 20 min in high-stringency solution (1× SSC, 0.1% SDS), and once for 30 min at 37 °C in a low-stringency solution containing 10 ug RNase A. Membranes are then rinsed in low-stringency solution and analyzed by incubation with X-ray film. We have successfully used this protocol for the assay of transcription of the mouse *hsp70.1*, c-*fms*, IL-1beta, and TNF-alpha genes [40].

(c) *Chromatin Immunoprecipitation-PCR (or -qPCR) to Measure HSF1 and RNA Polymerase II Occupancies on HSP Genes*

To understand transcriptional regulation and activation at *HSP* genes, it is important to monitor transcription factors including HSF1, histone markers, RNA polymerase II (Pol II), and more. Occupancy changes of certain protein factors and modifications indicate and correlate with transcriptional status, repression, or activation. For example, HSF1 recruitment on the promoter of *HSP* genes indicates gene activation, while a reduction of Pol II or phosphorylated serine 2 in the Pol II CTD in the gene body of HSP genes reliably suggests gene repression. ChIP is a powerful tool to probe and monitor the genetic occupancy of specific protein factors in the cell. The specificity of ChIP attributes to the utilization of antibodies targeting their antigens. The basic principle of ChIP includes the following: (1) cross-linking of DNA and DNA-associated proteins, (2) DNA fragmentation, (3) capturing antigens (target protein factors) using their antibodies, (4) pull-down of the antibody–antigen–DNA complex, and (5) reverse cross-linking and DNA purification. The purified DNA of interest can be quantified by PCR followed by gel electrophoresis or quantitative PCR (ChIP-qPCR) or the entire DNA can be both identified and quantified by sequencing (ChIP-seq). Using ChIP-seq, the genome-wide occupancy of a specific transcription factor can be monitored, which informs the genes that are likely to be regulated by the factor.

We have developed a ChIP method with mild modifications that has been tested in studying stress-inducible genes including *HSP70* and serum-activated genes [17, 41, 42]. In addition, this method has been utilized for various human and murine cell lines in combination with PCR, qPCR, and sequencing [17, 20, 43]. To perform the ChIP assay, the cells are grown to 80–90% confluence, approximately 10^7 adherent cells per each sample. Control and heat-shocked cells are prepared in the same manners, side by side, except for heat treatment. We typically heat shock the cells at 40–42 °C for 0.5–15 min immediately before cross-linking by directly supplementing 200 μL 37% formaldehyde to 10 mL cell culture and gently rocking the mixture for 10 min at room temperature (RT). Cross-linking is stopped by supplementing 500 μL of 2.5 M glycine and gently agitating the mixture for 5 min at RT. Then the cells are rinsed with cold PBS twice before scraping and are pelleted by centrifugation. The cell pellet is resuspended with 400 μL ChIP cell lysis buffer. At this stage, the mixture can be snap-frozen and stored at −70 °C for a while. The cells are again pelleted down and resuspended in 600 μL ChIP nuclei lysis buffer. Chromatin fragmentation, which is a crucial step for successful ChIP analysis, can be achieved enzymatically or by physical shearing using sonication. We have been using the latter method. The ideal sizes of fragmented chromatins are 200–1000 bp, and the frequency and intensity of sonication should be empirically determined (Fig. 2). After sonication, a brief centrifugation at 4 °C is recommended to remove any debris (14,000 rpm, 1 min) and 50 μL of fragmented chromatin solution is spared and stored at −70 °C to be used as the input DNA in qPCR and sequencing. For a ChIP reaction, 100 μL fragmented chromatin solution is diluted with 900 μL ChIP dilution buffer. An antibody to probe a desired protein of interest is added to the mixture (1–5 μg per a reaction). For ChIP-seq, antibodies validated for the method by manufacturers are recommended. An IgG control is also recommended as a negative control and experimental noise (used to calculate fold enrichment in ChIP-qPCR). Protein G magnetic beads are supplemented, typically 30–50 μL, according to the manufacturer's instruction. The mixture of fragmented chromatins, an antibody, and protein G beads is incubated on a rotator at 4 °C for 3 h overnight. Using a magnetic stand, the beads–proteins–DNA

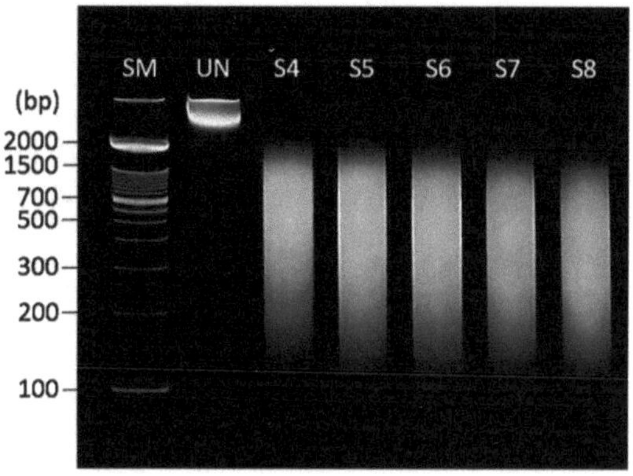

Fig. 2 An example of sonication condition setup. Approximately 10^7 HEK293 cells were sonicated using EpiShear Probe Sonicator (output 12%, active motif). After each sonication cycle (4–8 sonications), 25 μL of the sonicated sample was taken to be reverse-cross-linked, and the DNA was purified. The purified DNA was separated on an 8% DNA polyacrylamide gel. UN, intact chromatins before sonication; S4–S8, sonicated 4 times to 8 times; SM, size marker

complex is washed three times with ChIP wash buffer 1 (1 mL per each wash) and then once with ChIP wash buffer 2 (1 mL per each wash). Please note that the salt concentration in the ChIP wash buffer can be empirically adjusted. The pellet is resuspended with 120 μL ChIP elution buffer including 1.5 μg RNase A. At the same time, the chromatin input is thawed and resuspended with 80 μL ChIP elution buffer including 1 μg RNase A. These mixtures are incubated at RT for 15 min and then at 65 °C overnight for reverse cross-linking. The DNA is purified using a PCR purification kit (Qiagen) and stored at −70 °C and is used as the template in PCR, qPCR, or sequencing for qualitative and quantitative analyses.

3.1 ChIP Assay Buffers and Reagents

Protease inhibitors (1 mM DTT, 1 mM sodium metabisulfite, 1 mM benzamidine, 0.25 mM PMSF, and aprotinin [Sigma A6279, 1:1000])

ChIP cell lysis buffer (5 mM PIPES pH 8.0, 85 mM KCl, 0.5% NP-40, and freshly added protease inhibitors)

ChIP nuclei lysis buffer (50 mM Tris–HCl pH 8.0, 10 mM EDTA, 1% SDS, and freshly added protease inhibitors)

ChIP dilution buffer (50 mM Tris–HCl pH 8.0, 150 mM NaCl, 2 mM EDTA pH 8.0, 1% NP-40, 0.5% sodium deoxycholate, 0.1% SDS, and freshly added protease inhibitors)

ChIP-grade protein G magnetic beads (cell signaling)

ChIP washing buffers: *washing buffer 1* (0.1% SDS, 1% Triton X-100, 2 mM EDTA, 20 mM Tris–HCl pH 8.0, 150 mM NaCl, and freshly added protease inhibitors) *washing buffer 2* (0.1% SDS, 1% Triton X-100, 2 mM EDTA, 20 mM Tris–HCl pH 8.0, 500 mM NaCl, freshly added protease inhibitors)

ChIP elution buffer (1% SDS, 0.1 M $NaHCO_3$)

RNase A

ChIP-PCR *HSPA1B* primers: Primer sets targeting the promoter, transcription start site, and gene body of *HSPA1B* have been used, and their sequence information for both humans and mice is available [17]. Representative human *HSPA1B* primers for ChIP-PCR are as follows:

Promoter (−167 to +10; forward: GCGGCACCCTGCCCTC TGATTGGTCCAAGGAAGGC; reverse: GCCGTTTTCC GGACCGCGCGCCCCTCGGC)

Transcription start site (−19 to +103; forward: GCCGAGGGGCGCGCGGTCCGGAAAACG; reverse: CGGAACCGGGGAAACTCAACACGCCGGTGCC)

Gene body (+1861 to +2010; forward: GGTGTCAGCCAA-GAACGCCCTGGAGTCC; reverse: CTCCTTCCTC TTGTGCTCAAACTCGTCCTTCTC)

4 Notes

1. *The EMSA technique* has the advantages that it is rapid, sensitive, and straightforward to carry out. For assessing the significance of the transcription factor– response element interaction, it is, however, lacking in that response elements in chromatin are wound along nucleosomes and may not be available for binding. The EMSA reaction is carried out using naked DNA. In addition, as ChIP on CHIP and ChIP-seq studies begin to accumulate, it is evident that response elements for particular factors are more flexible than suspected from early studies.
2. Some of these problems can be avoided using the *ChIP assay* that measures HSF binding to chromosomal DNA in vivo. This technique is highly dependent on the availability of high-affinity and specific antibodies for transcription factors. This can be overcome by overexpression of the factor with a sequence tag and carrying out ChIP with an anti-TAG antibody. However, this can introduce potential artifacts involved with protein overexpression.
3. To assess the results of HSF–DNA binding, we have used two approaches. We have used transfection of *reporter constructs* containing either HSP promoters or HSE coupled to reporter genes CAT or luciferase. The assays have the advantages of being rapid and permitting the accumulation of plentiful data. The promoter portion of the construct can be tailored to assess the activity of a single transcription factor such as HSF1. The assay is however indirect and does not measure the

transcription of the native, chromosomally embedded gene. There are other potential complications, as reporters require to be translated and yield enzymatically active proteins.

4. Transcriptional rate of HSP genes can be assessed directly by run-on assay. This assay indicates the joint activities of all response elements in the *HSP* gene promoters. Although genes such as *HSP70B* respond only to HSF1 or heat shock, others such as *HSP70A* have more complex promoters.

References

1. Sorger PK, Pelham HRB (1987) Purification and characterization of a heat-shock element binding protein from yeast. EMBO J 6:3035–3041
2. Sorger PK, Nelson HCM (1989) Trimerization of a yeast transcriptional activator via a coiled-coil motif. Cell 59:807–813
3. Sorger PK, Pelham HRB (1988) Yeast heat shock factor is an essential DNA-binding protein that exhibits temperature-dependent phosphorylation. Cell 54:855–864
4. Rabindran SK et al (1993) Regulation of heat shock factor trimer formation: role of a conserved leucine zipper. Science 259:230–234
5. Wu C (1995) Heat shock transcription factors: structure and regulation. Annu Rev Cell Dev Biol 11:441–469
6. He H et al (2003) Elevated expression of heat shock factor (HSF) 2A stimulates HSF1-induced transcription during stress. J Biol Chem 278(37):35465–35475
7. Fujimoto M et al (2009) A novel mouse HSF3 has the potential to activate non-classical heat shock genes during heat shock. Mol Biol Cell 21(1):106–116
8. Tanabe M et al (1998) Disruption of the HSF3 gene results in the severe reduction of heat shock gene expression and loss of thermotolerance. EMBO J 17(6):1750–1758
9. Tanabe M et al (1999) The mammalian HSF4 gene generates both an activator and a repressor of heat shock genes by alternative splicing. J Biol Chem 274(39):27845–27856
10. Kumar M et al (2009) Heat shock factors HsfB1 and HsfB2b are involved in the regulation of Pdf1.2 expression and pathogen resistance in arabidopsis. Mol Plant 2(1):152–165
11. Scharf KD et al (1990) Three tomato genes code for heat stress transcription factors with a region of remarkable homology to the DNA-binding domain of the yeast HSF. EMBO J 9(13):4495–4501
12. McMillan DR et al (1998) Targeted disruption of heat shock transcription factor 1 abolishes thermotolerance and protection against heat-inducible apoptosis. J Biol Chem 273:7523–7528
13. Morange M (2006) HSFs in development. Handb Exp Pharmacol 172:153–169
14. Prince TL et al (2020) HSF1: primary factor in molecular chaperone expression and a major contributor to cancer morbidity. Cell 9(4): 1046
15. Abravaya K et al (1992) The human heat shock protein hsp70 interacts with HSF, the transcription factor that regulates heat shock protein expression. Genes Dev 6:1153–1164
16. Zou J et al (1998) Repression of heat shock transcription factor HSF1 activation by HSP90 (HSP90 complex) that forms a stress-sensitive complex with HSF1. Cell 94(4):471–480
17. Bunch H et al (2014) TRIM28 regulates RNA polymerase II promoter-proximal pausing and pause release. Nat Struct Mol Biol 21(10): 876–883
18. Guettouche T et al (2005) Analysis of phosphorylation of human heat shock factor 1 in cells experiencing a stress. BMC Biochem 6(1): 4
19. Shamovsky I et al (2006) RNA-mediated response to heat shock in mammalian cells. Nature 440(7083):556–560
20. Bunch H et al (2015) Transcriptional elongation requires DNA break-induced signalling. Nat Commun 6:10191
21. Mosser DD et al (1997) Role of the human heat shock protein hsp70 in protection against stress-induced apoptosis. Mol Cell Biol 17(9): 5317–5327
22. Price BD, Calderwood SK (1992) Heat-induced transcription from RNA polymerases II and III and HSF binding are co-ordinately regulated by the products of the heat shock genes. J Cell Physiol 153:392–401

23. Zhao M et al (2002) Double-stranded RNA--dependent protein kinase (pkr) is essential for thermotolerance, accumulation of HSP70, and stabilization of ARE-containing HSP70 mRNA during stress. J Biol Chem 277(46): 44539–44547
24. Subjeck JR, Sciandra JJ, Johnson RJ (1982) Heat shock proteins and thermotolerance; a comparison of induction kinetics. Br J Radiol 55(656):579–584
25. Wang X et al (2006) Phosphorylation of HSF1 by MAPK-activated protein kinase 2 on serine 121, inhibits transcriptional activity and promotes HSP90 binding. J Biol Chem 281(2): 782–791
26. Soncin F, Prevelige R, Calderwood SK (1997) Expression and purification of human heat-shock transcription factor 1. Protein Expr Purif 9(1):27–32
27. Schreiber E et al (1989) Rapid detection of octamer binding proteins with "mini-extracts" prepared from a small number of cells. Nucleic Acids Res 17:6419
28. Wu B, Hunt C, Morimoto RI (1985) Structure and expression of the human gene encoding the major heat shock protein HSP70. Mol Cell Biol 5:330–341
29. Bruce JL et al (1999) Activation of heat shock transcription factor 1 to a DNA binding form during the G(1)phase of the cell cycle. Cell Stress Chaperones 4(1):36–45
30. Cahill CM et al (1996) Transcriptional repression of the prointerleukin 1beta gene by heat shock factor 1. J Biol Chem 271(40): 24874–24879
31. Nunes SL, Calderwood SK (1995) Heat shock factor-1 and the heat shock cognate 70 protein associate in high molecular weight complexes in the cytoplasm of NIH-3T3 cells. Biochem Biophys Res Commun 213(1):1–6
32. Westwood T, Wu C (1993) Activation of drosophila heat shock factor: conformational changes associated with monomer-to-trimer transition. Mol Cell Biol 13:3481–3486
33. Xie Y et al (2003) Heat shock factor 1 contains two functional domains that mediate transcriptional repression of the c-fos and c-fms genes. J Biol Chem 278(7):4687–4698
34. Tang D et al (2005) Expression of heat shock proteins and heat shock protein messenger ribonucleic acid in human prostate carcinoma in vitro and in tumors in vivo. Cell Stress Chaperones 10(1):46–58
35. Rabindran SK et al (1991) Molecular cloning and expression of a human heat shock factor, HSF1. Proc Natl Acad Sci U S A 88:6906–6910
36. Oesterreich S et al (1996) Basal regulatory promoter elements in the hsp27 gene in human breast carcinoma cells. Biochem Biophys Res Commun 222:155–163
37. Chen C et al (1997) Heat shock factor 1 represses Ras-induced transcriptional activation of the c-fos gene. J Biol Chem 272(43): 26803–26806
38. Wang XZ, Asea A, Xie Y, Kabingu E, Stevenson MA, Calderwood SK (2000) RSK2 represses HSF1 activation during heat shock. Cell Stress Chaperones 5:432–437
39. Hunt C, Calderwood SK (1990) Characterization and sequence of a mouse HSP70 gene and its expression in mouse cell lines. Gene 87: 199–204
40. Xie Y et al (2002) Heat shock factor 1 represses transcription of the IL-1beta gene through physical interaction with the nuclear factor of interleukin 6. J Biol Chem 277(14): 11802–11810
41. Bunch H et al (2021) BRCA1-BARD1 regulates transcription through modulating topoisomerase IIbeta. Open Biol 11(10):210221
42. Bunch H et al (2019) P-TEFb regulates transcriptional activation in non-coding RNA genes. Front Genet 10:342
43. Jeong J et al (2021) Tetraarsenic oxide affects non-coding RNA transcriptome through deregulating polycomb complexes in MCF7 cells. Adv Biol Regul 80:100809

Chapter 4

A Workflow Guide to RNA-Seq Analysis of Chaperone Function and Beyond

Kristina M. Holton, Richard M. Giadone, Benjamin J. Lang, and Stuart K. Calderwood

Abstract

RNA sequencing (RNA-seq) is a powerful method of transcriptional analysis that allows for the sequence identification and quantification of cellular transcripts. RNA-seq can be used for differential gene expression (DGE) analysis, gene fusion detection, allele-specific expression, isoform and splice variant quantification, and identification of novel genes. These applications can be used for downstream systems biology analyses such as gene ontology or pathway analysis to provide insight into processes altered between biological conditions. Given the wide range of signaling pathways subject to chaperone activity as well as numerous chaperone functions in RNA metabolism, RNA-seq may provide a valuable tool for the study of chaperone proteins in biology and disease. This chapter outlines an example RNA-seq workflow to determine differentially expressed (DE) genes between two or more sample conditions and provides some considerations for RNA-seq experimental design.

Key words Chaperones, RNA-seq, Differential gene expression analysis, Overrepresentation analysis, Gene set enrichment analysis

1 Introduction

"RNA-seq" is a versatile method of transcriptional analysis that utilizes next-generation sequencing of RNA-derived cDNA libraries for various transcript analyses. In the past 15 years, the use of RNA-seq has expanded dramatically (Fig. 1). In this time, next-generation sequencing has become more financially accessible, commercial kits for library construction more readily available, and methods for downstream analysis have evolved and are continuing to do so. In addition to these factors, the increasing utilization of RNA-seq has been driven by the advantages it provides over previous transcriptomic methods such as tiling microarrays and cDNA Sanger sequencing, reviewed in [1, 2]. RNA-seq has a wide quantitative range and high sequencing accuracy, is high-throughput, is

Stuart K. Calderwood and Thomas L. Prince (eds.), *Chaperones: Methods and Protocols*, Methods in Molecular Biology, vol. 2693, https://doi.org/10.1007/978-1-0716-3342-7_4,

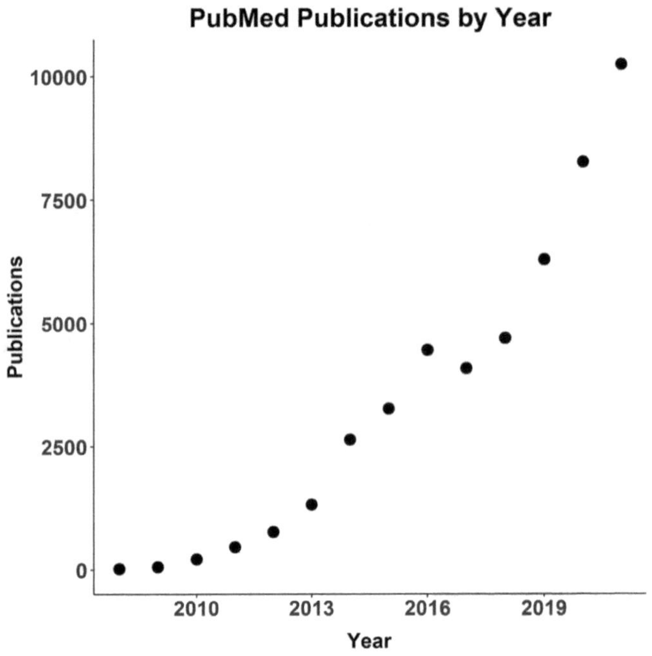

Fig. 1 Usage of RNA-seq has expanded in recent years. A PubMed search using the term "RNA-seq" returned the displayed number of studies by year. Since 2008 the number of studies utilizing RNA-seq has rapidly expanded

adaptable to large genomes and detection of lowly expressed genes, has low background noise, and is not completely dependent upon prior genome annotation. Unlike RT-qPCR, RNA-seq quantifies transcript expression across the entire gene body simultaneously and therefore has a greater capacity to distinguish between splice variants and isoforms. RNA-seq data typically return hundreds or thousands of changes in gene expression and can be analytically demanding to discern what is of biological significance. This can be aided by complementary functional assays to either validate findings within RNA-seq data or to provide some focus to questions to which RNA-seq is applied.

Chaperones represent a diverse class of proteins that serve to ensure the proper folding of proteins to limit their misfolding, aggregation, and resulting aberrant proteotoxicity leading to many diseases [3]. Chaperone proteins have been implicated in a wide range of biological processes including development, aging, and disease across diverse species [4, 5]. Through biophysical interactions with client proteins and co-chaperones, chaperones are critical components of complex protein homeostasis (proteostasis) networks [6–8]. Beyond their proteostatic functions, several studies have demonstrated chaperones to have various regulatory functions in transcription and mRNA metabolism. These have included roles for HSP90 members in the assembly of the RNA polymerase II

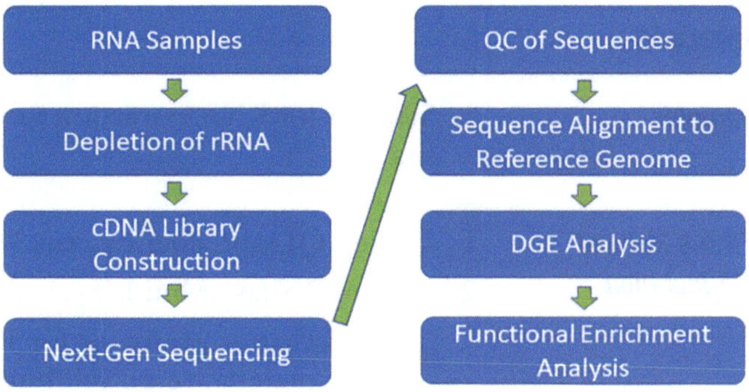

Fig. 2 A generalized workflow for an RNA-seq protocol examining differential gene expression (DGE) and the topics of this chapter

complex, facilitation of transcriptional pausing, mRNA splicing, and catalytic activity of the RNA-induced silencing complex (RISC) [9–12]. In addition, HSP70 family members (e.g., HSPA1A) bind to and mediate mRNA stability and decay [13–18], of note, these properties can be distinct between HSP paralogs and regulated by co-chaperones (e.g., BAG3 for HSP70 family members) [19, 20]. These studies indicate a broad potential for chaperones to have a significant impact on transcriptomic profiles; however, few have yet to assess chaperone functions at the transcriptomic level. Given its abovementioned advantages, RNA-seq is likely to be a central method used in future studies of RNA chaperone functional relationships [21].

The RNA-seq workflow begins with the experimental conditions from which RNA samples are collected, followed by total RNA extraction, construction of cDNA libraries, sequencing of cDNA libraries, and subsequent bioinformatic analyses (Fig. 2). These general steps are somewhat universal for studies using RNA-seq; however, the parameters and method within each step are typically selected for the chosen application. This is reflected by the large variation in published RNA-seq protocols [22]. Bioinformatic analysis of RNA-seq data typically involves performing sequence quality control, alignment of sequences to a reference genome, and generation of gene feature count data, followed by various assessments of gene transcription including DGE analysis, refinement and/or identification of gene models, splice variant analysis, isoform expression quantification, gene fusion detection, and allele-specific expression. This chapter is intended to provide a workflow protocol for differential gene expression analysis of coding genes. Investigators may adapt the protocol to their chosen application and we provide some guidance for making such changes.

2 Materials

2.1 Samples

1. Samples for RNA extraction from cell culture or animal-derived tissue

2.2 Tissue Harvest and RNA Isolation

1. Qiagen QIAshredder cat. no. 79654.
2. Qiagen RNeasy Mini Kit cat. no. 74106:
   (a) Alternative protocols for RNA isolation for smaller starting material (e.g., cell number or tissue) are available via commercial kits (e.g., RNeasy Micro Kit). Additionally, protocols may be optimized to isolate and purify different RNA species in addition to mRNA (e.g., miRNA, piRNA, etc.). If RNA quantity from tissue or cell lysates is limited, consider using extraction methods optimized for low RNA abundance (e.g., Qiagen RNeasy Micro Kit cat. no. 74004).
3. β-Mercaptoethanol.
4. RNAse-free microcentrifuge tubes.
5. RNase-away Molecular BioProducts cat. no. 7002.
6. DNase I Qiagen cat. no. 79254.
7. Cell culture plates.
8. 70% ethanol.

2.3 Assessment of the Concentration, Purity, and Integrity of the RNA Sample

1. Agilent Bioanalyzer or TapeStation.
2. Agilent RNA 6000 nano microfluidic chips or ScreenTape.
3. Reagents and consumables:
   (a) These instruments and services can commonly be accessed at core facilities.

2.4 cDNA Library Construction (Service Often Available at Next-Generation Sequencing Facility)

1. Solid-phase reversible immobilization (SPRI) nucleotide-binding paramagnetic beads, e.g., Agencourt®, AMPure® XP (Beckman Coulter cat no. A63880), or KAPA Pure Beads (KAPA Biosystems)
2. Illumina® and TruSeq® "forked" adapters
3. Thermocycler
4. Magnetic rack
5. Low-adhesion, RNAse-free, DNAse-free, 1.5 ml microcentrifuge tubes
6. Low-adhesion, RNAse-free, DNAse-free, PCR tubes sized for thermocycler to be used

7. Agilent Bioanalyzer/TapeStation
8. Filtered pipette tips
9. KAPA stranded RNA-Seq with RiboErase cat. no. KK8483

2.5 Next-Generation Sequencing

1. Illumina NextSeq 500, 1000, 2000, or NovaSeq 6000 sequencer

2.6 Sequence Processing and Analysis

1. High-performance computer cluster with UNIX command line for QC and alignment or powerful computer with UNIX command line for QC and pseudoalignment
2. R 4.X.X, Bioconductor

3 Methods

3.1 Experimental Design

1. Determine the number of samples for each condition needed for analysis. *See* Subheading 4.1 for further guidance on choosing the number of biological replicates.
2. Determine the sequencing platform to be used. For this protocol, an Illumina NextSeq 500 sequencer platform is used.
3. Decide how the cDNA libraries will be constructed (stranded or unstranded). This protocol uses the KAPA stranded RNA-Seq with RiboErase kit.
4. Determine the minimum read depth needed and budget. *See* Subheading 4.1.

3.2 RNA Collection from Cell Culture

1. Generate experimental samples ready for RNA collection.
2. Clean the working area and pipettes with ethanol and RNase away.
3. Remove media from culture dishes as much as possible:
   (a) Note that the maximum number of animal cells recommended for the use of the RNeasy Plus Mini Kit protocol is 1×10^7.
4. Following the RNeasy Plus Mini Handbook, use RLT buffer with β-mercaptoethanol to lyse the cells, and collect cell lysate in a new RNase-free microcentrifuge tube.
5. Homogenize the sample by passing it through a QIAshredder column.
   (b) In the event that cell lysates are used, QIAshredder columns are not needed.

6. Continue protocol with the inclusion of the optional DNase I digestion step.
7. Continue to **step 3.3** of this protocol, RNA integrity analysis, or store at −80 °C.
   (c) RNeasy Mini Kit buffers RLT and RW1 should not be mixed with bleach.

3.3 RNA Extraction from Mouse Tissue

1. Sacrifice mice using an approved protocol.
2. Excise the tissue of interest from the mouse.
3. Mince up to 30 mg of tissue into small pieces (1–3 mm) with scissors or an electric mechanical homogenizer, and transfer into a round bottom tube with an appropriate volume of cell lysis buffer RLT with β-mercaptoethanol per RNeasy Plus Mini Handbook.
4. Continue protocol with the inclusion of the optional DNase I digestion step.
5. Continue to **step 3.4** of this protocol, RNA integrity analysis, or store at −80 °C:
   (a) RNeasy Mini Kit buffers RLT and RW1 should not be mixed with bleach.

3.4 Assess the Concentration and Integrity of the RNA Sample

1. Quantify the concentration of RNA by absorbance spectrophotometry.
2. Assess RNA integrity using an Agilent Bioanalyzer. Submit 2 to 4 μl of RNA sample to the core facility for analysis, or if an instrument is available in-house, perform analysis per manufacturer's directions.

3.5 Stranded cDNA Library Construction Using KAPA Stranded RNA-Seq Kit with RiboErase for Illumina® Platforms

1. rRNA depletion to enrich coding genes.
2. Fragmentation of RNA.
3. Reverse transcription of fragmented RNA to produce first-strand cDNA.
4. Second strand synthesis, second strand marked with dUTP.
5. Ligation of dAMP to 3′ ends of dscDNA strands.
6. Ligation of indexed adaptors to each sample dscDNA.
7. PCR amplification of cDNA libraries:
   (a) Note that numerous SPRI purifications are involved throughout the above steps and are dependent upon the protocol used. When using Agencourt®, AMPure ® XP, or KAPA Pure Beads®, the RNA/DNA-binding steps must be performed at room temperature for the RNA/DNA to associate with the beads. RNA/DNA/Bead solutions that are not equilibrated to room temperature will result in loss of sample in washing steps.

3.6 Next-Generation Sequencing

1. Perform library QC and quantification of cDNA libraries.
2. Pool the cDNA libraries at equal proportions to create a master pool mix.
3. Run the cDNA library pool on an Illumina NextSeq sequencer flow cell or platform of choice.

3.7 Sequence Processing and Analysis

1. Export FASTQ files to the remote directory (these are the raw sequence files containing the sequences or "reads" for each indexed sample).
2. Quality control of reads to remove low-quality reads. Assess to remove or "trim" the indexed adaptors from the reads, and identify any batch effects or bias in read distribution.
3. Alignment to the reference genome using alignment software (e.g., HISAT2 [23], STAR [24], Kallisto [25], Salmon [26], Sailfish [27]).
4. Generate a count matrix using the count [28] or rsubread [29].
5. Differential expression analysis using edgeR [30] or DESeq2 [31].
6. Functional enrichment analysis using clusterProfiler [32].
7. Representation of data.

4 Notes

4.1 Experimental Design

Primary RNA-seq experimental design decisions include the number of biological replicates for each condition, whether the cDNA libraries will be stranded or unstranded, whether the sequencing run will be paired-end (PE) or single-end (SE), read depth, read length, and cost. Choosing the most cost and statistically effective combination of these parameters can be challenging as there is no current standard; however, recent studies have provided direction on how each of these parameters should be prioritized according to the experimental application [33–36].

One starting point is determining how many biological replicates the experiment will entail. Biological replicates are essential for encapsulating the biological variation within replicates and between conditions in the statistical analysis [36, 37]. For DGE analysis, the investigator may perform a power analysis to predict the number of replicates needed based on desired power, biological coefficient of variation (BCV) (the coefficient with which the abundance of a gene varies between replicates) [38], and effect size (like the relative expression of change to detect) [39]. This can be a challenge if the BCV for the model is not known. If a costly experiment is planned, it may pay to perform a pilot study to approximate the BCV for power analysis. A typical BCV for animal tissue-derived samples is 0.2–0.6 and therefore may require more replicates to achieve a given power and effect size than samples derived from cell lines (BCV closer to 0.1). The R package RNA-SeqPower allows for exploration of the number of biological replicates given sequencing depth, BCV, effect size, and the selected alpha and power [39]. NGS was shown to be highly reproducible, therefore prioritizing biological replication over technical replication in benefiting DGE analysis [40].

Increasing the number of sequence reads attributed to each biological sample, or "read depth," increases DGE power to a point of saturation [41]. A limiting factor for this parameter is commonly the cost of additional sequencing lanes and flow cells, and it is therefore balanced across an optimal number of biological replicates, the relationship of which was examined by Liu et al. [33]. For example, the Illumina NextSeq 500 platform yields approximately 220 M reads per sequencing lane, and if the cDNA library from one biological sample is applied to one flow cell lane, the read depth would be approximately 220 M reads. As a general principle, aiming for 10–20 million reads per sample is the most cost-efficient distribution of reads to gain maximum statistical power for DGE RNA-seq [33].

Strand-marked or "stranded" cDNA libraries provide additional mapping information over unstranded cDNA libraries. Unlike unstranded libraries, stranded cDNA libraries allow the RNA template strand to be recognized. This is an important distinction in regions where overlapping gene features exist on opposing DNA strands or when the antisense strand is also transcribed [42]. Levin et al. assessed different methods of strand marking protocols for RNA-seq and concluded dUTP marking of the second strand with paired-end sequencing is a preferred method using the Illumina platform [42]. Opting for paired-end (PE) sequencing is generally considered a cost-effective choice to yield greater information from the RNA-seq experiment. PE RNA-seq improves read mapping to features and is also beneficial for splice variant analysis. For DGE analysis, 75-150 nt sequence cycles/read length is sufficient to assign reads to gene features, with very few genes mapping to multiple features [43]. Longer read lengths will significantly increase the cost and provide greater benefits for other applications.

As much as possible, the handling of RNA samples and processing of all cDNA libraries should be kept consistent throughout the protocol. Differences in sample processing can influence gene calls downstream and lead to batch effects in the data. Factors in an RNA-seq experiment that may contribute to a batch effect include experiment date, isolation date, sample storage, sample age, sample sex, isolation area, isolation kit, researcher, and lab [34]. It is best practice to construct all cDNA libraries in concert and to sequence the samples as normalized-pooled cDNA libraries on the same flow cell lane if possible or equal numbers of each condition spread across multiple flow cell lanes.

4.2 RNA Collection

Careful RNA isolation, handling, and storage are critical for generating valid RNA-seq data. If using a commercial column-based extraction method, the column should be treated with DNase to reduce the chance of genomic contamination of the library preparations. Care should also be taken to prevent any buffer residue on the outside of the column from reaching the final elution and thereby avoiding unwanted impurities entering the sample. RNAses are ubiquitous, and steps to protect the RNA sample from RNases such as using clean gloves, wiping the working area, and pipettes with RNAse solution should be taken. Progressing through the RNA extraction protocol as promptly as possible and swift storage at −80 °C are also measures the investigator can take to limit RNAse-mediated sample degradation. We recommend either completing the RNA extraction protocol in its entirety once the sample is lysed or if commercial RNA preservation reagents are used, to closely follow the recommendations for their use. Repeated freeze-thaw cycles can lead to reduced RNA integrity and should be avoided.

4.3 RNA Integrity Analysis

Prior to the construction of cDNA libraries for sequencing, it is important to assess the RNA integrity of each sample. This step not only provides confidence going forward through the workflow but can save a great deal of time where poor sample quality leads to misleading or no results. Currently the best method to determine RNA integrity within a sample is to use the Agilent Bioanalyzer or TapeStation for RNA integrity number (RIN) analysis. The RIN is an arbitrary number derived from an algorithm that uses the fluorescent spectra generated by running an RNA sample through the Agilent Bioanalyzer microcapillary electrophoresis system [44, 45]. Up to 12 samples can be run on one chip at a time and the analysis simultaneously determines RNA concentration in the sample. An example of the fluorescent spectra generated is shown in Fig. 3, where large peaks can be observed in samples with high integrity that correspond to 18S and 28S ribosomal RNA (Fig. 3a) that are absent in degraded samples (Fig. 3b). The algorithm used to generate the RIN number encompasses multiple features of RNA degradation that provides a more robust and reproducible method of determining RNA integrity than assessments based upon the 18S:28S ratio alone [44, 45].

4.4 cDNA Library Construction

4.4.1

Ribosomal RNA (rRNA) makes up 80–90% of total RNA and for differential expression (DE) analysis of coding genes, and enrichment of non-rRNA is commonly employed to ensure that rRNA does not consume most sequence reads [36]. One method to enrich coding genes is the selection of polyA-mRNAs using poly-dT beads or poly-dT primers for reverse transcription [36]. This is a highly effective method of reducing downstream reads being

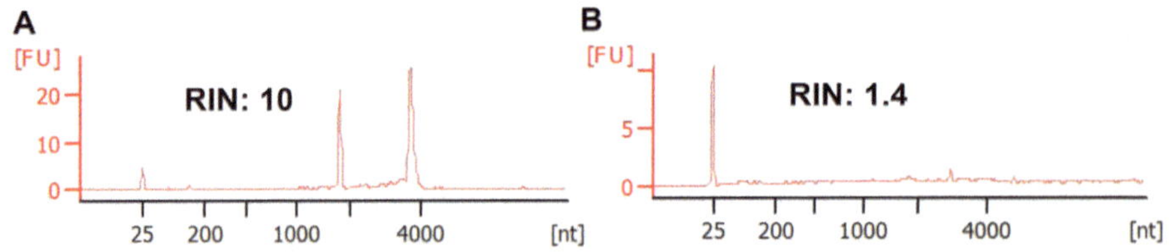

Fig. 3 Agilent Bioanalyzer analysis of RNA integrity. (**a**) An example fluorescence spectra generated by the Agilent Bioanalyzer from an intact RNA sample with a RIN value of 10 with distinct peaks corresponding to 18S and 28S rRNA. (**b**) A degraded RNA sample of low RIN value showing the absence of fluorescent signal corresponding to 18S and 28S rRNA and high level of small RNA molecules

allocated to rRNAs; however, be aware that poly-dT-based library construction can bias the representation of 3′ reads [1, 46]. A common alternative to polyA-based enrichment is rRNA-depletion followed by random hexamer-primed reverse transcription, which too is subject to some bias, albeit to a lesser degree [47]. The KAPA stranded RNA-seq Kit with RiboErase kit used in this protocol depletes rRNA by hybridizing rRNA-complementary DNA oligonucleotides to rRNA and degradation of RNA:DNA hybrids by RNase H while sparing the non-hybridized non-rRNA RNA molecules. The starting volume of total RNA used for the rRNA-depletion stage of the KAPA stranded RNA-seq with RiboErase protocol is 10 μl. Use the same mass of starting RNA across the samples. The amount that remains post-rRNA reduction will vary with variations in the proportion of rRNA across the samples; however, it is not necessary to re-quantify the rRNA-depleted RNA as variations in rRNA-depleted RNA input will be compensated for downstream when the cDNA libraries are pooled based upon cDNA concentration prior to sequencing.

4.4.2

For fragmentation of RNA, the KAPA stranded RNA-seq Kit with RiboErase kit uses heat treatment in the presence of magnesium. Varying the time and heat treatment will result in fragments of dependent length. For RNA-seq analysis of coding gene expression, we have typically used 6 min at 94 °C as designated by the kit's protocol to achieve fragments of approximately 200–300 bp. Optimal fragment length can be application-dependent and may also influence how many sequencing cycles one chooses. For example, some sequencing redundancy will occur if 100 bp fragments are generated, and more than 50 sequencing cycles PE are used at the sequencing stage as the same sequence information will be duplicated.

4.4.3

If using a strand-specific protocol such as the KAPA stranded RNA-seq Kit with RiboErase kit, it is important to be aware that the second strand marked with dUTP will not be amplified at the latter PCR stage (3.4.7). The complement of the coding gene region will therefore be amplified and sequenced. When using the sequence alignment software, be sure to designate the strandedness of the cDNA libraries. This also applies for counting reads to gene features at stage (htseq count, 3.6.4) where the --stranded = reverse option needs to be designated.

4.4.4
A unique indexed adapter is ligated to the cDNA libraries of each biological sample. The KAPA stranded RNA-seq Kit with RiboErase kit provides a guide to how many adaptors should be included in each ligation reaction and is proportionate to the amount of input RNA. The indexed adaptors serve two purposes. First, the ends of the adaptors allow for the cDNA libraries to bind the sequencing flow cell. Second, a unique 6-nucleotide sequence identifier or “index” allows for the cDNA libraries to be pooled together and run on the same flow cell lane as each sequence read is attributed to an individual biological sample based upon the unique index identifier. The adenylation of the 3′ ends of dscDNA allows for ligation of the adaptors at 3′ ends, and the indexed adaptors are also ligated to the 5′ ends. The inclusion of adaptors at both ends of dscDNA allows for paired-end sequencing (PE), where the final cDNA library fragments are sequenced from both directions. PE sequencing is desirable as it allows for sequences to be located in the genome with greater confidence and allows quantification of alternatively spliced transcripts.

4.4.5
PCR amplification of cDNA libraries is the final step of cDNA library construction. PCR primers complementary to the Illumina adaptors are added, and the cDNA libraries are amplified by PCR for 8–16 cycles. The cycle number recommended by the KAPA protocol is dependent upon the amount of input RNA. The goal of this amplification stage is to increase the cDNA library yield while maintaining a representative distribution of the transcript-derived cDNA molecules. Molecules of cDNA representing transcripts in high abundance will be enriched more rapidly than cDNA molecules representing transcripts of low abundance, and therefore choosing the optimum number of amplification cycles is a balance between increasing the cDNA library to a sufficient yield for sequencing while limiting overamplification of highly abundant cDNAs. Overamplification will lead to highly abundant cDNAs taking up a disproportionate number of binding sites on the flow cell leading to the underrepresentation of lowly expressed transcripts. After the post-PCR purification stage, cDNA libraries should be stored at −20 °C.

4.5 Next-Generation Sequencing

Once the cDNA libraries have been constructed, it is important to quantify the cDNA concentration and assess the size distribution of each cDNA library. This is commonly achieved using the Agilent TapeStation system to assess cDNA size distribution and by qPCR to determine cDNA library concentration [43]. The cDNA libraries are then normalized by concentration and combined or "pooled" to form a master/parent cDNA library that will be run on an Illumina flow cell lane.

As described in Subheading 4.1, the sequencing parameters the investigator must choose are application-dependent. These parameters include whether the sequencing run is to be paired-end (PE) or single-end (SE) and the number of sequencing cycles.

4.6 Sequence Processing and Analysis

4.6.1 Quality Control of Raw Reads

Quality control should always be performed as a first step to understanding the fastq sequence generated by RNA-seq. A fastq file is essentially a fasta (sequence) file, but with an added line, for the PHRED (quality) score, the scale of which depends on the sequencer generating it. Quality/degradation of input RNA, RNA library preparation, model and chemistry kit of the Next-Gen Sequencer, and length of sequence can all affect the consistency and quality of sequence throughout the read. The Java program FastQC generates an HTML report, including crucial graphical metrics including: per base sequence quality, average quality per read, sequence duplication levels, and overrepresented sequence. MulitQC can be used to aggregate the reports of many fastq files into one, easy-to-read dynamic analysis HTML report [48]. In some cases, barcodes/adapters have not been fully removed from the sequence and are flagged. While there is a myriad of trimming/grooming programs available (e.g., Cutadapt/Trim Galore [49], FASTX-Toolkit, Skewer [50], and Trimmomatic [51]), modern-day aligners are often able to softclip (i.e., successively ignore the 3′ tail until a match is hit), so that trimming is not always necessary. For more advanced QC metrics and visualization, RSeQC contains an expansive collection of quality control tools and sequencing metrics, along with format conversion [52].

4.6.2 Alignment to the Reference Genome

TopHat2 with its underlying dependency on the Bowtie2 aligner was once the standard for Illumina sequencing alignment [53], but this tool has been deprecated and replaced with HISAT2 from the same group at Johns Hopkins University [23]. Taking an indexed genome along with an optional genome annotation file (GAF/GTF), HISAT2 is able to discern reads that span splice junctions and detect novel isoforms. Given the high degree of conservation between many HSP genes, the accuracy of the alignment as opposed to novel splice junctions is of greater interest, so the parameters for mismatches and allowance of multimapping should be considered, to avoid read fragments mistakenly mapping to a different HSP isoform. HISAT2 is a command line Unix tool, and as such, is best suited to run in a high-performance compute (HPC) environment. Additionally, it can distribute its computation over multiple processors to speed up alignment, which is a workflow very amenable to HPC.

STAR aligner emerged as an ultrafast alignment alternative to HISAT2, equally capable of mapping reads across splice junctions and detecting novel isoforms [24]. STAR is a Unix tool with higher memory requirements, so STAR is best to run in an HPC environment. STAR affords greater options and specificity as to mismatches, multimappers, and output format, and includes verbose alignment statistics. It has built-in functionality for read counting, using the htseq-count algorithm [24]. Rsubread's featureCounts is another option to count reads [29].

Pseudo-aligners have gained popularity as well and differ in that they quantify the abundance of transcripts, but not per se the actual alignment coordinates of each read. These memory-efficient programs are capable of running on a regular desktop computer, benchmarked in minutes [54]. Kallisto [25], Salmon [26], and Sailfish [27] are three such pseudo-aligners that output transcript abundances. Preprocessing with tximport [55], counts can be generated from these abundances, along with adding gene annotation, to use as input for downstream differential gene expression testing in DESeq2. Similarly, edgeR includes functionality to import Kallisto and Salmon transcript abundances directly [30].

4.6.3 Differential Expression Analysis

The powerful R packages edgeR [30] and DESeq2 [31] are available through Bioconductor to perform DE analysis on an HSP-associated experiment to derive a list of genes that statistically differ between conditions [30]. The edgeR and DESeq2 packages are able to derive meaningful biological signal from a low number of replicates, in terms of the number of significant genes and precision, compared to other algorithms [35, 56].

The edgeR and DESeq2 packages perform DE analysis on raw counts data, as opposed to RPKM (reads per kilobase million) or FPKM (fragments per kilobase million), which have been shown to not effectively eliminate gene length bias [57]. Instead, by considering each gene and acknowledging that each gene length, and therefore bias, is the same between conditions, direct gene-wise comparisons can be made between conditions.

Lowly expressed genes (usually 1–10 counts per million (CPM)) are often considered to be noise and are conditionally omitted from the analysis, to prevent them from skewing statistics. This thresholding can be set to a determined number of samples (e.g., all the genes in condition 1 must have a CPM > 5 to be considered). Removing lowly expressed genes also shrinks the size of the input matrix to an easier-to-compute size.

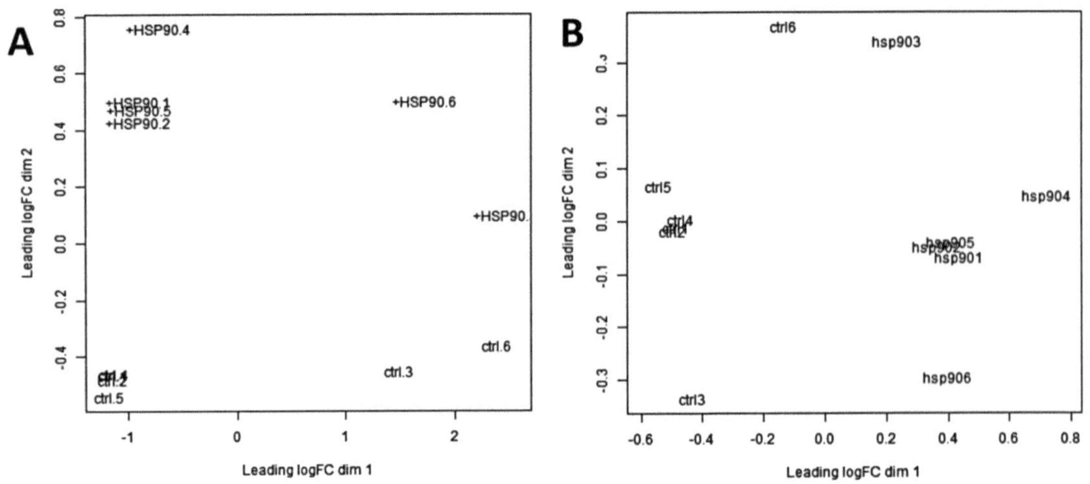

Fig. 4 Multidimensional scaling (MDS) plots. MDS plots are an informative visual representation of relatedness between samples. MDS plots may allow for batch effects to be identified (**a**). Ideally samples will be separated across dimension 1 based on the biological condition such as shown in (**b**)

While RPKM or FPKM normalization is not applied, the input data matrix still consists of varying library sizes (total read counts per sample) that must be normalized to allow for even-footing comparison. Highly expressed genes may dominate the total number of reads and need to be compensated for. The edgeR package achieves this using a trimmed means of M normalization (TMM) based method [57]. TMM minimizes the log-fold changes between samples and derives a scaling factor. A scaling factor below one scales the counts up and the library size down, indicating that the library size is monopolized by a small number of high-count genes [57]. Conversely, a scaling factor above one scales the counts down and the library size up. With the scaling factor, the count data has a new, effective library size. In DESeq2, normalization is based on the median of ratios, where the sample-specific size factor used to normalize the counts is calculated from the median ratio of gene counts relative to the geometric mean per gene [58].

It is often useful to visualize how samples cluster using unsupervised methods. Multidimensional scaling (MDS) plots the leading log-fold change in the highest two dimensions. Ideally, samples from conditions will separate along the primary x-axis (Fig. 4). Similarly, principal component analysis (PCA) should cluster and separate samples at least on the x-axis (PC1) as well. Hierarchical clustering on the top quartile varying genes, possibly in conjunction with a heatmap, can also identify any issues with sample consistency. When a batch effect is detected, algorithms like surrogate vector analysis (SVA) [59], as implemented through svaseq [60], can be used to add a surrogate variable to the design model in edgeR [30] and DESeq2 [31].

Like many popular R DE algorithms, edgeR and DESeq2 utilize a negative binomial (NB) model. A NB model is an over-dispersed Poisson model, that is, it has a variation term. That variation is how much the genes in samples within a condition differ from their mean. The BCV for an ideal model cell line is around 0.1 but is often as high as 0.6 for human samples (McCarthy et al. reported a BCV of 0.4 to be typical of human samples [38]. The edgeR algorithm encapsulates the BCV in two ways: common and gene-wise. The common dispersion assumes that genes all have the same mean–variance relationship. This dispersion is further refined on a tagwise (gene) level with empirical Bayesian shrinkage; genes more consistent between replicates are ranked higher [38]. DESeq2 also models the sample and gene-specific dispersion, but shrinkage is addressed after fitting the NB [31].

EdgeR has three major implementations: "classic," "likelihood ratio test" (LRT), and "quasi-likelihood F-test" (QLF). The classic mode is for experiments with only two conditions (one factor) and uses quantile-adjusted conditional maximum likelihood (qCML) to determine if a gene differs between conditions. The common and tagwise dispersions work on pseudo counts conditioned on the effective library size. The actual DE is an exact test using the negative binomial distribution. The resultant p-values are then corrected to q-values for multiple testing errors with a false discovery rate (FDR), using Benjamini–Hochberg (BH) correction [61].

For multiple factor experimental designs, the QLF approach is more appropriate. Based on a generalized linear model, the common and tagwise dispersions are estimated using Cox–Reid profile-adjusted likelihood (CR) by fitting GLMs to the design (experimental setup) matrix. The DE can be calculated between any pairwise factors using a quasi-likelihood F-test. p-values are corrected for multiple testing with BH to get resultant q-values (FDR). The LRT uses the same dispersion estimates, but the QLF model has been found to outperform the LRT for type I error control [30].

DESeq2 is also available in several implementations. The classic Wald test estimates the standard error of the logFC, to test if it is equal to 0. When more than one term is being tested, the likelihood ratio test (LRT) is more applicable, where there are two models: a full model with some additional terms and a reduced model where extra terms are removed. In the LRT, the likelihood tested is that the extra terms in the full model are more than if the extra terms are equal to 0. As an extension to estimating shrinkage of effect size, DESeq2 offers a number of options for moderating the logFC for gene ranking and visualization, like apeglm [62] and ashr [63].

Some investigators choose to threshold q-values with a combination of logFC or adjusted logFC (DESeq2) to filter out only the statistically significant genes with the highest differential signal.

Differentially expressed genes are most commonly represented as logFC to provide a scale upon which observed fold changes can be represented.

Recording procedures should include all commands with data inputs and analysis outputs, analysis date, version of the statistical package used for each stage, and any additional details needed to replicate the analysis. Many statistical packages and R libraries are subject to updates and future replication may be affected if a different version is used. Similarly, the commands entered in R may also be affected when a different version of the R software is used. Use .txt files for counts matrix data and result output to prevent inadvertent keystrokes that may change data records when using Microsoft Excel spreadsheets. In addition, be aware that Microsoft Excel will automatically change some gene names to dates (Oct2 to 2-Oct, etc).

4.6.4 Functional Enrichment Analysis Using clusterProfiler

Interrogating the biological context of a DE gene list is imperative to understanding the consequence of con-commitment upregulated and downregulated genes. Gene Ontology (GO) is controlled vocabulary to describe a gene by three features: biological process (BP), cellular component (CC), and molecular function (MF). The Gene Ontology Consortium maintains a constantly updated catalog of ontology for many model organisms and is manually curated and evidence-based [64]. While no longer maintained as a free resource, the Kyoto Encyclopedia of Genes and Genomes (KEGG) has curated pathways across thousands of species, mapping genes into functional relationships [65]. Alternatively, newer ontologies include: Reactome [66], WikiPathways [67], and Hallmark Pathways [68]. These and other ontologies are curated by the Broad Institute of Harvard and MIT as part of the MSigDB resource [68] and can be accessed by downloading files from the web portal or in R through the msigdbr tool [69].

There are two main methods for querying a term given significant genes or a list of ranked genes. In overrepresentation analysis (ORA), significant genes lists (q-value at alpha $\leq$0.05, etc) can be used as input to find enrichment of each term/pathway's membership using the hypergeometric distribution, generating a p-value and multiple testing corrected q-value [70]. Similarly, terms and pathways can be queried through gene set enrichment analysis (GSEA) [71]. In GSEA, all the genes from an experiment can be ranked, for example, by q-value or the -log10 of the p-value multiplied by the sign of their logFC, and assessed for term enrichment in their leading edges (tails) to generate a p-value, corrected q-value, and normalized enrichment score (NES) dictating the direction of the genes.

The clusterProfiler R package is a fully featured meta-analysis tool that enables querying significant gene lists or ranked gene lists

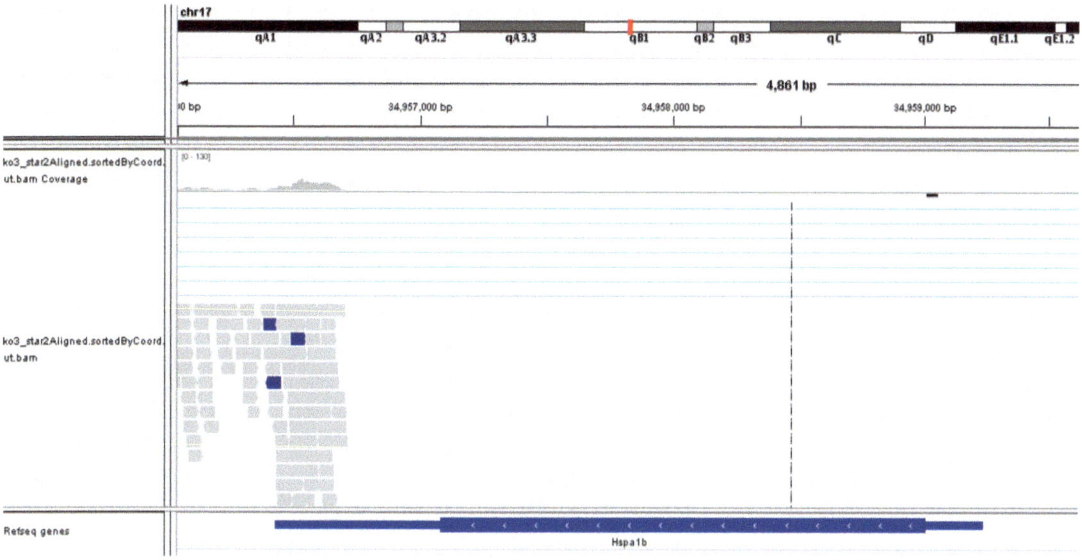

Fig. 5 Visualization of mapped reads using IGV. An example representation of reads mapped to the mouse *Hspa1b* gene expressed in MMT Hsp70$^{-/-}$ mice [76]. Here, the coding region of the Hspa1b gene has been knocked out; however, reads are still mapped to the *Hspa1b* gene. Closer inspection using IGV found these reads mapped to the 5'UTR and demonstrates one benefit of further examination of mapped reads

through ORA or GSEA [32]. ClusterProfiler is able to query GO and KEGG directly or to use gene sets/pathways pulled from MSigDB (msigdbr providing seamless functionality) [69]. clusterProfiler provides visualizations through inverse bar charts, dot plots, network graphs of genes and terms, heatmaps, upset plots, leading edge graphs, pathview representations of KEGG pathways, and more, with dplyr support [32]. The igraph tool, available in R, Python, and C/C++, is another popular alternative to visualizing the networks from Gene Ontology [72].

4.6.5 Representation of Data

Data visualization plays a crucial role in conveying the results of an RNA-seq experiment. Beginning with a cursory glance at the alignment, Integrative Genomics Viewer (IGV) can display how reads map to the reference genome (Fig. 5) [73, 74]. Volcano plots generated via ggplot2 [75] with additional functionality from ggrepel [76] are a common way to visualize data: the x-axis is typically the logFC, and the y-axis is either the -log10 of the p-value or q-value. Genes meeting a significance threshold (q-value below a set alpha such as 0.05, with or without logFC cutoff) are typically assigned a different color like red, giving the "volcano"-like appearance (Fig. 6). Heatmaps of the genes that are DGE, based on the z-scored normalized counts or the variance stabilization transformation (VST) of the counts, graphically present the sample and gene expression (Fig. 7). Enabling hierarchical clustering, with distances such as "Euclidean," "Manhattan," or "Pearson" and

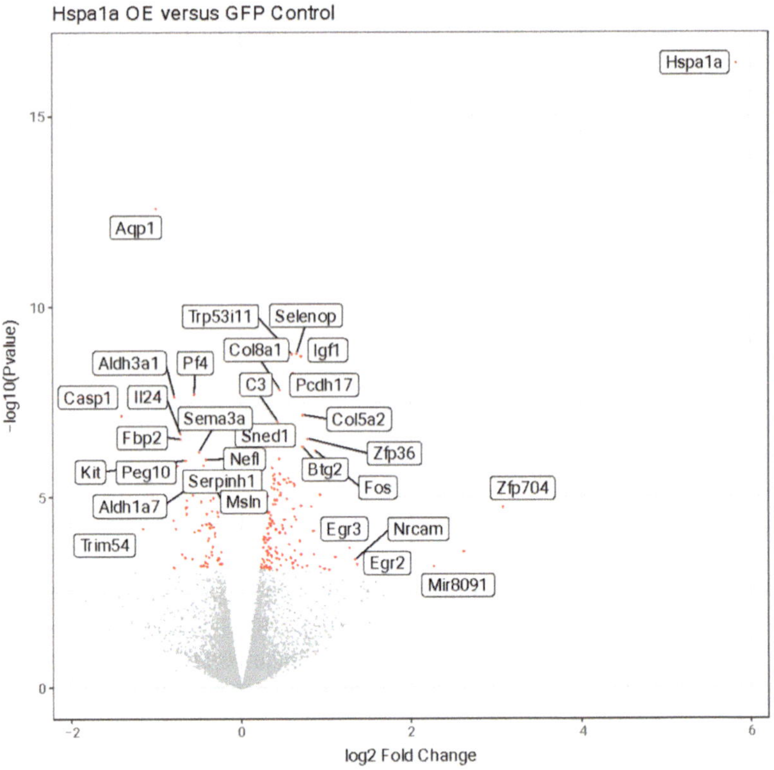

Fig. 6 Volcano plot depicting DGEs. The volcano plot highlights Hspa1a as the most upregulated gene in cells overexpressing Hspa1a compared to control cells. Red dots signify genes that are FDR < 0.05

linkage types like "ward.D2," "average," and "absolute," visually maximizes the DGE pattern of high/low expression encapsulated by DGEs and produces clusters. Samples of the same experimental factor should cluster together, and genes will cluster according to count or z-score in high/low blocks. The pheatmap package in R allows for additional annotation bars on the x- and y-axis to add metadata to genes and samples [77]. For ORA analysis, significant terms can be displayed as a bar graph with the x-axis as the inverse log10 p-value, or count of contributing genes, with a color scale reflective of the strength of the p-value (Fig. 8a). With GSEA analysis, the x-axis can again reflect the inverse log10 p-value, with color scale reflective of NES value (Fig. 8b). Either ggplot2 [75] or clusterProfiler [32] can create these types of visualizations for ORA and GSEA.

4.7 *Additional Notes*

Where institutional access to a high-performance computing cluster is not available, several commercial analysis services are available. These include the Amazon cloud computing service which uses a UNIX command line and many of the steps outlined in this protocol could be adapted to using the Amazon cloud; see Griffith et al., for reference [78]. The ACCESS project is also openly available

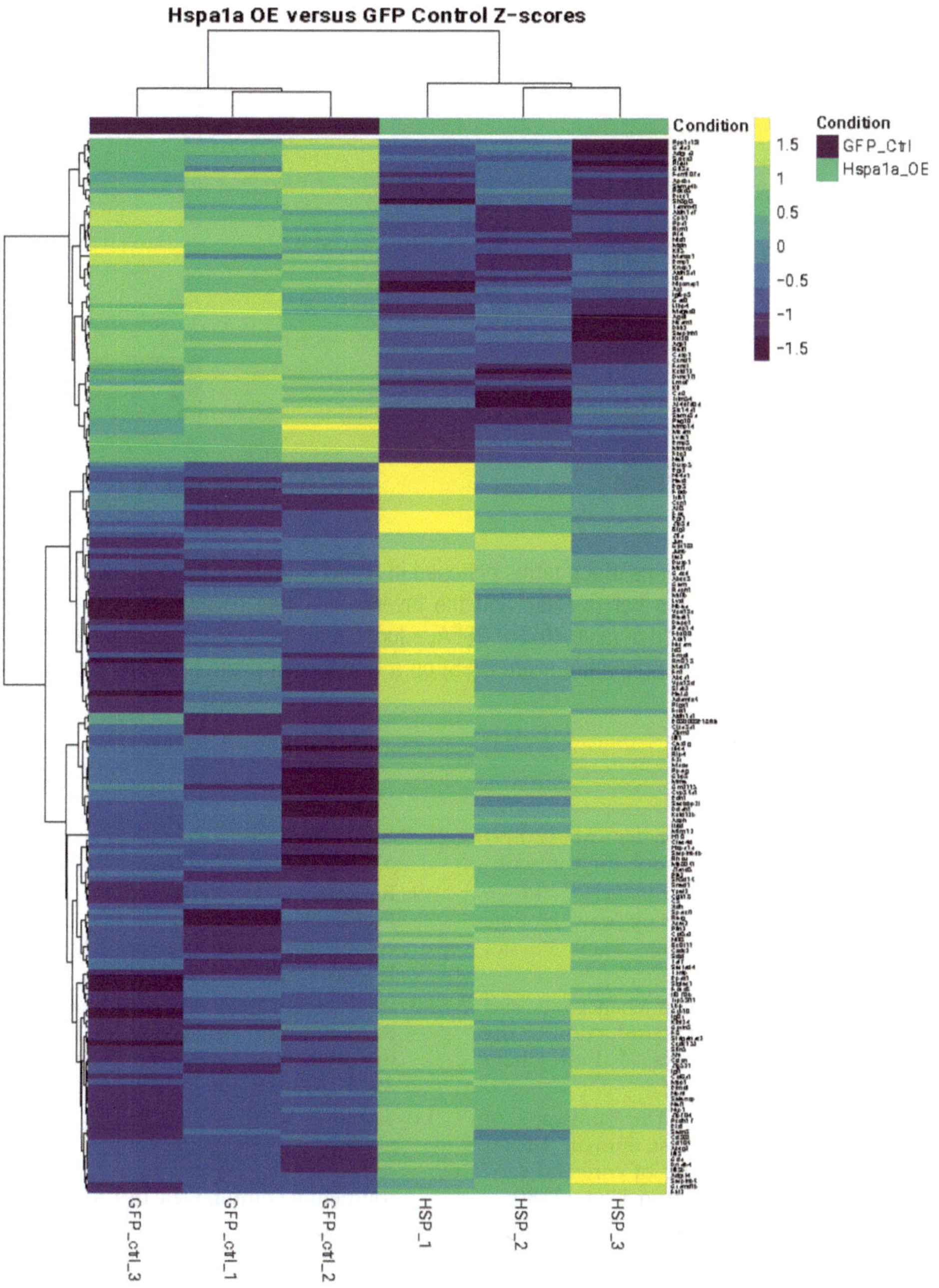

Fig. 7 Heatmap depicting DGEs. The heatmap displays significant genes (FDR < 0.05) in Hspa1a overexpression cells compared to GFP vector control cells, with genes z-scored and both genes and samples clustered by Euclidean distance, ward.D2 linkage

(https://access-ci.org/) and provides computing resources to perform RNA-seq analysis. Online discussion forums such as Biostars (https://www.biostars.org) and SEQanswers (http://seqanswers.com/) are valuable tools to help navigate issues commonly faced when working through an RNA-seq workflow.

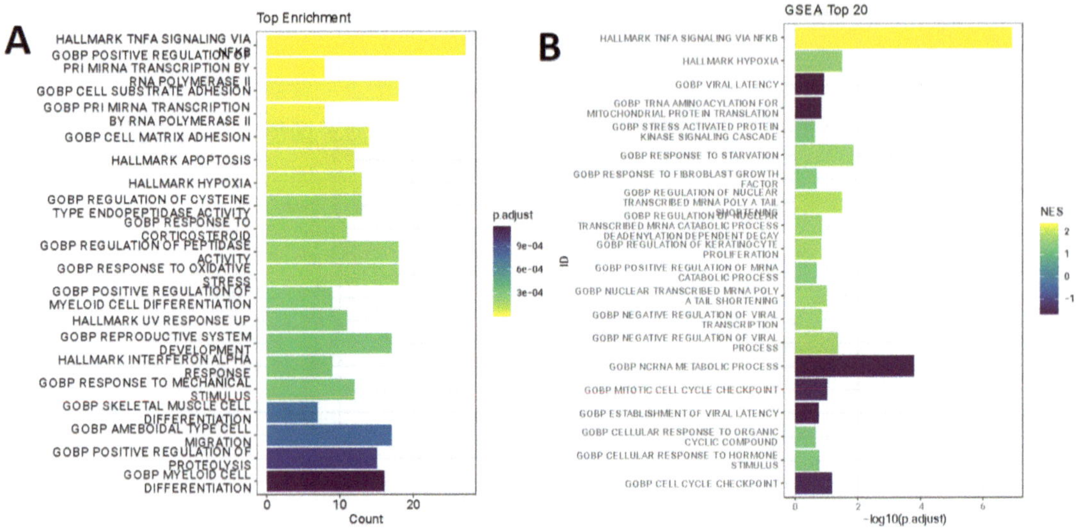

Fig. 8 ORA and GSEA representations of top terms enriched in Hspa1a. (**a**) For ORA, the number of significant genes determines the size of the bar (x-axis), and the inverse adjusted p-value on a viridis color scale indicates the degree of the adjusted p-value. (**b**) For GSEA, −log10 adjusted p-value determines the size of the bar (x-axis), and the normalized enrichment score (NES) on a viridis color scale indicates the direction

References

1. Wang Z, Gerstein M, Snyder M (2009) RNA-Seq: a revolutionary tool for transcriptomics. Nat Rev Genet 10(January):57–63
2. Weber AP (2015) Discovering new biology through sequencing of RNA. Plant Physiol 169(3):1524–1531
3. Buchner J (2019) Molecular chaperones and protein quality control: an introduction to the JBC reviews thematic series. J Biol Chem 294(6):2074–2075
4. Shemesh N et al (2021) The landscape of molecular chaperones across human tissues reveals a layered architecture of core and variable chaperones. Nat Commun 12(1):2180
5. Tittelmeier J, Nachman E, Nussbaum-Krammer C (2020) Molecular chaperones: a double-edged sword in neurodegenerative diseases. Front Aging Neurosci 12:581374
6. Sinnige T, Yu A, Morimoto RI (2020) Challenging proteostasis: role of the chaperone network to control aggregation-prone proteins in human disease. Adv Exp Med Biol 1243:53–68
7. Hipp MS, Kasturi P, Hartl FU (2019) The proteostasis network and its decline in ageing. Nat Rev Mol Cell Biol 20(7):421–435
8. Elsasser S et al. (2022) A comprehensive enumeration of the human proteostasis network. 1. components of translation, protein folding, and organelle-specific systems. bioRxiv, p. 2022.08.30.505920
9. Boulon S et al (2010) HSP90 and its R2TP/Prefoldin-like cochaperone are involved in the cytoplasmic assembly of RNA polymerase II. Mol Cell 39(6):912–924
10. Iki T et al (2010) In vitro assembly of plant RNA-induced silencing complexes facilitated by molecular chaperone HSP90. Mol Cell 39(2):282–291
11. Sawarkar R, Sievers C, Paro R (2012) Hsp90 globally targets paused RNA polymerase to regulate gene expression in response to environmental stimuli. Cell 149(4):807–818
12. Ferraldeschi R et al (2016) Second-generation HSP90 inhibitor onalespib blocks mRNA splicing of androgen receptor variant 7 in prostate cancer cells. Cancer Res 76(9):2731–2742
13. Duttagupta R et al (2003) A yeast homologue of Hsp70, Ssa1p, regulates turnover of the MFA2 transcript through its AU-rich 3' untranslated region. Mol Cell Biol 23(8):2623–2632
14. Henics T et al (1999) Mammalian Hsp70 and Hsp110 proteins bind to RNA motifs involved in mRNA stability. J Biol Chem 274(24):17318–17324

15. Kishor A et al (2013) Hsp70 is a novel post-transcriptional regulator of gene expression that binds and stabilizes selected mRNAs containing AU-rich elements. Mol Cell Biol 33(1): 71–84
16. Laroia G et al (1999) Control of mRNA decay by heat shock-ubiquitin-proteasome pathway. Science 284(5413):499–502
17. Marucci A et al (2009) The role of HSP70 on ENPP1 expression and insulin-receptor activation. J Mol Med (Berl) 87(2):139–144
18. Wilson GM et al (2001) Thermodynamics and kinetics of Hsp70 association with A + U-rich mRNA-destabilizing sequences. J Biol Chem 276(48):44450–44456
19. Matsui H, Asou H, Inaba T (2007) Cytokines direct the regulation of Bim mRNA stability by heat-shock cognate protein 70. Mol Cell 25(1):99–112
20. Zimmer C, Gabain AV, Henics T (2001) Analysis of sequence-specific binding of RNA to Hsp70 and its various homologs indicates the involvement of N- and C-terminal interactions. RNA 7:1628–1637
21. Okusha Y et al (2022) Extracellular Hsp90alpha stimulates a unique innate gene profile in microglial cells with simultaneous activation of Nrf2 and protection from oxidative stress. Cell Stress Chaperones 27(5):461–478
22. Conesa A et al (2016) A survey of best practices for RNA-seq data analysis. Genome Biol 17:13
23. Kim D, Langmead B, Salzberg SL (2015) HISAT: a fast spliced aligner with low memory requirements. Nat Methods 12:357–360
24. Dobin A et al (2013) STAR: ultrafast universal RNA-seq aligner. Bioinformatics 29(1):15–21
25. Bray NL et al (2016) Near-optimal probabilistic RNA-seq quantification. Nat Biotechnol 34(5):525–527
26. Patro R et al (2017) Salmon provides fast and bias-aware quantification of transcript expression. Nat Methods 14(4):417–419
27. Patro R, Mount SM, Kingsford C (2014) Sailfish enables alignment-free isoform quantification from RNA-seq reads using lightweight algorithms. Nat Biotechnol 32(5):462–464
28. Anders S, Pyl PT, Huber W (2015) HTSeq–a python framework to work with high-throughput sequencing data. Bioinformatics 31(2):166–169
29. Liao Y, Smyth GK, Shi W (2019) The R package Rsubread is easier, faster, cheaper and better for alignment and quantification of RNA sequencing reads. Nucleic Acids Res 47(8):e47
30. Robinson MD, McCarthy DJ, Smyth GK (2010) edgeR: a Bioconductor package for differential expression analysis of digital gene expression data. Bioinformatics 26(1): 139–140
31. Love MI, Huber W, Anders S (2014) Moderated estimation of fold change and dispersion for RNA-seq data with DESeq2. Genome Biol 15(12):550
32. Wu T et al (2021) ClusterProfiler 4.0: a universal enrichment tool for interpreting omics data. Innovation (Camb) **2**(3):100141
33. Liu Y, Zhou J, White KP (2014) RNA-seq differential expression studies: more sequence or more replication? Bioinformatics 30(3): 301–304
34. Fang Z, Cui X (2011) Design and validation issues in RNA-seq experiments. Brief Bioinform 12(3):280–287
35. Schurch NJ et al (2016) How many biological replicates are needed in an RNA-seq experiment and which differential expression tool should you use? RNA 22(6):839–851
36. Sims D et al (2014) Sequencing depth and coverage: key considerations in genomic analyses. Nat Rev Genet 15(2):121–132
37. Anders S, Reyes A, Huber W (2012) Detecting differential usage of exons from RNA-seq data. Genome Res 22(10):2008–2017
38. McCarthy DJ, Chen Y, Smyth GK (2012) Differential expression analysis of multifactor RNA-Seq experiments with respect to biological variation. Nucleic Acids Res 40(10):4288–4297
39. Hart SN et al (2013) Calculating sample size estimates for RNA sequencing data. J Comput Biol 20(12):970–978
40. Marioni JC et al (2008) RNA-seq: an assessment of technical reproducibility and comparison with gene expression arrays. Genome Res 18(9):1509–1517
41. Tarazona S et al (2011) Differential expression in RNA-seq: a matter of depth. Genome Res 21(12):2213–2223
42. Levin JZ et al (2010) Comprehensive comparative analysis of strand-specific RNA sequencing methods. Nat Methods 7(9):709–715
43. Korpelainen, E., et al., RNA-seq data analysis: a practical approach. Mathematical and computational biology series, ed. N.F. Britton, et al. 2015, Boca Raton, Florida: Chapman & Hall/ CRC Press
44. Schroeder A et al (2006) The RIN: an RNA integrity number for assigning integrity values to RNA measurements. BMC Mol Biol 7:3
45. Imbeaud S et al (2005) Towards standardization of RNA quality assessment using user-independent classifiers of microcapillary electrophoresis traces. Nucleic Acids Res 33(6):e56

46. Shanker S et al (2015) Evaluation of commercially available RNA amplification kits for RNA sequencing using very low input amounts of total RNA. J Biomol Tech 26(1):4–18
47. Hansen KD, Brenner SE, Dudoit S (2010) Biases in Illumina transcriptome sequencing caused by random hexamer priming. Nucleic Acids Res 38(12):e131
48. Ewels P et al (2016) MultiQC: summarize analysis results for multiple tools and samples in a single report. Bioinformatics 32(19): 3047–3048
49. Martin M (2011) Cutadapt removes adapter sequences from high-throughput sequencing reads. EMBnet J 17(1):10–12
50. Jiang H et al (2014) Skewer: a fast and accurate adapter trimmer for next-generation sequencing paired-end reads. BMC Bioinformatics 15: 182
51. Bolger AM, Lohse M, Usadel B (2014) Trimmomatic: a flexible trimmer for illumina sequence data. Bioinformatics 30(15): 2114–2120
52. Wang L, Wang S, Li W (2012) RSeQC: quality control of RNA-seq experiments. Bioinformatics 28(16):2184–2185
53. Trapnell C, Pachter L, Salzberg SL (2009) TopHat: discovering splice junctions with RNA-Seq. Bioinformatics 25(9):1105–1111
54. Schaarschmidt S et al (2020) Evaluation of seven different RNA-Seq alignment tools based on experimental data from the model plant arabidopsis thaliana. Int J Mol Sci **21**(5)
55. Soneson C, Love MI, Robinson MD (2015) Differential analyses for RNA-seq: transcript-level estimates improve gene-level inferences. F1000Res **4**:1521
56. Seyednasrollah F, Laiho A, Elo LL (2015) Comparison of software packages for detecting differential expression in RNA-seq studies. Brief Bioinform 16(1):59–70
57. Robinson MD, Oshlack A (2010) A scaling normalization method for differential expression analysis of RNA-seq data. Genome Biol 11(3):R25
58. Anders S, Huber W (2010) Differential expression analysis for sequence count data. Genome Biol 11(10):R106
59. Leek JT, Storey JD (2007) Capturing heterogeneity in Gene expression studies by surrogate variable analysis. PLoS Genet 3(9):e161
60. Leek JT (2014) Svaseq: removing batch effects and other unwanted noise from sequencing data. Nucleic Acids Res **42**(21):e161
61. Benjamini Y, Hochberg Y (1995) Controlling the false discovery rate: a practical and powerful approach to multiple testing. J R Stat Soc Ser B Methodol 57(1):289–300
62. Zhu A, Ibrahim JG, Love MI (2019) Heavy-tailed prior distributions for sequence count data: removing the noise and preserving large differences. Bioinformatics 35(12): 2084–2092
63. Stephens M (2017) False discovery rates: a new deal. Biostatistics 18(2):275–294
64. Gene Ontology C (2015) Gene Ontology consortium: going forward. Nucleic Acids Res 43 (Database issue):D1049–D1056
65. Kanehisa M et al (2017) KEGG: new perspectives on genomes, pathways, diseases and drugs. Nucleic Acids Res 45(D1):D353–D361
66. Gillespie M et al (2022) The reactome pathway knowledgebase 2022. Nucleic Acids Res 50 (D1):D687–D692
67. Slenter DN et al (2018) WikiPathways: a multifaceted pathway database bridging metabolomics to other omics research. Nucleic Acids Res 46(D1):D661–D667
68. Liberzon A et al (2015) The molecular signatures database (MSigDB) hallmark gene set collection. Cell Syst 1(6):417–425
69. Dolgalev I (2022) msigdbr: MSigDB Gene Sets for multiple organisms in a tidy data format.
70. Boyle EI et al (2004) GO::TermFinder--open source software for accessing Gene Ontology information and finding significantly enriched Gene Ontology terms associated with a list of genes. Bioinformatics 20(18):3710–3715
71. Subramanian A et al (2005) Gene set enrichment analysis: a knowledge-based approach for interpreting genome-wide expression profiles. Proc Natl Acad Sci U S A 102(43): 15545–15550
72. Csardi G, Nepusz T (2006) The igraph software package for complex network research. Int J Complex Syst 1695(5):1–9
73. Robinson JT et al (2011) Integrative genomics viewer. Nat Biotechnol 29:24–26
74. Thorvaldsdottir H, Robinson JT, Mesirov JP (2013) Integrative genomics viewer (IGV): high-performance genomics data visualization and exploration. Brief Bioinform 14(2): 178–192
75. Wickham H (2009) ggplot2: Elegant Graphics for Data Analysis. Springer-Verlag, New York
76. Slowikowski K (2021). ggrepel: automatically position non-overlapping text labels with 'ggplot2'. Available from: https://CRAN.R-project.org/package=ggrepel
77. Kolde R (2019). Pheatmap: pretty heatmaps
78. Griffith M et al (2015) Informatics for RNA sequencing: a web resource for analysis on the cloud. PLoS Comput Biol 11(6):e1004393

Chapter 5

Chromatin Immunoprecipitation (ChIP) of Heat Shock Protein 90 (Hsp90)

Ritwick Sawarkar

Abstract

Chromatin immunoprecipitation followed by sequencing (ChIP-seq) is a widely used technique for genome-wide mapping of protein–DNA interactions and epigenetic marks in vivo. Recent studies have suggested an important role of heat shock protein 90 (Hsp90) in chromatin. This molecular chaperone assists other proteins to acquire their mature and functional conformation and helps in the assembly of many complexes. In this chapter, we provide specific details on how to perform Hsp90 ChIP-seq from *Drosophila* Schneider (S2) cells. Briefly, cells are simultaneously lyzed and reversibly cross-linked to stabilize protein–DNA interactions. Chromatin is prepared from isolated nuclei and sheared by sonication. Hsp90-bound loci are immunoprecipitated and the corresponding DNA fragments are purified and sequenced. The described approach revealed that Hsp90 binds close to the transcriptional start site of around one-third of all *Drosophila* coding genes and characterized the role of the chaperone at chromatin.

Key words ChIP-seq, *Drosophila* Schneider (S2) cells, Hsp90

1 Introduction

Heat shock protein 90 (Hsp90) is a specialized molecular chaperone that facilitates complex formation, stability, and activity of many protein kinases and transcription factors. Predominantly localized and characterized in cellular cytosol Hsp90 has a well-established role in proteostasis, cell signaling, and carcinogenesis [1]. It has emerged as a promising target in cancer therapeutics with 13 Hsp90 inhibitors undergoing clinical evaluation [2].

Recent reports suggest an important function of Hsp90 in nuclear events [3]. Nuclear Hsp90 has been found to regulate the function of both transcription factors as well as the general transcription machinery thus contributing to the regulation of gene expression [4]. One of the first examples of Hsp90 regulating transcription factor function was shown in studies of steroid hormone receptors [5, 6] where it modulates their assembly,

Stuart K. Calderwood and Thomas L. Prince (eds.), *Chaperones: Methods and Protocols*, Methods in Molecular Biology, vol. 2693,
https://doi.org/10.1007/978-1-0716-3342-7_5,

disassembly, and activity [7]. In addition to hormone receptors, Hsp90 physically interacts with and affects the activity of a variety of TFs, such as p53, c-myc, STAT3, NF-kB, and HSF1 [4]. Another intriguing role for Hsp90 in the nucleus appears to be its interaction with the general transcription machinery. It not only mediates Pol II assembly in the cytoplasm and nuclear import of the fully assembled holoenzyme [8] but regulates Pol II pausing via stabilization of the negative elongation factor complex [9]. In addition, the chaperone is known to affect the packaging of DNA into chromatin and thus contributes to the epigenetic regulation of gene expression [10]. It induces chromatin accessibility changes through interactions with chromatin-modifying/remodeling complexes that could either activate (SMYD3, trithorax/MLL, RSC complex) [11–13] or repress (EZH2) [14] gene expression.

Thus, several studies link Hsp90 to transcription regulation at different levels. A comprehensive characterization of the nuclear clients of Hsp90 and the mechanism by which the chaperone controls gene expression has recently been done [22]. Such efforts have enhanced our understanding of the nuclear functions of Hsp90, traditionally considered to be a cytosolic chaperone. An important question remains largely unclear: does Hsp90 modulate the nuclear translocation and consequent activity of soluble, non-DNA-bound forms of clients, or does Hsp90 directly interact with chromatin-bound forms? Further studies employing genome-wide occupancy analysis will be required to reveal Hsp90-bound chromatin loci to clearly understand its role in chromatin.

Identification of genome-wide occupancy of transcription factors, components of the basal transcription machinery, and modified histones has been critical in understanding the mechanisms of transcriptional regulation. Several widely used methods contribute to the study of how proteins interact with chromatin, and some of the popular techniques are detailed below:

- Chromatin Immunoprecipitation [15]—it has become an indispensable tool for studying gene regulation and epigenetic mechanisms. The first step involves reversible cross-linking of cells in their growth media in order to stabilize protein–DNA interactions. Following chromatin preparation and sonication, DNA fragments bound to the protein of interest are enriched using an antigen-specific antibody. It allows the genome-wide localization of not only transcription (co)/factors but specific posttranslational modifications of proteins such as histones. The major issues here originate from the cross-linking and immunoprecipitation steps. Cross-linking is very likely to mask some of the antigen epitopes and thus prevent antibody binding. The choice of a specific ChIP-grade antibody of very good quality is critical.

The following three approaches offer an important advantage over ChIP—they allow mapping the binding site of proteins for which a good ChIP-seq-grade antibody is not available, as is the case with many chaperones. However, all of them have a significant disadvantage—they do not allow specific mapping of posttranslationally modified proteins.

- Chemical affinity capture followed by sequencing (Chem-seq) [16]—this recent technique relies on small chemical compounds binding with a high affinity to the protein of interest to enrich it from cross-linked or uncross-linked chromatin. The small size of chemical components makes antigen recognition easier and is not that much hampered by the cross-linking of protein–DNA complexes. It is the first method that allows researchers to directly determine the location of cellular factors targeted by small molecules throughout the genome. The successful outcome of the protocol however relies on the existence of a chemical that binds to an antigen with a high specificity and efficiency. Therefore it has been applied to a limited number of proteins. The increasing number of highly specific and potent Hsp90 inhibitors could soon allow the genome-wide Hsp90 localization through Chem-seq [17].
- Biotin-mediated ChIP (BioChIP) [18]—it involves endogenous expression of critical chromatin/DNA-binding factors tagged with biotin. This allows their affinity capture by streptavidin. Biotin–streptavidin interaction is highly specific and one of the strongest non-covalent interactions in nature. These features significantly reduce the nonspecific binding background and favor a highly efficient target pull-down. A major disadvantage here is the need for recombinant biotin-tagged protein expression which is more time-consuming, elaborate, and could introduce artifacts.
- DNA adenine methyltransferase identification (DamID) [19]—this alternative method involves the expression of the DNA-binding protein of interest as a fusion protein with DNA methyltransferase. This enzyme mediates in vivo DNA adenosine methylation in the region of the binding site. By using methylation-sensitive restriction enzymes to enrich methylated DNA, this technique allows the identification of protein-bound DNA independent of immunoprecipitation.

In the following protocol, we focus on ChIP and provide specific details on how to identify Hsp90-bound chromatin loci in *Drosophila* Schneider (S2) cells. It could also be applied to other cell types and (co)chaperones such as p23 [13].

2 Materials

Reagents

Prepare all solutions with molecular biology grade water (Milli-Q™) and supplement them with protease inhibitors.

2.1 Cells Preparation and Cross-Linking

1. Protease inhibitors at a final concentration of 3 μg/mL aprotinin, 10 μg/mL leupeptin, 1 μg/mL pepstatin, and 0.1 mM PMSF
2. Methanol-free formaldehyde 16% (Thermo Scientific, Prod. Nr. 28906)
3. 10X cell lysis buffer: 50 mM HEPES, 50 mM NaCl, 10 mM EDTA, 5 mM EGTA
4. Glycine 1 M
5. Wash buffer: 50 mM Tris-HCl, pH 8.0, 15 mM NaCl, 0.5 mM EGTA, 60 mM KCl

2.2 Chromatin Extraction and Sonication

1. Sonication buffer: 10 mM Tris–HCl, pH 8.0, 1 mM EDTA, 0.5 mM EGTA

2.3 Chromatin Quality Check

1. RNase A 10 mg/mL (DNase and protease-free; AppliChem, Prod. Nr. A3832)
2. Proteinase K 20 mg/mL (Sigma Aldrich, Prod. Nr. P2308)
3. TE buffer: 1 mM EDTA pH 8.0, 10 mM Tris-HCl pH 8.0
4. Phenol–chloroform–isoamyl alcohol 25:24:1 (ROTH, Prod. Nr. A156.3)
5. Glycogen 5 mg/mL (Roche, Prod. Nr. 13741729)
6. 100% ethanol
7. EB buffer: 10 mM Tris-HCl, pH 8.5
8. 1X TAE buffer: 40 mM Tris-acetate, 1 mM EDTA, pH 8.0
9. 2% agarose gel in TAE buffer
10. GELRed Nucleic Acid Gel Stain, 10,000X in water
11. 100 bp DNA ladder
12. DNA loading dye
13. High Sensitivity DNA Kit (Agilent)

2.4 Immunoprecipitation

1. ChIP dilution buffer: 10 mM Tris–HCl, pH 8.0, 1 mM EDTA, 1% Triton X-100, 0.1% sodium deoxycholate, 0.1% SDS, 140 mM NaCl
2. Specific antibody against a protein of interest (anti-Hsp90 polyclonal serum used in this case)

3. Protein A-Dynabeads (Novex by Life Technologies, Prod. Nr. 10002D)
4. LiCl buffer: 10 mM Tris–HCl, 250 mM LiCl, 1 mM EDTA, 0.5% NP-40, 0.5% sodium deoxycholate
5. Elution buffer: 1% SDS, 100 mM $NaHCO_3$

2.4.1 Equipment

1. End-over-end rotator
2. Refrigerated tabletop centrifuge
3. Protein low binding tubes
4. DNA low binding tubes
5. BioRuptor (Diagenode)
6. Magnetic rack
7. Thermomixer
8. Laboratory heating oven
9. Vortex mixer
10. Qubit fluorometer
11. Agarose gel electrophoresis equipment
12. UV transilluminator
13. 2100 Bioanalyzer

3 Methods

3.1 Cells Preparation and Cross-Linking

1. Use 50×10^6 exponentially growing cells per reaction in 10 mL of growth media. Add 10X cell lysis buffer and Triton X-100 directly in the media to reach 1X and 0.5% final concentration, respectively.
2. Immediately add formaldehyde to a final concentration of 1% to facilitate cell cross-linking (*see* **Notes 1** and **2**).
3. Incubate for 15 min with end-over-end rotation at room temperature (*see* **Note 3**).
4. Stop the cross-linking reaction by adding glycine to a final concentration of 100 mM. Incubate for 5 min with end-over-end rotation at room temperature.
5. Spin down the isolated and cross-linked nuclei at 750× *g* for 5 min at 4 °C and discard the cross-linking solution (*see* **Note 4**).
6. Wash twice with ice-cold wash buffer (*see* **Note 5**).

3.2 Chromatin Extraction and Sonication

All the following procedures should be performed at 4 °C with ice-cold buffers containing protease inhibitors.

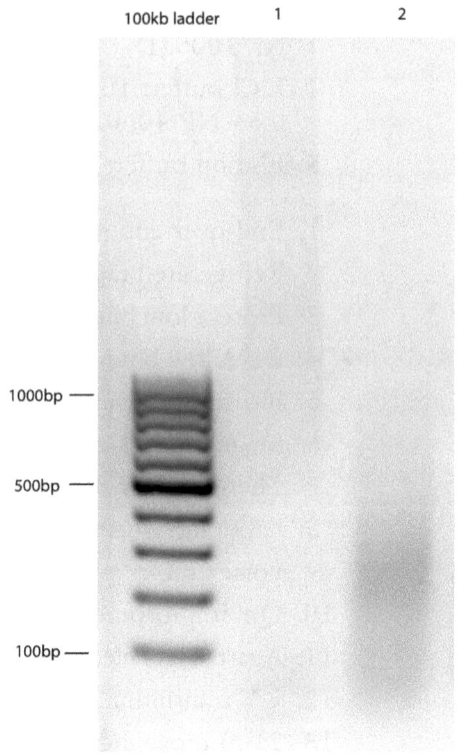

Fig. 1 Not sonicated [1] and sonicated [2] chromatin from *Drosophila* S2 cells

1. Resuspend the nuclei pellet thoroughly in 200 μL of sonication buffer and transfer to a protein low-binding tube (*see* **Notes 6** and 7).
2. Sonicate with Bioruptor (Diagenode) in cold water with the following settings: 30 cycles with intervals of 30 s ON/OFF at maximum power. The optimal size range of the acquired DNA fragments for ChIP-seq analysis should be between 200 and 600 base pairs appearing as a smear (Fig. 1) (*see* **Note 8**).
3. Spin down the sonicated material at 16,000× *g* for 10 min (*see* **Notes 9** and **10**).
4. Transfer the supernatant to a fresh protein low-binding tube (*see* **Note 11**).

3.3 Chromatin Quality Check

For optimal results confirm sonication efficiency and chromatin yield prior to IP. The following steps could be skipped once the protocol is optimized.

1. Take a 20 μL aliquot (1%) of the sheared chromatin. Add 1 μL of RNase A and 1 μL of proteinase K and incubate for 30 min at 37 °C.
2. Reverse cross-link overnight in a 65 °C oven (*see* **Note 12**).

3. Bring the volume of both IP and Input samples to 400 μL with TE buffer.
4. Add an equal volume of Phenol–chloroform–isoamyl alcohol (25:24:1) mixture and vortex (*see* **Note 13**).
5. Incubate at room temperature for 1 min.
6. Spin down at 20,000× *g* for 10 min at 4 °C.
7. Transfer the nucleic-acid-containing aqueous upper phase to a DNA low-binding tube. Add sodium chloride and glycogen to 200 mM and 50 μg/mL final concentration, respectively (*see* **Note 14**).
8. Incubate the mixture for 2 h at −20 °C for 1 h at −80 °C (*see* **Note 15**).
9. Spin down at 20,000× *g* for 10 min at 4 °C. Discard the supernatant.
10. Wash with 1 mL of 80% EtOH (precooled at −20 °C).
11. Spin down at 20,000× *g* for 10 min at 4 °C. Discard supernatant.
12. Air-dry the pellet for around 5 min.
13. Resuspend in 20 μL of EB buffer.
14. Measure DNA amount by qubit following manufacturer's protocol (expected yield around 1 μg of DNA per 1 × 106 cells).
15. Assess chromatin shearing efficiency either on a 2% agarose gel or on Bioanalyzer 2100 using a High Sensitivity DNA chip according to the manufacturer's protocol. The optimal size range of DNA for ChIP-seq analysis should be between 200 and 600 base pairs (Fig. 1).

3.4 Immunoprecipitation

1. Use 20–50 μg of chromatin for ChIP-seq and 5–10 μg for qPCR per reaction—one for the specific antibody (IP) and one for an appropriate negative control (mock-IP; preimmunization serum or IgA/IgG antibody could be used) (*see* **Note 16**).
2. Dilute it 1:10 in ChIP dilution buffer.
3. Take a 5% aliquot as input DNA. Freeze at −20 °C until needed.
4. Add 5 μg of Hsp90 or control antibody to the IP and mock-IP reactions, respectively (*see* **Note 17**).
5. Incubate overnight at 4 °C with end-over-end rotation.
6. Prepare 25 μL of Protein A-Dynabeads per reaction. Wash in ChIP dilution buffer (*see* **Note 18**).
7. Capture the antigen–antibody complexes by adding 25 μL of Protein A-Dynabeads per reaction and incubate for 4 h.

8. Separate the beads on a magnetic rack and discard the supernatant.
9. Transfer the beads in ChIP dilution buffer to a new tube.
10. Wash the beads five times per 5 min with 1 mL of ChIP dilution buffer.
11. Wash once per 5 min with 1 mL of LiCl buffer.
12. Wash twice per 5 min with 1 mL of TE buffer.
13. Elute by adding 50 μL of elution buffer and incubating in a thermomixer at 65 °C for 15 min under vigorous shaking (1400 rpm) (*see* **Note 19**).

    At this point, thaw the input DNA and process it along with the IP samples.
14. Reverse cross-link overnight at 65 °C in a laboratory oven (*see* **Note 20**).
15. Add 200 μL of TE buffer and 8 μL of RNaseA. Incubate for 2 h at 37 °C in a thermomixer.
16. Add 4 μL of Proteinase K, mix, and incubate overnight at 55 °C in a laboratory oven.
17. Purify the DNA using phenol–chloroform–isoamyl alcohol (25:24:1) as previously described (Subheading 3.3, **steps 4–12**). DNA purification kits are not recommended due to loss of material and size selection.
18. Resuspend the DNA pellet in 50 μL of EB buffer and transfer to a fresh tube—expected yield around 15–30 ng per 50×10^6 cells.

    IP DNA can then be sent for next-generation sequencing [20] (*see* **Note 21**).

4 Notes

1. Always use methanol-free formaldehyde for optimal and reproducible results (methanol reduces the fixing power of formaldehyde). Ampoule-sealed solutions are highly recommended as they are well-protected from both air oxidation and light. Formaldehyde is highly toxic and has to be handled carefully under a fume hood.
2. Simultaneous cell lysis and cross-linking achieved at this step generate isolated cross-linked nuclei—a much purer substrate for subsequent steps than whole cells. That initial cellular fractionation improves significantly the chromatin shearing process. A recently published protocol [21] offers an alternative—cross-linking of cells in growing media followed by ultrasound-based nuclear extraction. Nuclei isolation prior to chromatin

preparation offers an additional advantage for proteins that are highly abundant in the cytosol (like Hsp90). By eliminating the cytosolic pool, the efficiency of chromatin-bound Hsp90 pull-down is increased.

3. Cross-linking is a time- and temperature-dependent process. It should not exceed 15 min performed at room temperature. Extended cross-linking could lead to poor results due to reduced sonication efficiency, reduced antibody accessibility to antigen, and epitope masking.
4. Discard formaldehyde-containing solution according to the institutional safety rules.
5. *Pause point:* Cross-linked nuclei/cells could be snap frozen in liquid nitrogen and stored at −80 °C for at least 1 month.
6. Avoid foaming since it dramatically reduces cross-linking efficiency. If foaming occurs, centrifuge for 3 min at 20,000× *g*. Resuspend the material gently leaving no foam bubbles.
7. Use non-siliconized protein low-binding tubes to prevent proteins from sticking to tube walls and thus protein loss.
8. One could use alternative sonication devices, for example, the Branson tip sonicator or Covaris ultrasonicator. Sonication conditions however must be carefully optimized depending upon the specific sonicator to ensure the optimal quality of sheared chromatin. The optimal size range of the acquired DNA fragments for ChIP-seq analysis should be between 200 and 600 base pairs (Fig. 1). DNA fragments out of the range may not be sequenced later on.
9. Proper sonication will generate a clear solution. Turbidity could indicate insufficient sonication.
10. A minimal pellet should be visible at this point. If a larger pellet appears, the sonication did not work properly.
11. *Pause point:* Sheared chromatin could be stored at 4 °C for up to 1 week.
12. Use a laboratory oven for longer thermal incubations to prevent sample condensation on the cap of the tube and ensure uniform heating.
13. Phenol–chloroform is highly toxic and should be handled under the fume hood. Avoid inhalation and skin contact. Wear protective clothing and gloves.
14. Glycogen increases the quantitative recovery of DNA.
15. *Pause point:* Incubation at −20 °C or −80 °C could also be done overnight.
16. An isotype-matched control immunoglobulin (a negative control) is necessary to determine the nonspecific DNA enrichment (background).

17. The choice of antibody is crucial. To test whether an antibody is ChIP-grade, follow the ChIP protocol until the IP washes are done. At this point boil the beads in 1X SDS loading buffer for 30 min (a longer boiling time is necessary to reverse cross-link), and perform a western blot.

    The amount of antibody per IP and the incubation time depend mainly on the affinity properties of the antibody and the abundance of the protein of interest. Test each new antibody to determine optimal conditions.
18. Protein A/G agarose beads could also be used but magnetic beads give a lower background, do not require blocking, and are easier to handle.
19. Always use a fresh elution buffer. *Pause point:* The eluted material could be frozen at −20 °C and stored for up to a week.
20. Do not reverse cross-link for more than 18 h.
21. *Pause point:* The DNA could be stored frozen at −80 °C and stored for at least 1 month.

References

1. Verma S, Goyal S, Jamal S, Singh A, Grover A (2016) Hsp90: friends, clients and natural foes. Biochimie 127:227–240
2. Katerina S, Evangelia P (2014) HSP90 inhibitors: current development and potential in cancer therapy. Recent Pat Anticancer Drug Discov 9(1):1–20
3. Calderwood SK, Neckers L (2016) Chapter four – Hsp90 in cancer: transcriptional roles in the nucleus. In: Jennifer I, Luke W (eds) Advances in cancer research, vol 129. Academic, pp 89–106
4. Sawarkar R, Paro R (2013) Hsp90@chromatin. nucleus: an emerging hub of a networker. Trends Cell Biol 23(4):193–201
5. Pratt WB, Toft DO (1997) Steroid receptor interactions with heat shock protein and immunophilin chaperones. Endocr Rev 18(3): 306–360
6. Bennesch MA, Segala G, Wider D, Picard D (2016) LSD1 engages a corepressor complex for the activation of the estrogen receptor α by estrogen and cAMP. Nucleic Acids Res 44(18): 8655–8670
7. Freeman BC, Yamamoto KR (2002) Disassembly of transcriptional regulatory complexes by molecular chaperones. Science 296(5576): 2232–2235
8. Boulon S et al (2010) HSP90 and its R2TP/Prefoldin-like cochaperone are involved in the cytoplasmic assembly of RNA polymerase II. Mol Cell 39(6):912–924
9. Sawarkar R, Sievers C, Paro R (2012) Hsp90 globally targets paused RNA polymerase to regulate gene expression in response to environmental stimuli. Cell 149(4):807–818
10. Isaacs JS (2016) Chapter five – Hsp90 as a "chaperone" of the epigenome: insights and opportunities for cancer therapy. In: Jennifer I, Luke W (eds) Advances in cancer research, vol 129. Academic, pp 107–140
11. Brown MA et al (2015) C-terminal domain of SMYD3 serves as a unique HSP90-regulated motif in oncogenesis. Oncotarget 6(6): 4005–4019
12. Tariq M, Nussbaumer U, Chen Y, Beisel C, Paro R (2009) Trithorax requires Hsp90 for maintenance of active chromatin at sites of gene expression. Proc Natl Acad Sci U S A 106(4):1157–1162
13. Echtenkamp Frank J et al (2016) Hsp90 and p23 molecular chaperones control chromatin architecture by maintaining the functional pool of the RSC chromatin remodeler. Mol Cell 64(5):888–899
14. Fiskus W et al (2009) Panobinostat treatment depletes EZH2 and DNMT1 levels and enhances decitabine mediated de-repression of JunB and loss of survival of human acute leukemia cells. Cancer Biol Ther 8(10):939–950

15. Park PJ (2009) ChIP-seq: advantages and challenges of a maturing technology. Nat Rev Genet 10(10):669–680
16. Anders L et al (2014) Genome-wide determination of drug localization. Nat Biotechnol 32(1):92–96
17. Moulick K et al (2011) Affinity-based proteomics reveal cancer-specific networks coordinated by Hsp90. Nat Chem Biol 7(11): 818–826
18. Baubec T, Ivánek R, Lienert F, Schübeler D (2013) Methylation-dependent and -independent genomic targeting principles of the MBD protein family. Cell 153(2):480–492
19. Steensel BV, Henikoff S (2000) Identification of in vivo DNA targets of chromatin proteins using tethered dam methyltransferase. Nat Biotechnol 18(4):424–428
20. Bardet AF, He Q, Zeitlinger J, Stark A (2012) A computational pipeline for comparative ChIP-seq analyses. Nat Protoc 7(1):45–61
21. Arrigoni L et al (2016) Standardizing chromatin research: a simple and universal method for ChIP-seq. Nucleic Acids Res 44(7):e67–e67
22. Antonova A et al (2019) Heat-shock protein 90 controls the expression of cell-cycle genes by stabilizing metazoan-specific host-cell factor HCFC1. Cell Rep 29(6):1645–1659.e9

Chapter 6

Transfection and Thermotolerance Methods for Analysis of miR-570 Targeting the HSP Chaperone Network

Yuka Okusha and Stuart K. Calderwood

Abstract

Heat shock proteins (HSPs) are key stress proteins induced in cells exposed to proteotoxic insult and are critical for thermotolerance. The dynamic network of chaperone interactions, known as the chaperome, contributes significantly to the proteotoxic cell response and the malignant phenotype in cancer. We identified a potent microRNA, miR-570 that could bind the 3′untranslated regions of multiple HSP mRNAs and inhibit HSP synthesis. Here, we will introduce the transfection and thermotolerance methods for analysis of miR-570 targeting the HSP chaperone network.

Key words MicroRNA570, Heat shock protein, BAG3, Thermotolerance, Breast cancer

1 Introduction

The canonical role played by HSPs is as molecular chaperones, folding nascent polypeptides, refolding denatured proteins, and building multiprotein complexes [1–3]. The HSPs are also implicated in the pathophysiology of multiple diseases including cancer where elevated chaperone levels promote tumorigenesis [4]. MicroRNAs are short, single-stranded RNA species typically around 22 bp in length, found in cells of many types that can silence the expression of intracellular target mRNA typically by binding to 3′ regulatory sequences (3′ UTR) found in many mRNA species [5–7]. A previous report had shown miR-570 to inhibit cell proliferation [8]. In addition, we have investigated the potential use of a microRNA approach to simultaneously target multiple members of the cancer chaperome [9]. We determined a significant role for miR-570 in regulating markers of mammary tumor progression, including cell motility and invasion. As overexpression of miR-570 elicited tumor suppressive effects and mediated resistance to proteotoxic stress, we inferred that this miR could play a potential role

Stuart K. Calderwood and Thomas L. Prince (eds.), *Chaperones: Methods and Protocols*, Methods in Molecular Biology, vol. 2693, https://doi.org/10.1007/978-1-0716-3342-7_6,

in inhibiting tumorigenesis and cancer cell growth [9]. Here, we will introduce the transfection and thermotolerance methods for analysis of miR-570 targeting the HSP chaperone network.

2 Materials

2.1 Transfection

1. Cells of interest, in which you aim to transfect (*see* **Note 1**)
2. Medium for the cells, for example, DMEM or RPMI 1640
3. 12-well plates
4. mirVana miR-570-3p or nontargeting control miRNA (Thermo Fisher)
5. Lipofectamine RNAiMAX reagent (Thermo Fisher)

2.2 Heat Shock/ Thermotolerance Conditions

1. Parafilm or plastic material to seal the lids of 12-well plates (*see* **Note 2**)
2. Water bath

2.3 Protein Assay

2.3.1 Sample Preparation for Western Blotting

1. RIPA buffer
2. 25G needle syringe
3. BCA protein assay reagent (Thermo Fisher)
4. Microplate reader
5. 4× SDS sample buffer
6. β-Mercaptoethanol (β-ME) (*see* **Notes 3** and **4**)
7. Protein marker
8. Peltier heating block incubator
9. Parafilm or plastic material to seal the lids of tubes
10. Gel loading tips
11. Mini centrifuge

2.3.2 SDS-PAGE

1. Polyacrylamide gel (PAG): 4–12% gradient (*see* **Note 5**)
2. Electrophoresis tank
3. Power source
4. SDS running buffer

2.3.3 Protein Transfer

1. Transfer buffer
2. Protein transfer apparatus
3. Power source
4. 100% methanol
5. Polyvinylidene fluoride (PVDF) membrane
6. Whatman filter papers

7. Plastic bat
8. Ice bag

2.3.4 Immunoblotting

1. TBS-T
2. Blocking agent, for example, 5% skim milk
3. Primary antibodies for detection of HSPs and BAG3 proteins and loading control (GAPDH or β-actin)
4. Secondly antibodies

2.3.5 Imaging and Data Analysis

1. An imaging system, for example, Odyssey Imaging System (LI-COR)
2. Image Studio Lite Ver. 5.2 (LI-COR)

2.4 Cell Proliferation Assay

1. 96-well plates
2. Hemocytometer
3. Medium for the cells

2.5 Colony Formation Assay

1. 6-cm dish plates
2. Medium for the cells
3. 4% paraformaldehyde (thermo fisher)
4. 0.5% crystal violet
5. Methanol
6. Distilled water
7. ImageJ software Flow Chart (Fig. 1)

Day -1 Seed cells

↓

Day 0 Transfection (3.1.)

↓

Day 1 Change medium

↓

Day 2 Heat Shock stress (3.2.)

↓

Day 3 Protein assay (3.3.), cell proliferation assay (3.4.) and colony formation assay (3.5.)

Fig. 1 Flowchart of transfection and thermotolerance methods for analysis of miR-570 targeting the HSP chaperone network

3 Methods

3.1 Transfection

1. Seed 1 × 10^5 cells per well in 12-well plates and incubate it overnight.
2. Transfect cells with mirVana miR-570-3p or nontargeting control miRNA (25–50 nM) and lipofectamine RNAiMAX reagent.
3. Replace medium on day 1 after transfection.

3.2 Heat Shock/Thermotolerance Conditions

1. Heat water bath at 37 °C and 43 °C.
2. Place transfected cells with mirVana miR-570-3p and nontargeting control miRNA in each water bath at 37 °C and 43 °C for 30 min.
3. Maintain the cells in the incubator at 37 °C for 6 h to recovery.
4. Place the cells in each water bath at 37 °C and 43 °C for 90 min.
5. Incubate the cells in the incubator at 37 °C overnight.
6. Proceed to the protein assay (Subheading 3.3), cell proliferation assay (Subheading 3.4), and colony formation assay (Subheading 3.5).

3.3 Protein Assay

3.3.1 RIPA Buffer and Homogenization Protocol

1. Add 1× RIPA buffer to cover the cells, for example, 2 mL for a 10-cm dish.
2. Scrape and collect the cells using a cell scraper.
3. Homogenize samples with a 25G needle syringe for 10 strokes.
4. Incubate samples for 30 min on ice.
5. Centrifuge at 15,000× *g* at 4 °C for 20 min to remove cell debris.
6. Transfer the supernatant (as a protein sample) to another tube (*see* **Note 6**).
7. Snap freeze in liquid nitrogen and store at −80 °C.
8. Dilute the WCL 1:10 with water.
9. Perform protein assay using a BCA protein assay kit, according to the manufacturer's protocol.

3.3.2 Sample Preparation for SDS-PAGE

1. Prepare 4× SDS sample buffer. If required, add β-ME to the sample buffer. The final concentration of β-ME should be 5%.
2. Mix the 4× SDS sample buffer, protein samples, and water.
3. Seal the lids of tubes with plastic materials or parafilm to avoid the lids opening by heating.
4. If you added β-ME, boil the samples at 70 °C for 10 min.
5. Spin down the samples at 10,000 rpm for 30 min.

3.3.3 SDS-PAGE

1. Set a PAGE in an electrophoresis tank.
2. Pore the running buffer in the tank.
3. Load equal volumes of samples to each well. A blank lane may be important.
4. Run at 85 V for an initial 25 min and then 180 V for the next 50 min (*see* **Note 7**).
5. Stop the running before the leading dye reaches the bottom of the gel.

3.3.4 Wet Transfer

1. Soak a PVDF membrane in 100% methanol for 30 s for hydrophilizing.
2. Soak the membrane in the transfer buffer and gently shake for 5 min for equilibrating.
3. "Sandwich" wet system sequentially using filter paper and foam
4. Transfer at 100 V for 60 min at 4 °C.

3.3.5 Immunoblotting

1. Soak the membrane in 5% skim milk in TBS-T for blocking.
2. Gently swirl for 60 min at RT.
3. Add anti-HSPs or BAG3 primary antibodies to 5% skim milk in TBS-T and mix well.
4. Soak the membrane in the anti-HSPs or BAG3 antibodies solution and gently swirl for overnight at 4 °C.
5. Wash the membrane with TBS-T for 5 min four times.
6. Prepare secondary antibodies in 5% skim milk in TBS-T and mix well.
7. Soak the membrane with the secondary antibodies and gently swirl at RT for 40 min.
8. Wash the membrane with TBS-T for 5 min four times (*see* **Note 8**).
9. Wash the membrane with TBS for 5 min two times.
10. Visualize the immunoreactive bands by the imaging system (*see* **Note 9**).
11. Quantify the densitometric analysis using Image Studio Lite Ver. 5.2 (*see* **Note 10**).

3.4 Cell Proliferation Assay

1. Seed 2×10^3 cells per well in 96-well plates.
2. Count the number of cells at 1–8 days post-seeding by hemocytometer.
3. Replace the medium every 3 days during this analysis.

3.5 Colony Formation Assay

1. Seed 1×10^3 cells per well in 6 cm dishes.
2. Discard the supernatants and fix them with 4% paraformaldehyde for 10 min on day 15 after seeding.
3. Stain with 0.5% crystal violet in 25% methanol for 10 min.
4. Rinse with water and dry overnight.
5. Calculate the percentage of colonies using ImageJ software.
6. Replace the medium every 3 days during this analysis.

4 Notes

1. Choose a cell type that expresses the gene of interest at a high level. In our case, HSPs-high and BAG3-high cells were useful for demonstrating the effects of miR-570.
2. Seal the lids tightly to avoid contamination.
3. β-ME is a powerful reductant that cuts bisulfate bonds between cysteines in proteins. The reduction of proteins is needed for many antibody types to detect the proteins. However, another antibody type favors 3D structures of the protein, for which β-ME is useless. See user manuals of antibodies carefully to know if the β-ME is needed or unnecessary.
4. Confirm a strong odor of β-ME right before use. If it does not smell strong, it is expired.
5. Make or buy gels with an appropriate polyacrylamide concentration for your proteins of interest.
6. The pellet can be analyzed as an insoluble fraction that contains chromatin.
7. These sequential steps are effective in preventing too much heating that triggers proteolysis.
8. Avoid light to protect the membrane.
9. Take photographs with short, medium, and long exposure periods.
10. Some reviewers or journals request relative quantification of band intensities, reproducibility, and statistics in Western blotting.

Acknowledgements

We wish to thank the Department of Radiation Oncology for their continued support. This work was supported by the National Institutes of Health grant R01CA176326 (SKC).

References

1. Lang BJ et al (2021) The functions and regulation of heat shock proteins; key orchestrators of proteostasis and the heat shock response. Arch Toxicol. https://doi.org/10.1007/s00204-021-03070-8
2. Ellis RJ (2007) Protein misassembly: macromolecular crowding and molecular chaperones. Adv Exp Med Biol 594:1–13. https://doi.org/10.1007/978-0-387-39975-1_1
3. Hartl FU (1996) Molecular chaperones in cellular protein folding. Nature 381:571–579. https://doi.org/10.1038/381571a0
4. Ciocca DR, Calderwood SK (2005) Heat shock proteins in cancer: diagnostic, prognostic, predictive, and treatment implications. Cell Stress Chaperones 10:86–103. https://doi.org/10.1379/csc-99r.1
5. Bartel DP (2009) MicroRNAs: target recognition and regulatory functions. Cell 136:215–233. https://doi.org/10.1016/j.cell.2009.01.002
6. Bartel DP (2018) Metazoan MicroRNAs. Cell 173:20–51. https://doi.org/10.1016/j.cell.2018.03.006
7. Fabian MR, Sonenberg N, Filipowicz W (2010) Regulation of mRNA translation and stability by microRNAs. Annu Rev Biochem 79:351–379. https://doi.org/10.1146/annurev-biochem-060308-103103
8. Baker JR et al (2019) MicroRNA-570 is a novel regulator of cellular senescence and inflammaging. FASEB J 33:1605–1616. https://doi.org/10.1096/fj.201800965R
9. Okusha Y et al (2022) MicroRNA-570 targets the HSP chaperone network, increases proteotoxic stress and inhibits mammary tumor cell migration. Sci Rep 12:15582. https://doi.org/10.1038/s41598-022-19533-6

Chapter 7

Targeted Replacement of HSF1 Phosphorylation Sites at S303/S307 with Alanine Residues in Mice Increases Cell Proliferation and Drug Resistance

Xiongjie Jin, Demetrius Moskophidis, and Nahid F. Mivechi

Abstract

Mammalian heat shock factor HSF1 transcriptional activity is controlled by a multitude of phosphorylations that occur under physiological conditions or following exposure of cells to a variety of stresses. One set of HSF1 phosphorylation is on serine 303 and serine 307 (S303/S307). These HSF1 phosphorylation sites are known to repress its transcriptional activity. Here, we describe a knock-in mouse model where these two serine residues were replaced by alanine residues and have determined the impact of these mutations on cellular proliferation and drug resistance. Our previous study using this mouse model indicated the susceptibility of the mutant mice to become obese with age due to an increase in basal levels of heat shock proteins (HSPs) and chronic inflammation. Since HSF1 transcriptional activity is increased in many tumor types, this mouse model may be a useful tool for studies related to cellular transformation and cancer.

Key words $Hsf1^{303A/307A}$, Targeting vector, Knock-in mice, Cell proliferation, Drug resistance, Tumorigenesis

1 Introduction

Heat shock transcription factor 1 (HSF1) is an evolutionarily conserved gene and is widely expressed in most tissues in humans and other mammals. HSF1 responds to a variety of stressors and regulates the expression of a wide range of genes at the transcriptional level to maintain cellular proteostasis and support cell survival [1–4]. Under non-stressed physiological conditions, HSF1 exists as an inactive monomeric form and is distributed mainly in the cytoplasm. Upon exposure of cells to heat shock or other stresses, HSF1 is activated through a multistep process, including trimerization, nuclear translocation, binding to the heat shock element (HSE) on the promoter of target genes, and initiating gene transcription [1, 4–6]. HSF1 activation and inactivation are regulated by multiple factors. For example, heat shock proteins (HSPs),

Stuart K. Calderwood and Thomas L. Prince (eds.), *Chaperones: Methods and Protocols*, Methods in Molecular Biology, vol. 2693, https://doi.org/10.1007/978-1-0716-3342-7_7,

whose transcription is upregulated following HSF1 activation in stressed cells, provide negative feedback to inactivate HSF1. Studies show that both HSP70 and HSP90 bind to HSF1 and inhibit its DNA binding capacity [7–9]. In addition, the posttranslational modification of HSF1 protein, including phosphorylation, acetylation, and sumoylation, is an important mechanism for the regulation of its activation, inactivation, and stability.

Phosphorylation of HSF1 plays an important role in the regulation of its activity and has been widely investigated. Phosphorylation of HSF1 is strongly induced following exposure of cells to heat shock, and an increase in its phosphorylation correlates with its transcriptional activity. So far, 15 serine residues (S121, S216, S230, S292, S303, S307, S314, S319, S320, S326, S344, S363, S368, S419, S444) and 4 threonine residues (T142, T323, T367, and T369) of HSF1 protein, which can be phosphorylated under stress and/or unstress physiological conditions, have been reported [6, 10–12]. Notably, while phosphorylation of most of these serine or threonine residues does not affect HSF1 trans-activation capacity, it has been demonstrated that several serine residues (S230, S326, S363, S303, and S307) contribute to HSF1 activity. Phosphorylation of S326 plays a positive stimulatory role in HSF1 activation by facilitating the association between HSF1 and the coactivator Daxx and has been widely used as a marker for HSF1 activation [13]. Additionally, it has been reported that phosphorylation of S230 and S419 of HSF1 confers stimulatory function toward HSF1 activation [14, 15]. On the other hand, constitutive phosphorylation of S303, S307, and S363 represses the activation capacity of HSF1 [16–18]. It has been proposed that phosphorylation of S303 and S307 promotes the binding of the scaffolding protein 14–3-3ε to the HSF1 and subsequently sequesters HSF1 in the cytoplasm [19, 20]. Along this line, phosphorylation of S303/S307 residues is required for HSF1 degradation by facilitating its binding to FBXW7 ubiquitin ligase [21].

The functions of the endogenous phosphorylation of S303/S307 residues on cellular processes under physiological and stressed conditions in vivo remain elusive. Here, we generated a knock-in mouse model in which the S303 and S307 of HSF1 were mutated to alanine residues (named $Hsf1^{303A/307A}$). Our previous studies show that lack of S303 and S307 phosphorylation reduces the threshold of HSF1 activation and that HSF1 is mildly activated in mouse embryo fibroblasts (MEFs), as well as in multiple tissues in vivo under non-stressed physiological conditions [22]. In the current study, we describe in detail the strategy and method for the generation of the $Hsf1^{303A/307A}$ mouse line. We further provide evidence on the effects of HSF1 activation on cellular proliferation, tumorigenesis, and drug sensitivity using MEFs and $Hsf1^{303A/307A}$ mouse lines. The $Hsf1^{303A/307A}$ knock-in MEFs and mice are ideal cellular and animal models for studying the role of S303 and S307 on HSF1 function in vitro and in vivo.

2 Materials

2.1 Reagents

2.1.1 Molecular Biology

LB Broth (Fisher Scientific, cat no.AAH2667636)

Agar powder (Fisher Scientific, cat no.AAA1075236)

Chloroform (Fisher Scientific, cat no. J67241.AP)

Isopropanol (Fisher Scientific, cat no. T036181000)

Ethanol (Fisher Scientific, cat no. T038181000)

Agarose (Invitrogen, cat no. 16500500)

Sodium chloride (Fisher Scientific, cat no. 447302500)

dNTP (Fisher Scientific, cat no. FERR0181)

Tris (Fisher Scientific, cat no. 17926)

EDTA (Fisher Scientific, cat no. 17892)

Sodium dodecyl sulfate (SDS) (Fisher Scientific, cat no. 15525017)

Polyvinylidene fluoride membrane (Fisher Scientific, cat no. 88518)

Sodium deoxycholate (Fisher Scientific, cat no. 89905)

NP-40 (Fisher Scientific, cat no. J61055.AE)

Phosphatase inhibitor cocktail (Fisher Scientific, cat no. 78420)

TRIzol (Invitrogen, cat no. 15596026)

Diethylnitroseamine (Sigma, cat no. 73861)

2.1.2 Southern Blot

SSPE (20X) (Fisher Scientific, cat no. AM9767)

Denhardt's solution (Fisher Scientific, cat no. 750018)

Sodium hydroxide (Fisher Scientific, cat no. A16037.36)

32P-ATP (Perkin Elmer, NEG002A100UC)

Nylon membrane (for Southern blot) (Fisher Scientific, cat no. P177016)

2.1.3 Cell Culture and Histology

Dulbecco's Minimal Essential Medium (Corning, cat no. MT10102CV).

Fetal bovine serum (FBS) (Gibco, cat no. 26140079)

Crystal violet (Fisher Scientific, cat no. 405835000)

37% formaldehyde (Fisher Scientific, cat no. 119690250)

Hematoxylin and eosin staining kit (Abcam, cat no. ab245880)

2.1.4 Drug for Selection of Clones and Treatment

Ganciclovir (Fisher Scientific, cat no.AC461710250)

G418 sulfate (Corning, cat no.30234CR)

Puromycin (MP Biomedicals, cat no. 0219453910)

Blasticidin (MP Biomedicals, cat no. MP215047783)

Doxorubicin (Fisher Scientific, cat no. J64000.MF)

Etoposide (Fisher Scientific, cat no. J63651.MC)

2.2 Enzymes

T4 DNA ligase (Fisher Scientific, cat no. 15224017).

Taq DNA polymerase (Fisher Scientific, cat no. FEREP0406).

pfu Turbo DNA polymerase (Agilent, cat no. 600250).

XhoI Restriction Endonuclease (New England Biolabs, cat no. R0146L).

ClaI Restriction Endonuclease (New England Biolabs, cat no. R0197L).

SacI Restriction Endonuclease (New England Biolabs, cat no. R3156L).

BglII Restriction Endonuclease (New England Biolabs, cat no. R0144L).

NotI Restriction Endonuclease (New England Biolabs, cat no. R0189L).

2.3 Plasmids

BAC clone PR23-266H9 (BACPAC Resource Center).

pBluescript II KS plasmid (pBS) (Agilent, cat no. 212207).

pKT1 plasmid (gift from Dr. Gail Martin (27)).

phage DNA vector λDASHII-254-2TK (gift from Dr. N. R Manley, University of Georgia, Athens, GA).

2.4 Commercial Kits

PCR-based site-directed mutagenesis (Agilent, cat no. 200523).

Gigapack II Packaging Extract (Agilent, cat no. 200201).

DecaLabel DNA labeling kit (Fisher Scientific, cat no. K0622).

DH5alpha competent bacteria (Fisher Scientific, cat no. 18265017).

iScript cDNA synthesis kit (Bio-Rad, cat no. 170–8891).

iQ SYBR Green Supermix (Bio-Rad, cat no. 708880).

3 Methods

1. **Generation of Hsf1$^{303A/307A}$ Knock-in Mice**

To replace the HSF1 wild-type (WT) allele S303/S307 with 303A/307A mutant allele, we constructed a targeting vector in which a 3007 bp proximal, 1252 bp neomycin-Lox P site (for positive selection of embryonic stem (ES) cells), and a 3809 bp distal DNA fragments were ligated into the targeting vector (Fig. 1a). The proximal and distal DNA fragments were amplified by PCR with high fidelity DNA polymerase (*pfu* Tubor DNA polymerase (Agilent) and using a BAC clone PR23-266H9 (BAC-PAC Resource Center), which contained the entire HSF1 gene as a template. The sequences of the primers used for the PCR reactions were as follows: (i) For the proximal DNA fragment, forward primer: 5'-CCGCTCGAG CAAGGAGCTGAAGGGATCTGC-3′ (contains a Xho I site, highlighted with an underline), reverse pri mer: 5'-CCGGAGCTCGCTCTGCCCAAGG-TCATAAAGG-3′ (contains a Sac I site); and (ii) for the distal fragment, forward primer: 5'-CCGATCGATTGAAAATGGGTGGAACTAACTT-3′ (contains a Cla I site), reverse primer: 5'-CCGCTCGAGGCACA CTCCTCTTAGTCTGGG-3′ (contains a Xho I site). The PCR fr agments were digested with the indicated restriction enzymes and inserted into the pBluescript II KS plasmid (pBS). The plasmids w ere introduced into DH5alpha competent bacteria (18,265,017, Fisher Scientific), and a single bacterial clone was selected and am plified. The DNA sequence of the inserted HSF1 fragment was ve rified by DNA sequencing. To generate the neomycin-Lox P fragment, we cloned the neomycin cassette and Lox P (sequence: 5'-ATAACTTCGTATAGCATACATTATACGAAGTTAT-3′) frag ment into pKT1 plasmid (gift from Dr. Gail Martin [23]). The neomycin-Lox P fragment was released by Sac I and Cla I restric tion enzymes. The 303A and 307A mutations were introduced by PCR-based site-directed mutagenesis (200,523, Agilent) using pri mers containing point mutations (underlined), forward primer:5'-AAGCAAGAGCCCCCCGCCCCACCTCACGCCCCTCGGGT ACTGGAG-3′, reverse primer: 5'-CTCCAGTACCCGAGGGG CGTGAGGTGGGGCGGGGGGCT-CTTGCTT-3′ (using the p Bluescript plasmid containing the distal arm fragment as a substrate).

The proximal, distal, and neomycin-Lox P DNA fragments were released from relevant plasmids using indicated restriction enzymes. In the following steps, these DNA fragments were introduced into the Xho I site of the phage DNA vector λDASHII-254-2TK that was flanked by two thymidine kinase (TK) genes for drug negative selection (ganciclovir) of the ES cell clones. The phage DNA vectors were packaged using Gigapack II Packaging Extract. The single phage clone whose genome contained all 3 DNA fragments in the correct orientation was identified and verified by PCR and DNA sequencing. The final targeting vector, containing 2 TK genes and proximal, distal, and neomycin-Lox P fragments, was released from the phage DNA by Not I restriction enzyme digestion.

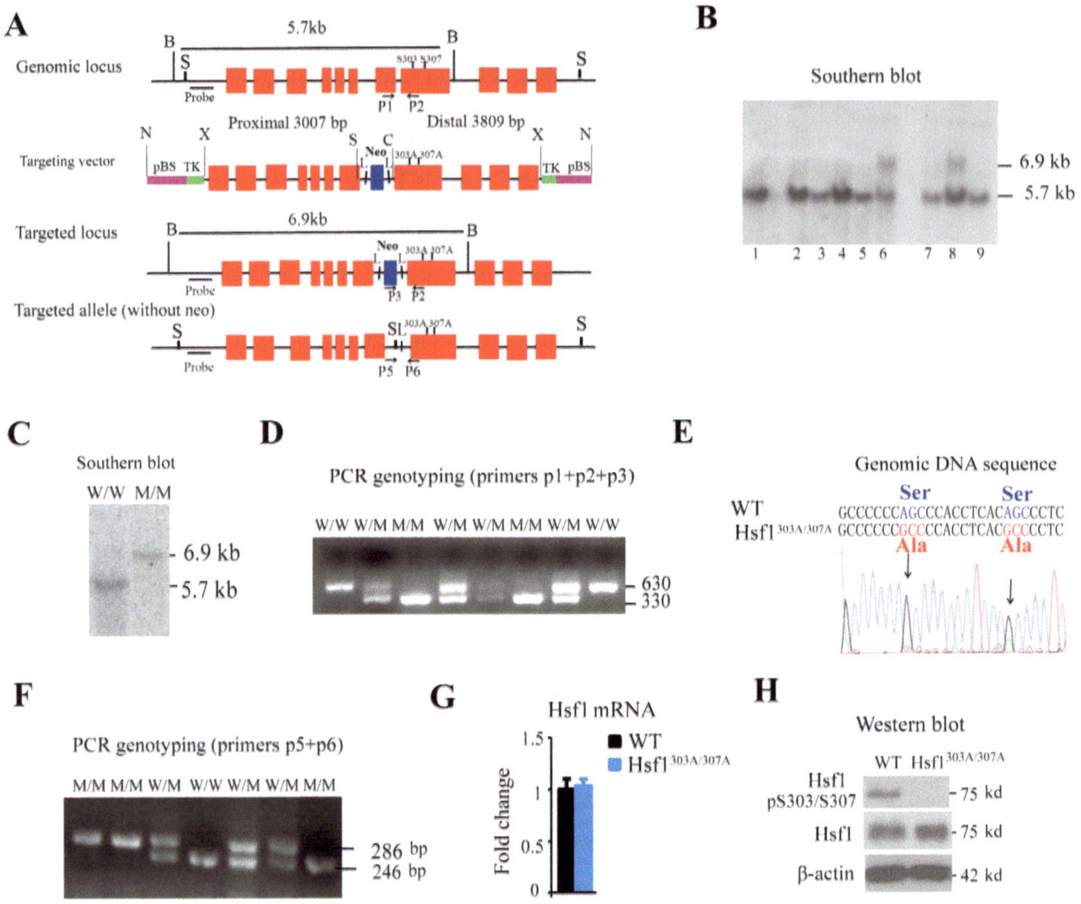

Fig. 1 Generation of HSF1 (303A/307A) knock-in mice. (**a**) Schematic illustrating the targeting strategy for generation of HSF1 (303A/307A) (Hsf1$^{303A/307A}$) mutant mice (modified from published Figure (Mol Cell Biol. 2018 Aug 28;38(18). pii: MCB.00095–18) [22]. The genomic HSF1 locus, targeting vector, targeted locus (with mutant HSF1 (303A/307A) and the neomycin (neo) gene in the genome of ES cells and mouse), and the targeted locus with mutant HSF1 (303A/307A) (without neo gene) in the genome of mutant mice are presented. The neo gene is presented in blue and exons are in red. Locations of S303 and S307 sites in the genomic HSF1 locus, probes for Southern blot analysis, and genotyping primers (P1 to P6) for PCR are indicated. Notably, Neo is flanked by Lox P sites (L) and can be removed after crosses with the Cre recombinase transgenic mice. Note: distances are not in proportion. B, Bgl II; S, Sac I; X, Xho I; N, Not I; C, Cla I restriction enzyme sites. TK is thymidine kinase, and pBS is pBluescript KS plasmid. (**b**) Southern blot analysis to select correctly targeted ES cell clones. The genomic DNA was isolated from ES cell clones (obtained after double drug selection, *see* Materials and Methods), digested with Bgl II restriction enzyme, fractionated on agarose gel, and transferred to the nitrocellulose membrane. The fractionated genomic DNA was hybridized with the external probe to yield a 5.7 kb (for WT) or 6.9 kb (for mutant) DNA band. Lanes 1–5, 7, and 9 represent WT ES cell clones; lanes 6 and 8 represent heterozygous ES cell clones. (**c**) Southern blot analysis to verify WT and homozygous mutant mice. Southern blot analysis was performed as described in (**b**) using genomic DNA isolated from the tails of WT and homozygous mutant mice. (**d**) PCR-based genotyping and verification of WT (W/W), Hsf1$^{303A/307A/WT}$ (W/M), and Hsf1$^{303A/307A}$ (M/M) mice containing neo gene cassette in the HSF1 allele. The PCR analysis was performed with a primer combination of P1, P2, and P3 (locations of the primers are indicated in panel A) using tail genomic DNA as templates. A 630 bp fragment for the mutant allele and a 330 bp fragment for the WT allele were amplified by the above primers. (**e**) DNA sequencing analysis to confirm the 303A/307A mutation in the HSF1 gene. The sequence of DNA fragments including the

2. **Electroporation of Targeting Vector into ES Cells and Identification of Positively Targeted ES Cell Clones**

The targeting vector (plasmid) was amplified and linearized at the Not I site to allow transfection into ES cells. The ES cells were electroporated with the targeting vector. The genomic DNA of ES cell clones doubly selected using G418 (200 μg/ml) (positive selection) and ganciclovir (2 μM) (negative selection) were subjected to Southern blotting analysis to identify correctly targeted ES cell clones.

To identify the clones containing homologous recombination of mutant HSF1, an external probe was designed. The external probe hybridizes to the DNA region that is close to, but not included in the targeting vector. According to the targeting strategy, we designed a 600 bp probe that predictably hybridizes to the 5′ region of the targeting vector as indicated in Fig. 1a. The external probe for Southern blotting was prepared by PCR using specific primers, forward primer: 5'-GGGAAGAATGGGGA CTAAAC-3′, reverse primer: 5'-TACATGTGGCACTCACATAT-3′. The DNA probe was labeled with ^{32}P-ATP (NEG002A100UC, Perkin Elmer) using commercially available kits (Decalabel DNA labeling kit). The genomic DNA isolated from ES cells was digested with restriction enzyme Bgl II, fractionated by agarose gel electrophoresis, transferred to a nylon membrane, and hybridized with ^{32}P-ATP labeled external probe. The targeted (mutant) allele yielded a 6.9 kb band, while the WT allele shows a 5.7 kb band in Southern blotting analysis (Fig. 1b). Three ES cell clones (out of 129 clones obtained after double drug selection) containing homologous recombination (HR) of HSF1 mutant allele were identified by Southern blotting. The representative Southern blot result is presented in Fig. 1b.

3. **Microinjection of Positive ES Cell Clones into Blastocysts, Generation of Chimeras, and Selection of Mice with Germline Transmission**

Fig. 1 (continued) S303 and S307 coding sequence was amplified by PCR using genomic DNA from WT and homozygous Hsf1$^{303A/307A}$ mice. The coding substitutions replacing serine with alanine (AGC to GCC) are indicated in the mutant genomic locus. (**f**) PCR-based genotyping of WT (W/W), heterozygous Hsf1$^{303A/307A/WT}$ (W/M), and homozygous Hsf1$^{303A/307A}$ (M/M) mice following crossing Hsf1$^{303A/307A}$ homozygous mice with transgenic mice expressing Cre recombinase to remove the neo cassette. The PCR analysis was performed with a primer combination of P5 and P6 (locations of the primers are indicated in panel A) using tail genomic DNA as templates. A 286 bp fragment for the mutant allele and 246 bp fragment for the WT allele were amplified by these primers. (**g**) Relative mRNA levels of HSF1 in WT and Hsf1$^{303A/307A}$ MEFs determined by RT-PCR. Bars are means +/− SD (n = 5 per group). (**h**) Immunoblot analysis using antibodies to detect total HSF1 or the phosphorylated form of HSF1 at S303/S307 in MEF cell lysates prepared from WT and Hsf1$^{303A/307A}$ (M/M). β-Actin represents loading control

Two correctly targeted ES cell clones were microinjected into mouse blastocysts to generate germline-transmitting chimeric mice. The chimeric mice were intercrossed with C57BL/6 mice. The heterozygous mice (Hsf1$^{WT/303A/307A}$) with germline transmission of mutant allele were identified by PCR and Southern blot analysis.

The heterozygous mice with germline transmission of the mutant allele were intercrossed to generate homozygous (Hsf1$^{303A/307A}$) and WT mice. The genotype of the mice was detected by Southern blotting as described above and by PCR. The Southern blotting results of DNA isolated from homozygous mutant and WT mice are presented in Fig. 1c. When the genotype of HSF1 knock-in mutant mice was confirmed by Southern blot, thereon mice were routinely genotyped using PCR of tail DNA. To identify the mutant and WT HSF1 alleles, the following primers were used in the PCR reactions: primers P1 (5′- CCATGG-GACTGCCAGTAAGT-3′) and P2 (5'-CTTGTCTAGGCA GGCTACGC-3′) were used to generate a 630 bp DNA fragment for WT allele, and primers P2 and P3 (5′- CTCGACATCGGAAGA TCCAT-3′) were used to produce 330 bp DNA fragment for the mutant allele (Fig. 1a and d). To remove the neomycin (neo) gene from the genome of Hsf1$^{303A/307A}$ mice, the homozygous mice were crossed with Cre transgenic female mice (Jackson Laboratory). After the neo gene cassette was removed, a 40 bp Lox P sequence remained in the intron 8 of the HSF1 gene and thereby in Hsf1$^{303A/307A}$ cells, which was used in genotyping using primers P5 and P6, resulting in the mutant band to being 40 bp longer. The removal of the neo gene and genotype of mice were identified by PCR using primers P5 (5'-GCAGGACCTTTATCCCTCCT-3′) and P6 (5'-GTTGGGGACAAAGGGGTATC-3′) and DNA sequencing. The P5 and P6 primers produced a 246 bp band for the WT allele and a 286 bp band for the mutant (Fig. 1f).

Note that all the work using mice was approved by the Institutional Animal Use and Care Committee of Augusta University.

4. **Characterization of HSF1 Expression in Hsf1$^{303A/307A}$ Knock-in Mice**

To investigate whether our knock-in strategy affects the transcription of HSF1, we performed RT-PCR to determine the HSF1 mRNA level using the following primers: forward primer: 5'-TCTCACTGGTGCAGTCGAAC-3′ and reverse primer: 5'-GT AGGCTGGAGATGGAGCTG-3′. The data indicate that the mRNA level of HSF1 in Hsf1$^{303A/307A}$ cells is comparable to that of WT cells (Fig. 1g). The DNA point mutation of S303/S307 residues in HSF1 was verified by DNA sequencing using DNA isolated from homozygous Hsf1$^{303A/307A}$ mouse tissue (Fig. 1e). The elimination of phosphorylation at residues S303/S307 in

HSF1 was further confirmed by immunoblot analyses using S303/S307 phospho-specific antibody (gift from Dr. L. Sistonen, Abo Akademi University, Finland) (Fig. 1h).

5. **Generation and Transformation of Mouse Embryo Fibroblasts (MEF)**

Primary MEFs were prepared from embryonic day 13.5 (E13.5) following timed pregnancies as described previously [24, 25]. MEFs were immortalized following transfection with plasmids encoding SV40 large T antigen or transformed by stably expressing oncogenes (E1A and RAS) using a retroviral vector system. Transformed MEFs were selected using puromycin (2 μg/ml for E1A) and blasticidin (3 μg/ml, for RAS) and cultured in Dulbecco's Minimal Essential Medium (DMEM) supplemented with 10% heat-inactivated fetal calf serum (FCS).

6. **Drug Treatment and Colony Formation Assay**

Cellular survival using colony formation assays was determined as previously described [24, 25]. Briefly, cells were treated with chemotherapeutic agents (doxorubicin at 0.5 μg/ml or etoposide at 5 μg/ml concentrations for 16 h at 37 °C). An appropriate number of cells were plated and incubated for 10 days at 37 °C and 5% CO2. Colonies were stained with crystal violet and those containing more than 50 cells were counted. The plating efficiency (PE) of untreated cells was also determined. The surviving fraction was determined as the number of colonies for treated cells divided by the number of cells plated and divided by PE for each group.

7. **Hematoxylin and Eosin (H & E) Staining**

Liver and tumor tissues were fixed in 10% formalin and embedded in paraffin, and 7 μm tissue sections were prepared as described previously [26, 27]. For the H & E staining, the tissue sections were deparaffinized in xylene and rehydrated in a series of alcohol/water mixtures before staining with H & E sequentially.

8. **Western Blotting**

Cell lysates were prepared from cultured cells using RIPA buffer (150 mM NaCl, 1% NP-40, 0.5% deoxycholate, 0.1% SDS, 50 mM Tris, pH 7.5) containing protease and phosphatase inhibitor cocktail (Thermo Scientific, IL). The cellular protein lysates were fractionated on SDS-PAGE gels, transferred onto a polyvinylidene fluoride membrane (PVDF) and probed with the indicated primary followed by secondary antibody [26, 27].

9. **Quantitative RT-PCR**

Quantitative RT-PCR was performed as described previously [26, 27]. Briefly, total RNA was isolated from tissues or cultured cells using TRIzol and reverse transcribed using the Iscript cDNA synthesis kit. Quantitative RT-PCR was performed using iQ SYBR green Supermix on a CFX Opus 96 Real-time PCR System (Bio-Rad Laboratories).

10. **Diethylnitroseamine (DEN)-Induced Hepatocellular Carcinoma (HCC)**

DEN-induced HCC was generated using Hsf1$^{303A/307A}$ and WT mice as described previously [26, 27]. A single dose of DEN (25 mg/kg body weight) was injected into 14-day-old male WT and Hsf1$^{303A/307A}$ mice intraperitoneally (i.p.) to initiate tumor formation. DEN-treated mice were observed for evidence of sickness twice a week. At 10 months post-DEN injection, mice were euthanized and livers were removed and analyzed for the presence of HCCs using a dissecting microscope. The number of all tumors and the number of tumors larger than 0.2 cm in diameter were quantified. The maximum size of the tumors in each liver was measured using a caliper. Tumor volume was calculated using the following formula: V (volume) = W^2 x L/2, where W represents the width of the tumor and L represents the length of the tumor.

4 Physiological Effects of HSF1 Following S303/S307 Mutations to Alanine Residues

4.1 Loss of HSF1 Phosphorylation at S303/S307 Promotes Cell Proliferation, Drug Resistance, and Tumorigenesis

A number of studies have shown that HSF1 is required for malignant transformation and is essential for tumor cell survival. It has been demonstrated that deletion or inhibition of HSF1 leads to cell death in various tumor types [27–30]. In vivo, we have shown that loss of HSF1 blocks DEN-induced HCC in an animal model [27]. As mentioned above, loss of HSF1 S303/S307 phosphorylation sites increases its transcriptional activity and the expression levels of its downstream target genes, such as HSP90, HSP70, and HSP25 [22]. To examine whether increasing HSF1 activity facilitates the process of tumorigenesis, we first determined the cellular proliferation of transformed Hsf1$^{303A/307A}$ MEFs. WT and Hsf1$^{303A/307A}$ primary MEFs were transformed by overexpression of oncogene E1A and RAS. The data presented in Fig. 2a show significantly higher growth rate of transformed Hsf1$^{303A/307A}$ MEFs compared to WT MEFs. The doubling time of transformed Hsf1$^{303A/307A}$ MEFs was reduced by 35% compared to that of WT MEFs (1.42 days in WT versus 0.94 day in Hsf1$^{303A/307A}$). Consistently, colony formation assay of Hsf1$^{303A/307A}$ MEFs shows a significantly higher number of colonies observed than for WT MEFs (Fig. 2b). Lack of HSF1 phosphorylation at S303/S307 significantly increased cell survival when cells were exposed to chemotherapeutic reagents (Fig, 2c-d). Taken together, the above results indicate that activation of HSF1 by mutations of S303/

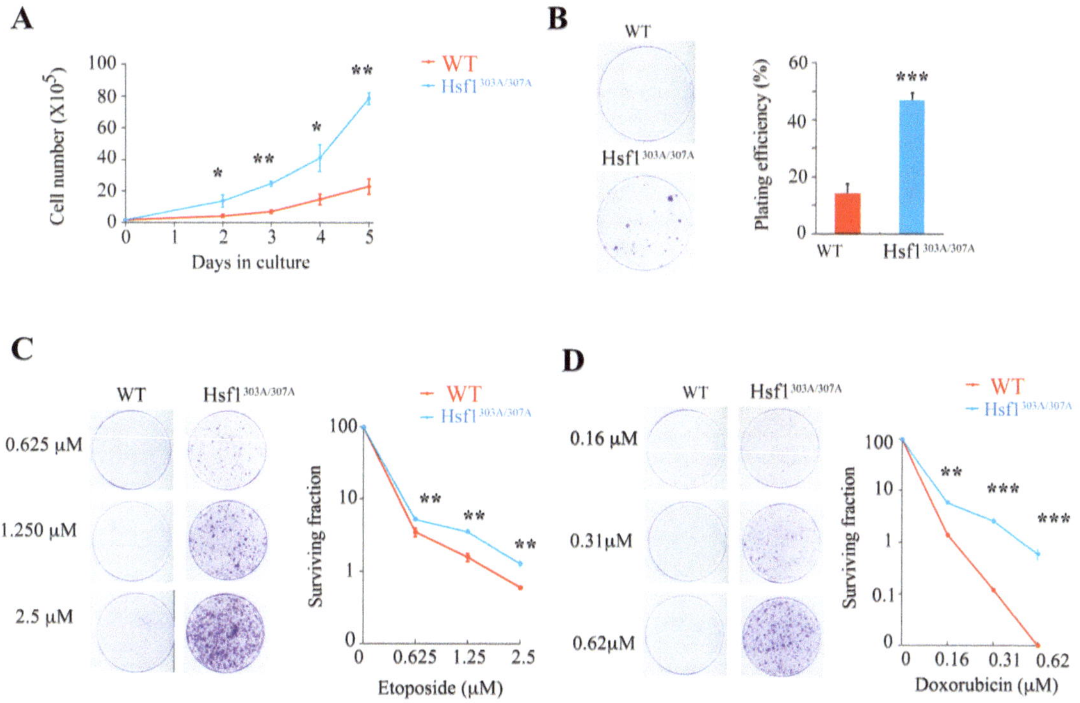

Fig. 2 Lack of phosphorylation of HSF1 at 303A/307A increases cell proliferation and promotes survival of cells exposed to chemotherapeutic drugs. The Hsf1$^{303A/307A}$ and WT primary MEFs were transformed by transfection of plasmids encoding E1A and RAS as described in Materials and Methods. (**a**) The growth rate of E1A and RAS transformed Hsf1$^{303A/307A}$ and WT MEFs. The proliferation rate of Hsf1$^{303A/307A}$ MEFs was significantly higher than that observed in WT MEFs. (**b**) Plating efficiency (PE) using colony formation assay. Untreated WT and Hsf1$^{303A/307A}$ E1A and RAS transformed MEFs were plated and cultured for 10 days. The number of colonies and the PE were calculated, as described in Materials and Methods. The left panel shows a representative image of the plates containing colonies. The right panel is the quantification of the PE. (**c–d**) The E1A and RAS transformed MEFs were treated with etoposide (**c**) or doxorubicin (**d**) at the indicated concentrations for 16 h. Determination of surviving fraction using colony formation assay was performed as described in Materials and Methods. The left panels show representative images of plates containing colonies. The right panels show the quantification of the surviving fractions. In all relevant panels, statistical significance is indicated (*$p < 0.05$, **$p < 0.01$, ***$p < 0.001$). Note that all experiments repeated at least two times with multiple samples

S307 phosphorylation sites promotes cell proliferation and their response to drug treatment.

Our previous study shows that loss of HSF1 S303/S307 phosphorylation in mice leads to age-associated obesity and obesity-related chronic inflammation [22]. It is well established that this metabolic state (obesity) and chronic inflammation are risk factors for cancer initiation and development [31, 32]. To investigate whether Hsf1$^{303A/307A}$ mice respond differently to tumor induction, we generated a DEN-induced HCC model using Hsf1$^{303A/307A}$ and WT mice (Fig. 3a–f). The total number of tumors at 10 months post-DEN treatment in Hsf1$^{303A/307A}$ mice slightly increased compared to that of WT mice (Fig. 3c). However,

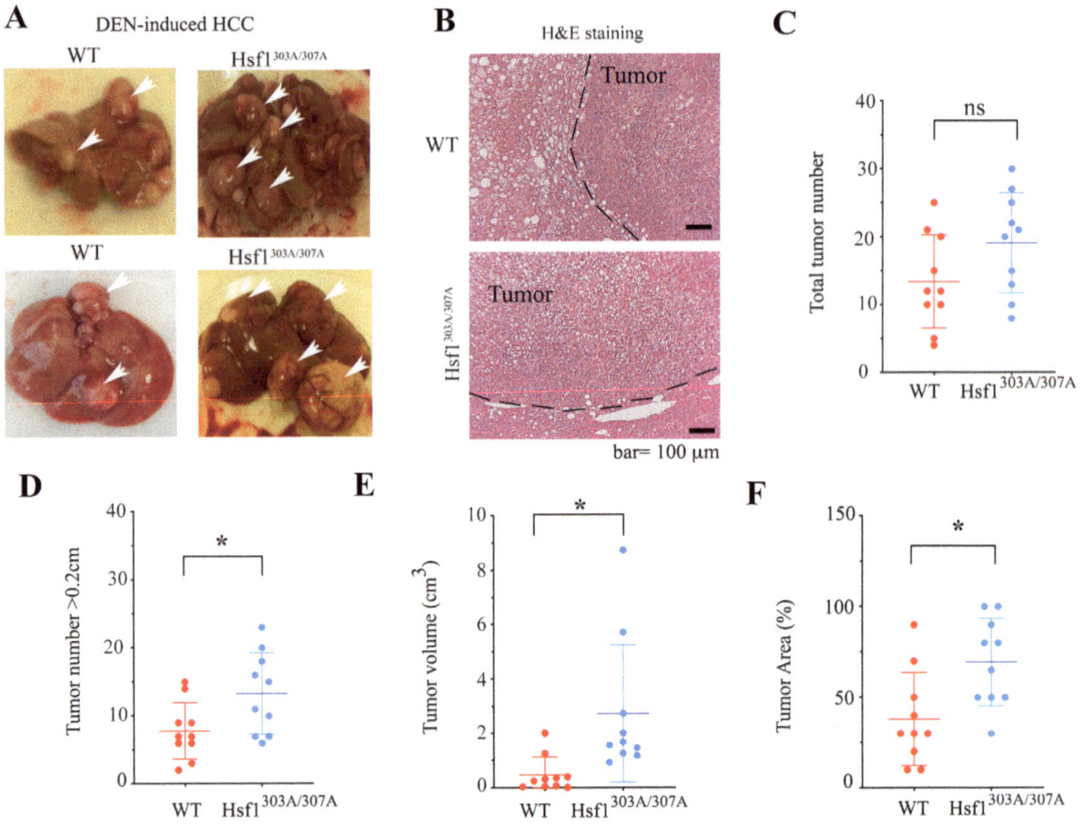

Fig. 3 Mice harboring HSF1 303A/307A exhibit accelerated tumorigenesis in vivo. (**a**) Two sets of representative macroscopic images of DEN-induced HCC in WT and Hsf1$^{303A/307A}$ mice. HCCs were induced by DEN as described in the Materials and Methods section. Arrows indicate tumor nodules. (**b**) Histological analysis of HCC (H & E staining). (**c–f**) Quantification of DEN-induced liver tumors observed in WT and Hsf1$^{303A/307A}$ mice. Total tumor number (**c**), tumor number larger than 0.2 cm in size (**d**), tumor volume (**e**), and tumor area (**f**) in livers were quantified. Scale bars are mean ± SD (n = 10 mice per group). For all relevant panels, statistical significance is indicated (*$p < 0.05$). ns, not significant

Hsf1$^{303A/307A}$ mice show significantly increased number of larger tumors (> 0.2 cm) compared to WT mice. In addition, the maximum tumor size and the tumor area covered in Hsf1$^{303A/307A}$ mice were also greater than that observed in WT mice (Fig. 3e–f). Taken together, the above experiments indicate that endogenous activation of HSF1 through lack of S303/S307 phosphorylation promotes tumorigenesis.

5 Conclusions

The above data indicate that mutations of only two of HSF1 serine phosphorylation residues to alanine leads to significant physiological consequences to the organism in vivo and to cells in vitro. Since most tumors exhibit higher levels of HSF1 activity, the above mouse line is a good model to study cellular proliferation, mechanisms of drug resistance, and cancer.

Acknowledgments

This work was supported by the NIH CA06213 grant.

References

1. Lindquist S (1986) The heat-shock response. Annu Rev Biochem 55:1151–1191
2. Westerheide SD, Morimoto RI (2005) Heat shock response modulators as therapeutic tools for diseases of protein conformation. J Biol Chem 280(39):33097–33100
3. Dai C, Sampson SB (2016) HSF1: guardian of proteostasis in cancer. Trends Cell Biol 26(1): 17–28
4. Mendillo ML et al (2012) HSF1 drives a transcriptional program distinct from heat shock to support highly malignant human cancers. Cell 150(3):549–562
5. Akerfelt M, Morimoto RI, Sistonen L (2010) Heat shock factors: integrators of cell stress, development and lifespan. Nat Rev Mol Cell Biol 11(8):545–555
6. Anckar J, Sistonen L (2011) Regulation of HSF1 function in the heat stress response: implications in aging and disease. Annu Rev Biochem 80:1089–1115
7. Ali A et al (1998) HSP90 interacts with and regulates the activity of heat shock factor 1 in Xenopus oocytes. Mol Cell Biol 18(9): 4949–4960
8. Zou J et al (1998) Repression of heat shock transcription factor HSF1 activation by HSP90 (HSP90 complex) that forms a stress-sensitive complex with HSF1. Cell 94(4):471–480
9. Abravaya K et al (1992) The human heat shock protein hsp70 interacts with HSF, the transcription factor that regulates heat shock gene expression. Genes Dev 6(7):1153–1164
10. Guettouche T et al (2005) Analysis of phosphorylation of human heat shock factor 1 in cells experiencing a stress. BMC Biochem 6:4
11. Soncin F et al (2003) Transcriptional activity and DNA binding of heat shock factor-1 involve phosphorylation on threonine 142 by CK2. Biochem Biophys Res Commun 303(2): 700–706
12. Olsen JV et al (2006) Global, in vivo, and site-specific phosphorylation dynamics in signaling networks. Cell 127(3):635–648
13. Boellmann F et al (2004) DAXX interacts with heat shock factor 1 during stress activation and enhances its transcriptional activity. Proc Natl Acad Sci U S A 101(12):4100–4105
14. Holmberg CI et al (2001) Phosphorylation of serine 230 promotes inducible transcriptional activity of heat shock factor 1. EMBO J 20(14): 3800–3810
15. Kim SA et al (2005) Polo-like kinase 1 phosphorylates heat shock transcription factor 1 and mediates its nuclear translocation during heat stress. J Biol Chem 280(13):12653–12657
16. Dai R et al (2000) C-Jun NH2-terminal kinase targeting and phosphorylation of heat shock factor-1 suppress its transcriptional activity. J Biol Chem 275(24):18210–18218
17. Kline MP, Morimoto RI (1997) Repression of the heat shock factor 1 transcriptional activation domain is modulated by constitutive phosphorylation. Mol Cell Biol 17(4):2107–2115
18. Chu B et al (1996) Sequential phosphorylation by mitogen-activated protein kinase and glycogen synthase kinase 3 represses transcriptional activation by heat shock factor-1. J Biol Chem 271(48):30847–30857
19. Wang X et al (2003) Regulation of molecular chaperone gene transcription involves the serine phosphorylation, 14-3-3 epsilon binding, and cytoplasmic sequestration of heat shock factor 1. Mol Cell Biol 23(17):6013–6026
20. Wang X et al (2004) Interactions between extracellular signal-regulated protein kinase 1, 14-3-3epsilon, and heat shock factor 1 during stress. J Biol Chem 279(47):49460–49469
21. Kourtis N et al (2015) FBXW7 modulates cellular stress response and metastatic potential through HSF1 post-translational modification. Nat Cell Biol 17(3):322–332
22. Jin X et al (2018) Modulation of heat shock factor 1 Activity through silencing of Ser303/Ser307 phosphorylation supports a metabolic program leading to age-related obesity and insulin resistance. Mol Cell Biol **38**(18): e00095–e00018

23. Meyers EN, Lewandoski M, Martin GR (1998) An Fgf8 mutant allelic series generated by Cre- and Flp-mediated recombination. Nat Genet 18(2):136–141
24. Jin X et al (2009) Heat shock factor 1 deficiency via its downstream target gene alphaB-crystallin (Hspb5) impairs p53 degradation. J Cell Biochem 107(3):504–515
25. Huang L et al (2007) Insights into function and regulation of small heat shock protein 25 (HSPB1) in a mouse model with targeted gene disruption. Genesis 45(8):487–501
26. Cho W et al (2019) The molecular chaperone heat shock protein 70 controls liver cancer initiation and progression by regulating adaptive DNA damage and mitogen-activated protein kinase/extracellular signal-regulated kinase signaling pathways. Mol Cell Biol **39**(9): e00391–e00318
27. Jin X, Moskophidis D, Mivechi NF (2011) Heat shock transcription factor 1 is a key determinant of HCC development by regulating hepatic steatosis and metabolic syndrome. Cell Metab 14(1):91–103
28. Dai C et al (2012) Loss of tumor suppressor NF1 activates HSF1 to promote carcinogenesis. J Clin Invest 122(10):3742–3754
29. Xi C et al (2012) Heat shock factor Hsf1 cooperates with ErbB2 (Her2/Neu) protein to promote mammary tumorigenesis and metastasis. J Biol Chem 287(42):35646–35657
30. Santagata S et al (2013) Tight coordination of protein translation and HSF1 activation supports the anabolic malignant state. Science 341(6143):1238303
31. Klil-Drori AJ, Azoulay L, Pollak MN (2017) Cancer, obesity, diabetes, and antidiabetic drugs: is the fog clearing? Nat Rev Clin Oncol 14(2):85–99
32. Hoy AJ, Nagarajan SR, Butler LM (2021) Tumour fatty acid metabolism in the context of therapy resistance and obesity. Nat Rev Cancer 21(12):753–766

Chapter 8

Bimolecular Fluorescence Complementation Assay to Evaluate HSP90-Client Protein Interactions in Cells

Abir Chakraborty, Gregory L. Blatch, and Adrienne L. Edkins

Abstract

Protein–protein interactions (PPI) in cells play a pivotal role in cellular function and dynamics. Cellular proteostasis is maintained by PPI networks between molecular chaperones, co-chaperones, and client proteins. Consequently, strategies to visualize and analyze PPI in cells are useful in understanding protein homeostasis regulation. The Bimolecular Fluorescence Complementation (BiFC) assay has emerged as a useful tool for studying PPI between proteins in live or fixed cells. BiFC is based on the detection of fluorescence generated when interacting protein pairs, produced as fusion proteins with either the N- or C-terminal fragment of a fluorescent protein, are in sufficient proximity to permit reconstitution of the split fluorophore. Here, we describe the application of the BiFC assay to a model of chaperone–client interactions using Hsp90 and the validated client protein CDK4. This assay allows for the distribution and spatiotemporal analysis of HSP90-CDK4 complexes in live or fixed cells and is amenable to studying the effects of inhibitors and mutations on chaperone–client protein networks.

Key words BiFC, Protein–protein interactions, PPI, HSP90, CDK4

1 Introduction

Protein–protein interactions (PPI) underpin many important biological processes in cells, including signal transduction, trafficking, metabolism, and transcriptional activation. In addition, PPIs are gaining popularity as drug targets, and hence efficient assays to screen for PPI inhibitors using small molecule inhibitors are required [1, 2]. However, visualizing the PPI in living cells can be challenging. Over the years, multiple strategies have been developed to observe cellular PPIs, including the Bimolecular Fluorescence Complementation (BiFC) assay [3–5]. The analysis of PPIs using BiFC is based on the structural complementation of two nonfluorescent fragments of a fluorescence protein, such as green fluorescence protein (GFP), resulting in this assay also being known as the "split GFP" assay. The two nonfluorescent fragments are fused to putative interacting proteins, and if these proteins assemble

Stuart K. Calderwood and Thomas L. Prince (eds.), *Chaperones: Methods and Protocols*, Methods in Molecular Biology, vol. 2693,
https://doi.org/10.1007/978-1-0716-3342-7_8,

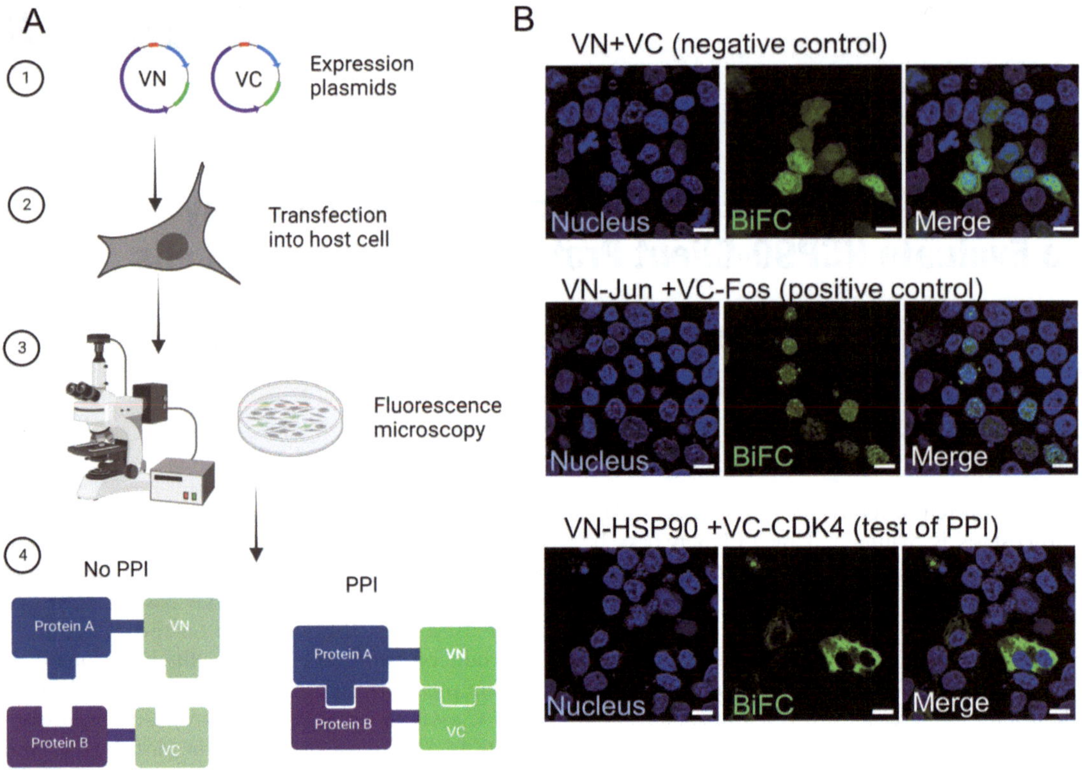

Fig. 1 Analysis of protein–protein interactions by bimolecular fluorescence complementation assay. (**a**) Schematic diagram of the BiFC strategy. (1) Putative interacting partner proteins are cloned into expression plasmids in a frame with either the C or N terminus of mVenus. (2) Expression plasmids are transfected into a suitable host cell. (3) The fluorescence intensity and localization detected by fluorescence microscopy. (4) If there is no interaction, the two domains of the fusion protein remain separate and have a diffuse, low-intensity signal, while a PPI will result in a high-intensity signal with specific subcellular localization. Image created with BioRender. (**b**) Example data from BiFC analysis of the interaction between HSP90 and CDK4 and minimum positive and negative controls. The scale bar is 10 μm. The negative control produces diffuse lower intensity signal that is distributed throughout the nucleus and cytoplasm (top panel), while the positive control and test reactions produce higher intensity signals that are localized to distinct positions in the cell depending on the PPI (middle and bottom panels)

to form a macromolecular complex, the two fragments are bought into close enough proximity to reconstitute the fluorophore [6, 7] (Fig. 1). BiFC was initially developed using YGFP. Subsequent to this, the rational modification and mutation of YGFP gave rise to a new version of super-enhanced yellow fluorescence GFP, called Venus fluorescent protein [3, 8, 9]. In BiFC, the N-terminal and C-terminal Venus fragments (VN173 and VC155, respectively) are fused to putative interacting partner proteins [3]. BiFC assays have been used to evaluate and identify numerous PPI in several organisms, including human cell line models, worms, plants, and yeast [10–13]. This assay can be utilized for live-cell imaging and in fixed cells [14] and can be applied to the screening of PPI inhibitors and to understand the contribution of specific amino acid residues to established PPIs [15]. The intensity of the fluorescence signal can be correlated with the strength of the PPI of the protein pairs [16]. Additionally, BiFC improves the stability of PPIs, making it an attractive option for assessing transient or weak PPIs in cells without compromising the native cellular state [16–18]. We have successfully applied this technique to investigate HSP90–client interactions in cells. HSP90 is a ubiquitous molecular chaperone that plays an important role in multiple cellular functions [19–21]. HSP90 function is controlled by intrinsic ATPase activity and participation in a network of PPIs with client proteins (viz., Hsp90-interacting proteins that rely on the chaperone for stability or function) and/or co-chaperones (non-client regulatory proteins that interact with HSP90 to modulate its function) [22, 23]. Here we utilized this assay to evaluate the subcellular location and interaction of the HSP90-CDK4 complex in cells as an indicator of HSP90–client protein interactions [24].

2 Materials

1. Empty backbone vector controls containing coding regions for only the mVenus C-terminal region (VC) and N-terminal region (VN) (pBiFC-HA-VC155 and pBiFC-FLAG-VN173) are available from Addgene [25].
2. Plasmid vectors pBiFC-HSP90α-VN173 and pBiFC-CDK4-VC155 that contain the coding regions for HSP90α and CDK4 within the pBiFC-HA-VC155 and pBiFC-FLAG-VN173 backbones, respectively. The coding regions for HSP90α and CDK4 are ligated in the frame with the FLAG and HA tags, respectively [24]. These plasmids are available on request.
3. BiFC-positive control plasmids containing the coding regions of interacting bFOS and bJUN transcription factors (pBiFC-bFOSVC155 and pBiFC-bJUNVN173) are available from Addgene [25].
4. Cell line model of choice (e.g., HEK293T).
5. Materials for cell culture (e.g., culture medium, tissue culture and plasticware).
6. Materials for cell transfection (e.g., X-tremeGENE HP transfection reagent).
7. Sterilized coverslips, microscope slides.
8. Clear nail varnish.

9. Ice-cold phosphate buffered saline (PBS; 137 mM NaCl, 2.7 mM KCl, 4.3 mM Na_2HPO_4, 1.47 mM KH_2PO_4, pH 7.4).
10. 3.7% (w/v) paraformaldehyde solution in 1 × PBS (freshly prepared).
11. Hoechst 33342 solution (1 μg/mL in distilled water).
12. Aluminum foil.
13. Anti-fade mounting medium (e.g., DAKO).

2.1 Equipment

1. Tissue culture hood
2. 37 °C cell culture incubator
3. Confocal fluorescence microscope or widefield fluorescence microscope equipped with filter sets for Hoechst 33358 and GFP/FITC (excitation 457–487 nm and emission 502–538 nm)

3 Methods

1. The day before transfection, prepare a 12-well tissue culture plate by placing a single sterile coverslip on each well (depending on the number of samples). Add 9×10^4 cells into each well from a log-phase growing culture in a 1 mL culture volume (cell density of 9×10^4 cells/mL), and allow cells to adhere overnight at 37 °C in a humidified cell culture incubator with the appropriate CO_2 concentration (*see* **Note 1**). The coverslips can be sterilized by overnight incubation in ethanol or by autoclaving.
2. Co-transfect cells with the appropriate pair of plasmids using the appropriate transfection reagent according to standard protocols (*see* **Note 2**). Note that transfection with both test and control plasmid pairs is necessary to ensure confidence in the interpretation of the signal (*see* **Note 3** and Table 1).
3. Incubate cells for 36 h at 37 °C in a cell culture incubator (*see* **Note 4**).
4. Remove the cell culture medium and wash the cell monolayer gently with 1 mL of 1 × PBS.
5. Fix cells with 1 mL of 3.7% (w/v) of freshly prepared paraformaldehyde in 1 × PBS for 10 min at room temperature in the dark by wrapping the plate in aluminum foil (*see* **Note 5**).
6. Discard the paraformaldehyde solution, and wash fixed cells twice with 1 mL of 1 × PBS.

Table 1
Recommended control and test plasmid pairs for BiFC

Set number	Plasmid 1	Plasmid 2	Purpose of experiment
1	pBiFC-VC155	pBiFC-VN173	Negative control
2	pBiFC-VC155	pBiFC-HSP90α-VN173	Negative control
3	pBiFC-CDK4-VC155	pBiFC-VN173	Negative control
4	pBiFC-bFOS-VC155	pBiFC-bJUN-VN173	Positive control
5	pBiFC-CDK4-VC155	pBiFC-HSP90α-VN173	Test of PPI

7. Add 1 mL of Hoechst 33342 (1 μg/mL in distilled water) to each well to stain nuclear DNA, and incubate for 3 min at room temperature in the dark (*see* **Note 6**).
8. Remove the Hoechst 33342 staining solution, and wash with 1 mL of 1 × PBS solution.
9. Using a spatula and forceps, carefully lift the coverslips from the bottom of the cell culture plate and place them on a paper towel on the bench with the cell-side surface facing upwards.
10. Add 10 μL of anti-fade mounting medium to the center of the coverslip.
11. Gently place a microscope slide over the coverslip with the mounting medium. This should result in the coverslip becoming attached to the microscope slide and equal spreading of the mounting medium across the surface. The surface of the coverslip containing the cells should now face the microscope slide (*see* **Note 7**)
12. Seal the edges of the attached coverslips with transparent nail varnish to prevent movement and for long-term storage (*see* **Note 7**).
13. At this stage, the slide can be stored in a microscope slide box at 4 °C in the dark until imaging.
14. A confocal or fluorescence microscope can be used to capture the image using standard FITC or GFP filter sets for the mVenus and standard DAPI filter sets for the Hoechst 33342. Use a 20× objective to obtain a broad overview of a large number of cells, while analysis using the 60× or 100× objective will be necessary to analyze the subcellular localization of the signal (*see* **Notes 8** and **9**).

4 Notes

1. *Selection and Preparation of Cells*

   The selection of an appropriate cell culture system is crucial for efficient investigation. Adherent cells that grow as a monolayer are convenient for high-quality imaging. The BiFC assay has been documented in a variety of cells, including COS-1, HeLa, Hep3B, NIH3T3, αTN4, HEK293, and HEK293T [4, 8, 26–30]. For efficient co-transfection, cell confluence plays a vital role. We observed the maximum transfection efficiency at 60–70% confluency with a minimum number of overlapping cells.

2. *Transfection and Confirmation of Protein Production*

   A range of transfection protocols can be used. We routinely use the X-tremeGENE HP transfection protocol from Roche. Briefly, a total of 1 μg of DNA at a 1:1 ratio was mixed with a 100 μL serum-free medium. Next, a 1 μL of X-tremeGENE HP transfection reagent was gently added to the middle of the solution and incubated at 24 °C for 15 min. The total transfection reaction complex was added dropwise in the respective well and incubated in the cell culture incubator for 24–36 h before further processing. The equivalent amount of all plasmids should be used in both test and control sets. Ideally, the expression of proteins from the plasmid pairs should also be confirmed by western blot (using established methods). This is particularly important to validate protein production in the event that no fluorescent signal is observed in BiFC. We recommend using either protein-specific antibodies (e.g., against HSP90α or CDK4) or antibodies against the HA and FLAG tags found in the fusion proteins produced from the pBiFC-HA-VC155 and pBiFC FLAG-VN173 plasmids, respectively.

3. *Control and Test Plasmid Pairs*

   Interpretation of the BiFC signal requires appropriate controls to identify nonspecific background fluorescence. We suggest that at the start of the experiment, at least three negative controls and one positive control be run alongside the test assay (Table 1). Usually, some diffuse background staining can be observed in the negative controls, while a *bona fide* PPI produces an intense fluorescence at the correct emission wavelength that shows a specific subcellular localization.

4. *Incubation Time Post-transfection*

   The incubation period post-co-transfection could be varied depending on the expression level of your protein of interest. Ideally, 24–36 h of incubation is sufficient for good-quality images. Avoid extended incubations as in our experience this results in higher background nonspecific fluorescence.

5. *Cell Fixation*

   One should always use freshly prepared paraformaldehyde solution. Dissolve 0.37 g of paraformaldehyde powder in 10 mL of 1 × PBS and heat to 60 °C. Vortex occasionally to dissolve the powder. Once the paraformaldehyde is completely dissolved, filter-sterilize with a 0.2 μm filter and store at 4 °C for a maximum of 2 weeks.

6. *Hoechst 33342 Staining*

   Hoechst 33342 reagent will stain live and fixed cells. DAPI or propidium iodide can be used as alternative nuclear dyes on fixed cell samples.

7. *Sample Handling*

   While removing the coverslip from the cell culture plate, avoid making contact with the cell surface as this will damage the cell morphology and render the results unusable.

   Avoid adding excess mounting medium as it can cause significant movement of coverslips after mounting. Gently press the middle of the coverslip to eliminate any bubbles if present. Use filter paper to remove the excess mounting medium around the edges, seal the edges with nail varnish, and dry to stop further movement and to seal for long-term storage.

8. *Antibody-Mediated Staining*

   Our analysis uses endogenous fluorescence of the recombined mVenus protein to detect the PPI. Antibody-mediated fluorescence detection using an antibody against GFP has been reported for this BiFC protocol. However, in our hands, we observed significant background fluorescence from the controls when using antibody-mediated fluorescence detection, which we determined to be due to the ability of all the antibodies we tested to detect the individual GFP domains. Therefore, we do not recommend using antibody-mediated detection unless the GFP antibody has been validated to only detect the recombined protein and when including all the relevant negative controls.

9. *Quantitation of BiFc Intensity*

   The protocol described herein gives a qualitative assessment of the PPI. However, if quantitative data are required, this can be achieved by co-transfection of a third plasmid producing RFP and quantifying the ratio of the BiFC signal intensity to this RFP signal. This is useful if comparing the effect of inhibitors on a PPI or if monitoring the effects of amino acid mutations on the strength of the PPI.

References

1. Delahunty CM, Yates JR (2019) Protein-protein interactions. Proteomics Biol Discov:125–144. https://doi.org/10.1002/9781119081661.ch5
2. Berggård T, Linse S, James P (2007) Methods for the detection and analysis of protein-protein interactions. Proteomics 7:2833–2842. https://doi.org/10.1002/pmic.200700131
3. Shyu YJ, Liu H, Deng X, Hu CD (2006) Identification of new fluorescent protein fragments for bimolecular fluorescence complementation analysis under physiological conditions. Biotechniques 40:61–66. https://doi.org/10.2144/000112036
4. Miller KE, Kim Y, Huh WK, Park HO (2015) Bimolecular fluorescence complementation (BiFC) analysis: advances and recent applications for genome-wide interaction studies. J Mol Biol 427:2039–2055. https://doi.org/10.1016/j.jmb.2015.03.005
5. Kerppola TK (2006) Design and implementation of bimolecular fluorescence complementation (BiFC) assays for the visualization of protein interactions in living cells. Nat Protoc 1:1278–1286. https://doi.org/10.1038/nprot.2006.201
6. Ventura S (2012) Bimolecular fluorescence complementation: illuminating cellular protein interactions. Curr Mol Med 11:582–598. https://doi.org/10.2174/156652411800615117
7. Morell M, Czihal P, Hoffmann R, Otvos L, Avilés FX, Ventura S (2008) Monitoring the interference of protein-protein interactions in vivo by bimolecular fluorescence complementation: the DnaK case. Proteomics 8: 3433–3442. https://doi.org/10.1002/pmic.200700739
8. Hu CD, Chinenov Y, Kerppola TK (2002) Visualization of interactions among bZIP and Rel family proteins in living cells using bimolecular fluorescence complementation. Mol Cell 9:789–798. https://doi.org/10.1016/S1097-2765(02)00496-3
9. Nagai T, Ibata K, Park ES, Kubota M, Mikoshiba K, Miyawaki A (2002) A variant of yellow fluorescent protein with fast and efficient maturation for cell-biological applications. Nat Biotechnol 20:87–90. https://doi.org/10.1038/nbt0102-87
10. Walter M, Chaban C, Schütze K, Batistic O, Weckermann K, Näke C, Blazevic D, Grafen C, Schumacher K, Oecking C, Harter K, Kudla J (2004) Visualization of protein interactions in living plant cells using bimolecular fluorescence complementation. Plant J 40:428–438. https://doi.org/10.1111/j.1365-313X.2004.02219.x
11. Bracha-Drori K, Shichrur K, Katz A, Oliva M, Angelovici R, Yalovsky S, Ohad N (2004) Detection of protein-protein interactions in plants using bimolecular fluorescence complementation. Plant J 40:419–427. https://doi.org/10.1111/j.1365-313X.2004.02206.x
12. Skarp KP, Zhao X, Weber M, Jäntti J (2008) Use of bimolecular fluorescence complementation in yeast saccharomyces cerevisiae. Methods Mol Biol 457:165–175. https://doi.org/10.1007/978-1-59745-261-8_12
13. Subotić A, Swinnen E, Demuyser L, De Keersmaecker H, Mizuno H, Tournu H, Van Dijck P (2017) A bimolecular fluorescence complementation tool for identification of protein-protein interactions in Candida albicans. G3 Genes Genomes Genet 7:3509–3520. https://doi.org/10.1534/g3.117.300149
14. Lai H-T, Chiang C-M (2013) Bimolecular fluorescence complementation (BiFC) assay for direct visualization of protein-protein interaction in vivo. Bio-Protocol 3:10.21769/bioprotoc.935
15. Wang Y, Wang Z, Liu J, Wang Y, Wu R, Sheng R, Hou T (2021) Discovery of novel HBV capsid assembly modulators by structure-based virtual screening and bioassays. Bioorg Med Chem 36. https://doi.org/10.1016/j.bmc.2021.116096
16. Morell M, Espargaró A, Avilés FX, Ventura S (2007) Detection of transient protein-protein interactions by bimolecular fluorescence complementation: the Abl-SH3 case. Proteomics 7: 1023–1036. https://doi.org/10.1002/pmic.200600966
17. Robida AM, Kerppola TK (2009) Bimolecular fluorescence complementation analysis of inducible protein interactions: effects of factors affecting protein folding on fluorescent protein fragment association. J Mol Biol 394:391–409. https://doi.org/10.1016/j.jmb.2009.08.069
18. Chu J, Zhang Z, Zheng Y, Yang J, Qin L, Lu J, Huang ZL, Zeng S, Luo Q (2009) A novel far-red bimolecular fluorescence complementation system that allows for efficient visualization of protein interactions under physiological conditions. Biosens Bioelectron 25:234–239. https://doi.org/10.1016/j.bios.2009.06.008
19. Edkins AL (2016) Hsp90 co-chaperones as drug targets in cancer: current perspectives.

Topics Med Chem 19:21–54. https://doi.org/10.1007/7355_2015_99

20. Chakraborty A, Edkins AL (2021) HSP90 as a regulator of extracellular matrix dynamics. Biochem Soc Trans 49:2611–2625. https://doi.org/10.1042/BST20210374
21. Jackson SE, Queitsch C, Toft D (2004) Hsp90: from structure to phenotype. Nat Struct Mol Biol 11:1152–1155. https://doi.org/10.1038/nsmb1204-1152
22. Schopf FH, Biebl MM, Buchner J (2017) The HSP90 chaperone machinery. Nat Rev Mol Cell Biol 18:345–360. https://doi.org/10.1038/nrm.2017.20
23. Prodromou C (1823) The "active life" of Hsp90 complexes. Biochim Biophys Acta, Mol Cell Res 2012:614–623. https://doi.org/10.1016/j.bbamcr.2011.07.020
24. Chakraborty A, Boel NME, Edkins AL (2020) HSP90 interacts with the fibronectin N-terminal domains and increases matrix formation. Cell 9:20–24. https://doi.org/10.3390/cells9020272
25. Hiatt SM, Shyu YJ, Duren HM, Hu CD (2008) Bimolecular fluorescence complementation (BiFC) analysis of protein interactions in Caenorhabditis elegans. Methods 45:185–191. https://doi.org/10.1016/j.ymeth.2008.06.003
26. Grinberg AV, Hu C-D, Kerppola TK (2004) Visualization of Myc/Max/Mad family dimers and the competition for dimerization in living cells. Mol Cell Biol 24:4294–4308. https://doi.org/10.1128/mcb.24.10.4294-4308.2004
27. Fang D, Kerppola TK (2004) Ubiquitin-mediated fluorescence complementation reveals that Jun ubiquitinated by Itch/AIP4 is localized to lysosomes. Proc Natl Acad Sci U S A 101:14782–14787. https://doi.org/10.1073/pnas.0404445101
28. Schmidt C, Peng B, Li Z, Sclabas GM, Fujioka S, Niu J, Schmidt-Supprian M, Evans DB, Abbruzzese JL, Chiao PJ (2003) Mechanisms of proinflammatory cytokine-induced biphasic NF-κB activation. Mol Cell 12:1287–1300. https://doi.org/10.1016/S1097-2765(03)00390-3
29. Tao RH, Maruyama IN (2008) All EGF(ErbB) receptors have preformed homo- and heterodimeric structures in living cells. J Cell Sci 121:3207–3217. https://doi.org/10.1242/jcs.033399
30. Kerppola TK (2008) Bimolecular fluorescence complementation (BiFC) analysis as a probe of protein interactions in living cells. Annu Rev Biophys 37:465–487. https://doi.org/10.1146/annurev.biophys.37.032807.125842

Chapter 9

Complementation Assays for Co-chaperone Function

Adrienne L. Edkins and Gregory L. Blatch

Abstract

The development of mutant microorganisms lacking J domain proteins (JDPs; formerly called Hsp40s) has enabled the development of complementation assays for testing the co-chaperone function of JDPs. In these assays, an exogenously expressed novel JDP is tested for its ability to functionally substitute for a non-expressed or nonfunctional endogenous JDP(s) by reversing a stress phenotype. For example, the in vivo functionality of prokaryotic JDPs can be tested on the basis of their ability to reverse the thermosensitivity of a *dnaJ cbpA* mutant strain of the bacterium *Escherichia coli* (OD259). Similarly, the in vivo functionality of eukaryotic JDPs can be assessed in a thermosensitive *ydj1* mutant strain of the yeast *Saccharomyces cerevisiae* (JJ160). Here we outline the use of these thermosensitive microorganisms in complementation assays to functionally characterize a JDP from the bacterium, *Agrobacterium tumefaciens* (AgtDnaJ), and a JDP from the trypanosomal parasite, *Trypanosoma cruzi* (TcJ2).

Key words Heat shock proteins, DnaJ, Hsp40, J domain proteins, Molecular chaperones, Protein folding

1 Introduction

Complementation assays are a convenient and medium-throughput method of assessing the cellular functionality of molecular chaperones and co-chaperones. The approach is advantageous in that it provides confirmation of chaperone activity in vivo and insight into similarity with well-characterized chaperones that can be used to infer function [1, 2], particularly where the parent organism may be pathogenic or less amenable to genetic manipulation [3–5]. A novel Hsp70 chaperone or Hsp40 co-chaperone (now called J domain proteins, JDPs [6]) is expressed within a prokaryotic (e.g., mutant strains of *Escherichia coli*) or eukaryotic model host (e.g., mutant strains of *Saccharomyces cerevisiae*), to assay for its ability to functionally replace a non-expressed or nonfunctional Hsp70 or JDP. Certain mutant *E. coli* strains with disrupted chromosomal genes encoding JDPs (such as the *dnaJ cbpA* mutant strain *E. coli* OD259 [1, 7]) have been routinely used to assess whether exogenously

Stuart K. Calderwood and Thomas L. Prince (eds.), *Chaperones: Methods and Protocols*, Methods in Molecular Biology, vol. 2693, https://doi.org/10.1007/978-1-0716-3342-7_9,

expressed prokaryotic JDPs (plasmid-encoded) are able to functionally compensate for the missing JDPs (DnaJ and CbpA) and reverse the thermosensitivity phenotype. Similarly, thermosensitive *S. cerevisiae* strains lacking certain JDPs (such as the *ydj1* mutant strain *S. cerevisiae* JJ160 [2, 8]) have been used in complementation assays for testing the functionality of eukaryotic JDP. In this methods paper, we outline the use of thermosensitive microorganisms in complementation assays to functionally characterize a JDP from the bacterium, *Agrobacterium tumefaciens* (AgtDnaJ; [9]), and the trypanosomal parasite, *Trypanosoma cruzi* (TcJ2; [4]).

2 Materials

2.1 Prokaryotic Complementation Assay Materials

1. The *dnaJ and cbpA* mutant strain, *E. coli* OD 259 (MC4100 *araD139* Δ*ara714* Δ*cbpA::kan dnaJ::Tn10-42*), was initially described as *E. coli* WKG190 [1, 7]. At temperatures below 16 °C and above 37 °C, this strain is thermosensitive.
2. The plasmids pBAD22A (Amp^R; *colE1 ori*) and pWKG90 (pBAD22A*dnaJ*+) are used as negative and positive controls, respectively, with the *E. coli dnaJ* coding region under the control of the arabinose promoter.
3. The commercially available empty vector pQE30 (Qiagen, USA) was used to construct the plasmid encoding the prokaryotic JDP from *Agrobacterium tumefaciens* (AgtDnaJ), pRJ30 [9]. A derivative called pRJ30-H33Q was developed encoding AgtDnaJ-H33Q, which contains a mutation in the HDP motif of the J domain (H33Q) and renders it nonfunctional [9].
4. All plasmids are transformed into *E. coli* OD259 using well-established methods.

2.2 Eukaryotic Complementation Assay Materials

1. *Saccharomyces cerevisiae* JJ160 is a temperature-sensitive *ydj1* mutant strain [genotype *mat a trp1-1 ura3-1 leu2-3,112 his3-11,15 ade2-1 can1-100 GAL2+ met2-*Δ*1 lys2-*Δ*2 ydj1::HIS3*] displaying reduced growth at temperatures (30 and 34 °C) above the permissive growth temperature (23 °C) [2]. The strain is cultured in yeast minimal medium lacking histidine for selection.
2. The coding region for the *Trypanosoma cruzi* JDP, Tcj2, was inserted via *Eco*RI and *Xho*I sites into the yeast expression plasmid pKG6 [4]. The low copy number pKG6 contains an ampicillin resistance gene for selection in bacteria, the URA3 marker for selection in *S. cerevisiae*, and a *gal* promoter for regulated expression in the presence of galactose. The resultant pKG6-Tcj2 plasmid enables the production of Tcj2 in *S. cerevisiae* JJ160, which is induced by culture in a medium containing galactose as the sole carbon source.

3. The plasmid pRS317-*YDJ1* which contains the *YDJ1* coding region and the *LYS2* selectable marker is used as a positive control to rescue the temperature-sensitive phenotype of *S. cerevisiae* JJ160. The pKG6 empty plasmid is used as the negative control.
4. Plasmids are transformed into the *S. cerevisiae* JJ160 strain using the Frozen-EZ Transformation II™ Kit, according to the manufacturer's instructions.

3 Methods

3.1 Prokaryotic Complementation Assay Protocol

1. Pre-cultures (5 ml) of *E. coli* OD259 [pBAD22A], *E. coli* OD259 [pWKG90], *E. coli* OD259 [pQE30], *E. coli* OD259 [pRJ30], and *E. coli* OD259 [pRJ30-H33Q] are prepared by inoculation with fresh transformants and overnight growth at 30 °C in yeast tryptone (YT) broth containing 100 μg/ml ampicillin for plasmid selection and 25 μg/ml kanamycin for strain selection.
2. The overnight pre-cultures are diluted 1:10 into fresh YT broth and then grown at 30 °C for 4 h.
3. The 4 h cultures are diluted to an A_{600} of 0.3 to ensure that they are all at the same cell density before diluting further for plating purposes (1 in 100, 1 in 10^4, 1 in 10^6, 1 in 10^8, and 1 in 10^{10}) (*see* **Note 1**).
4. An aliquot of each dilution (3 μl) is spotted onto four YT agar plates (*see* **Note 2**), two with and two without the appropriate inducer: 6.6 mM arabinose for *E. coli* OD259 [pBAD22A] and *E. coli* OD259 [pWKG90] and 50 μM isopropyl-β-D-thiogalactoside (IPTG) for *E. coli* OD259 [pQE30], *E. coli* OD259 [pRJ30], and *E. coli* OD259 [pRJ30-H33Q].
5. The YT agar plates are grown at 30 °C and at 40 °C (*see* **Notes 3** and **4**) to ascertain whether AgtDnaJ can complement the lack of *E. coli* DnaJ and CbpA. AgtDnaJ is able to reverse the thermosensitivity of *E. coli* OD259, while a modified derivative of AgtDnaJ in which the highly conserved HPD motif is mutated (AgtDnaJ-H33Q) is unable to reverse the thermosensitivity (Fig. 1) (*see* **Notes 6** and **8**).
6. Western blotting is conducted to confirm that AgtDnaJ has been produced in *E. coli* OD259 [pRJ30] and *E. coli* OD259 [pRJ30-H33Q], with *E. coli* OD259 [pQE30] serving as a negative control (*see* **Note 7**). The AgtDnaJ proteins encoded by the pQE30-derived plasmids are His-tagged and hence can be readily detected using an anti-His antibody (Amersham Pharmacia) and chemiluminescence-based western blotting using the BM Chemiluminescence Western Blotting Kit

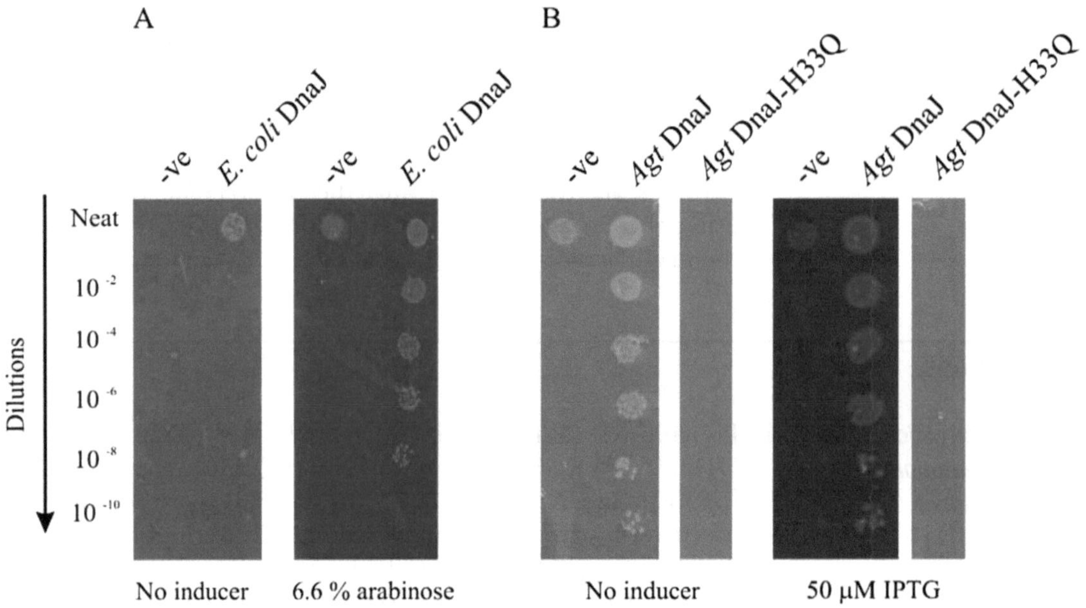

Fig. 1 AgtDnaJ can complement the lack of *E. coli* DnaJ and CbpA. Plasmids encoding *E. coli* DnaJ and AgtDnaJ as well as plasmids lacking the open reading frames for these proteins were transformed into *E. coli* OD259 [1]. The ability of these proteins to replace the lack of chromosomally expressed *E. coli* DnaJ and CbpA was investigated under stress conditions at 40 °C. The dilution series is indicated on the left. (**a**) Plasmid-encoded *E. coli* DnaJ can complement the lack of chromosomally encoded *E. coli* DnaJ and CbpA. The lane marked negative indicates *E. coli* OD259 [pBAD22A], and marked *E. coli* DnaJ indicates *E. coli* OD259 [pWKG90]; pWKG90 is pBAD22A containing the coding region for *E. coli* DnaJ. The inducer used was 6.6% arabinose. (**b**) Plasmid-encoded AgtDnaJ can complement the lack of chromosomally encoded *E. coli* DnaJ and CbpA. The lane marked negative indicates *E. coli* OD259 [pQE30], and marked AgtDnaJ indicates *E. coli* OD259 [pRJ30]; pRJ30 is pQE30 containing the coding region for AgtDnaJ; pRJ30-H33Q is pQE30 containing the coding region for AgtDnaJ-H33Q. The inducer used was 50 μM IPTG. These experiments were performed at least three times. (Reproduced from [9] with permission from Elsevier)

(Mouse/Rabbit) from Roche (Germany), as per the manufacturer's instructions.

3.2 Eukaryotic Complementation Assay Protocol

1. Competent *S. cerevisiae* JJ160 cells are prepared with the Frozen-EZ Transformation II™ Kit. Briefly, a yeast culture (10 ml) is grown with shaking at 23 °C in a yeast minimal medium containing glucose but lacking histidine ([YMM] glucose [HIS$^-$]) to select for the *S. cerevisiae* JJ160 strain. When an A_{600} of between 0.8 and 1.0 is reached, the cells are collected by centrifugation (500 *g*; 10 min) and washed with 10 ml EZ 1 solution. The mixture is centrifuged (500 *g*; 10 min), and the cell pellet is resuspended in 1 ml EZ 2 solution and stored in 60 μl aliquots at –80 °C until required.
2. The competent *S. cerevisiae* JJ160 cells are transformed with pKG6/Tcj2, pKG6 (negative control), or pRS317-*YDJ1*

(positive control) using the Frozen-EZ Transformation II™ Kit. A total of 100 ng of plasmids is incubated at room temperature with 15 µl competent *S. cerevisiae* JJ160 and 100 µL EZ 3 solution (100 µL) and vortexed every 15 min. The entire transformation mixture is plated onto yeast minimal agar (YMA) selective for each plasmid. YMA glucose [URA$^-$ HIS$^-$] was used for *S. cerevisiae* JJ160 containing pKG6 constructs; and YMA glucose [LYS$^-$ HIS$^-$] was used for *S. cerevisiae* JJ160 containing pRS317-*YDJ1*. Plates were incubated at 23 °C for 5 days.

3. A single yeast transformant for each plasmid is selected to inoculate 5 ml of the corresponding yeast minimal medium to induce protein production. YMM galactose [URA$^-$ HIS$^-$] is used for *S. cerevisiae* JJ160 transformed with pKG6 constructs and YMM galactose [LYS$^-$ HIS$^-$] for *S. cerevisiae* JJ160 transformed with pRS317-YDJ1. Cultures are grown at 23 °C with shaking for 3 days, the density corrected to equivalent A_{600}, and tenfold serial dilutions (10^0 to 10^{-5}) prepared (*see* **Note 1**).
4. A 10 µl aliquot of each dilution is spotted onto the appropriate selective minimal agar plate for growth and protein expression (YMA galactose [URA$^-$ HIS$^-$] for *S. cerevisiae* JJ160 containing pKG6 constructs and YMA galactose [LYS$^-$ HIS$^-$] for pRS317-*YDJ1*) (*see* **Note 2**). Separate plates are prepared for incubation at 23 °C, 30 °C, and 34 °C for 5 days (*see* **Notes 3–5**). The growth of yeast colonies at different temperatures was photographed.
5. The production of Tcj2 in *S. cerevisiae* JJ160 lysates is confirmed by western blot analysis using a peptide-directed antibody generated against a C-terminal peptide in the Tcj2 sequence [4]. Yeast protein extracts are prepared for SDS-PAGE by harvesting cells from 5 ml of undiluted liquid culture (prepared in step 3) by centrifugation (12,000 *g*; 5 min), resuspended in 30 µl of ESB (80 mM Tris-HCl, pH 6.8; 2% w/v SDS; 10% v/v glycerol; 1.5% w/v DTT; 0.1 mg/mL bromophenol blue), and boiled at 100 °C for 3 min. An equivalent volume of glass beads is added and the samples vortexed vigorously for 2 min. Another aliquot (30 µl) of ESB is added, and the samples vortexed briefly and boiled at 100 °C for 1 min before SDS-PAGE followed by western blot analysis using well-established methods. Protein extracts from untransformed or pKG6-transformed *S. cerevisiae* JJ160 cells can be prepared similarly and used as negative controls (*see* **Notes 6–8**).

4 Notes

1. The dilution regime is important to ensure that a dilution is achieved where single colonies derived from a single cell are produced at the highest dilutions; unless this is achieved, it cannot be interpreted that the thermosensitivity has been reversed down to a single-cell level.
2. The YT and YM agar plates should be properly dried before use to drive off any excess liquid; wet plates result in smeared growth which is difficult to interpret.
3. The temperature control of the plate incubation is critical to the success of the assays, and therefore the plates should be placed in a dedicated incubator that keeps a constant temperature (i.e., a temperature variation above or below the setting of less than 0.5 °C).
4. Consistent results will be achieved by applying a constant uninterrupted overnight incubation of 12–15 h for *E. coli* cultures and 60–72 h for yeast cultures.
5. For the *S. cerevisiae* JJ160 strain, growth at higher temperatures up to 37 °C can be done to increase the stringency of the assay [10].
6. The assays can be easily adapted to assess the contribution of specific domains or amino acids to the function of JDP by substitution of the wild-type sequences for modified protein-coding regions [2, 9–11].
7. Western blotting to show production of the novel JDP protein is particularly important, especially when testing mutant derivatives that are unable to reverse thermosensitivity. This lack of reversal cannot be interpreted accurately unless protein production is confirmed, as the lack of reversal of thermosensitivity could be due to lack of protein production.
8. The complementation assay can be adapted for screening of inhibitors of chaperones and JDP by culturing the transformed strains in the presence of selective inhibitors [5, 12]. In this case, the appropriate negative control to assess for general toxicity (i.e., inhibition of growth at permissive temperatures) and positive control of cultures grown at restrictive temperatures in the presence of a compound known to inhibit the Hsp70 or JDP will increase confidence in the results.

Acknowledgments

We gratefully acknowledge Dr. O Deloche (Geneva, Switzerland) for supplying the *E. coli* OD259 strain and plasmids pBAD22A and pWKG90. The *S. cerevisiae* JJ160 strain and PKG6 and pRS317-

YDJ1 plasmids were a kind gift of Elizabeth Craig (University of Madison—Wisconsin). G.L.B. acknowledges the financial support of Higher Colleges of Technology, UAE (Interdisciplinary Research Grant, IRG; Grant No. 213471), and Rhodes University, South Africa (Rated Researcher Grant, RRG; project number IFRR100006). Research in the lab of A.L.E is supported by a Newton Advanced Fellowships from the Academy of Medical Sciences (UK) and grants from the Resilient Futures Challenge-Led Initiative from the Royal Society (UK) (Grant No. CHL\R1 \180142), the South African Research Chairs Initiative of the Department of Science and Technology (DST) and the NRF (Grant No. 98566), Poliovirus Research Foundation (PRF, South Africa) (Grant No. 18/06), and Rhodes University and the Grand Challenges Africa Drug Discovery Programme (which is a partnership between the African Academy of Sciences [AAS], the Bill & Melinda Gates Foundation, Medicines for Malaria Venture [MMV], and the University of Cape Town Drug Discovery and Development Centre [H3D]) (Grant No. GCA/DD/rnd3/043).

References

1. Deloche O, Kelley WL, Georgopoulos C (1997) Structure-function analyses of the Ssc1p, Mdj1p, and Mge1p *Saccharomyces cerevisiae* mitochondrial proteins in *Escherichia coli*. J Bacteriol 179:6066–6075
2. Johnson JL, Craig EA (2001) An essential role for the substrate-binding region of Hsp40s in *Saccharomyces cerevisiae*. J Cell Biol 152:851–856
3. Shonhai A, Boshoff A, Blatch GL (2005) *Plasmodium falciparum* heat shock protein 70 is able to suppress the thermosensitivity of an *Escherichia coli* DnaK mutant strain. Mol Gen Genomics 274:70–78
4. Edkins AL, Ludewig MH, Blatch GL (2004) A *Trypanosoma cruzi* heat shock protein 40 is able to stimulate the adenosine triphosphate hydrolysis activity of heat shock protein 70 and can substitute for a yeast heat shock protein 40. Int J Biochem Cell Biol 36:1585–1598
5. Wang T, Bisson WH, Mäser P, Scapozza L, Picard D (2014) Differences in conformational dynamics between *Plasmodium falciparum* and human Hsp90 orthologues enable the structure-based discovery of pathogen-selective inhibitors. J Med Chem 57:2524–2535
6. Kampinga HH, Hageman J, Vos MJ, Kubota H, Tanguay RM, Bruford EA, Cheetham ME, Chen B, Hightower LE (2009) Guidelines for the nomenclature of the human heat shock proteins. Cell Stress Chaperones 14:105–111
7. Kelley WL, Georgopoulos C (1997) The T/t common exon of simian virus 40, JC, and BK polyomavirus T antigens can functionally replace the J-domain of the *Escherichia coli* DnaJ molecular chaperone. Proc Natl Acad Sci U S A 94:3679–3684
8. Yan W, Craig EA (1999) The glycine-phenylalanine-rich region determines the specificity of the yeast Hsp40 Sis1. Mol Cell Biol 19:7751–7758
9. Hennessy F, Boshoff A, Blatch GL (2005) Rational mutagenesis of a 40 kDa heat shock protein from *Agrobacterium tumefaciens* identifies amino acid residues critical to its *in vivo* function. Int J Biochem Cell Biol 37:177–191
10. Johnson JL, Craig EA (2000) A role for the Hsp40 Ydj1 in repression of basal steroid receptor activity in yeast. Mol Cell Biol 20: 3027–3036
11. Boshoff A, Hennessy F, Blatch GL (2004) The *in vivo* and *in vitro* characterization of DnaK from *Agrobacterium tumefaciens* RUOR. Protein Expr Purif 38:161–169
12. Mbaba M, de la Mare JA, Sterrenberg JN, Kajewole D, Maharaj S, Edkins AL, Isaacs M, Hoppe HC, Khanye SD (2019) Novobiocin-ferrocene conjugates possessing anticancer and antiplasmodial activity independent of HSP90 inhibition. J Biol Inorg Chem 24:139–149

Chapter 10

Optimized Microscale Protein Aggregation Suppression Assay: A Method for Evaluating the Holdase Activity of Chaperones

Ronald Tonui, Ruth O. John, and Adrienne L. Edkins

Abstract

Many molecular chaperones act as holdases by binding hydrophobic regions of substrates to prevent aggregation. Therefore, measuring holdase activity is an amenable method to determine chaperone activity. The holdase function is reliably and easily achieved by monitoring the suppression of heat-induced aggregation of well-characterized model protein substrates. However, the standard assay format requires large amounts of protein and hence is not applicable to all proteins. Using DnaK from *Escherichia coli* and heat-induced aggregation of malate dehydrogenase, we describe a protocol for absorbance and fluorescence-based miniaturized versions of the standard aggregation suppression assay that are affordable and have wide application for low abundance holdases. The assay can be used for both fundamental characterization of holdase function in proteins and screening of inhibitors of holdase activity.

Key words Holdase, Aggregation suppression, Malate dehydrogenase, DnaK, SYPRO Orange

1 Introduction

Chaperones have the intrinsic ability to recognize and bind to specific motifs in client proteins to prevent aggregation, promote protein folding, and assist with degradation. Selected chaperones, including Hsp70, Hsp83, DnaJ, and Hsp90, can bind or "hold" client proteins and are categorized under the holdase family [1–3]. Some holdases, including Hsp70, GRP70, Hsp90, and DnaJ, are potential drug targets for treating various diseases such as cancer [4–6] and targeting pathogenic microbes [7]. For example, the bacterium *Mycobacterium tuberculosis*, which causes tuberculosis, experiences several stresses that impact protein stability inside the host cell and hence an increase in reliance on protein homeostasis systems. This suggests that inhibiting the holdase family of proteins, such as DnaK, DnaJ, and GroES/GroEL chaperones, may be a therapeutic strategy [8].

Stuart K. Calderwood and Thomas L. Prince (eds.), *Chaperones: Methods and Protocols*, Methods in Molecular Biology, vol. 2693, https://doi.org/10.1007/978-1-0716-3342-7_10,

Screening for small inhibitors of holdases based on protein aggregation suppression assays using model substrates is a quick and easy in vitro pre-screening strategy applied in drug discovery. L-malate dehydrogenase (MDH), insulin, and luciferase have been used as heat-labile substrate models for screening the holdase activity of various proteins [2, 5, 7, 9–11]. The assay is based on the ability of the holdases to recognize exposed hydrophobic patches in unfolding proteins and bind, thereby occluding intermolecular interactions between exposed hydrophobic residues and preventing protein aggregation (Fig. 1). While this assay is easy, fast, and reproducible, the standard format consumes substantial amounts of reagents, including the substrate client, the holdase itself, and the ligand, were applicable. This is costly regarding the resources required to produce, purchase, and run the entire experiment, especially in screening many ligands or holdases. Here, using *Escherichia coli* DnaK as the holdase and MDH as the substrate, we describe an optimized, reproducible, small-scale protein aggregation suppression assay that considerably reduces the amount of input reagents. The assay can be implemented on a 96-well plate and read using an absorbance spectrophotometer (Fig. 1a). Alternatively, this assay can be coupled with a fluorescent dye that binds to aggregated protein, and measurements can be done using a quantitative thermocycler (Fig. 1b). Our assay provides a reproducible, cost-effective method for evaluating the holdase activity of various proteins and identifying small molecule modulators of holdases.

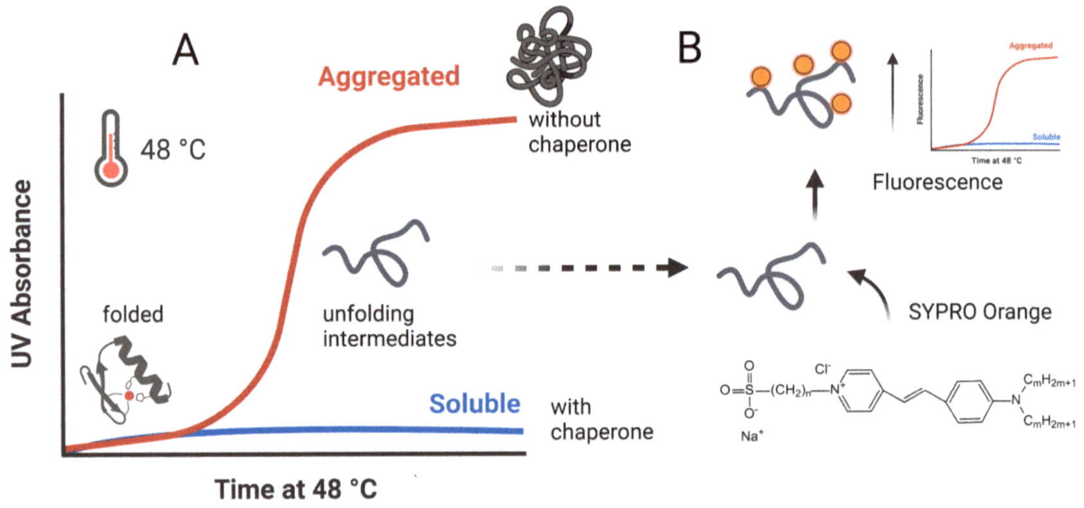

Fig. 1 Schematic diagram of aggregation suppression assay principle. An aggregation-prone model substrate (e.g., MDH) is incubated at 48 °C, and the transition from a folded native protein to an aggregate via unfolded intermediates is monitored over time by (**a**) increasing absorbance at 340 nm or (**b**) by increased fluorescence due to binding of a dye that binds specifically to hydrophobic patches in unfolded proteins. Image created using BioRender

2 Materials

2.1 Plasmids

The pQE80L-EcDnaK plasmid contains the coding sequence for *E. coli* DnaK in frame with an N-terminal hexahistidine tag. The plasmid carries ampicillin resistance and can be used to produce His-EcDnaK in a suitable *E. coli* T7-based protein production strain (e.g., we use BL21-Codonplus). This plasmid will be made available on request.

2.2 Buffers and Stock Solutions

1. **2XYT broth**: 16 g/L tryptone powder, 10 g/L yeast extract, 5 g/L NaCl. Autoclave.
2. **2XYT agar**: 2XYT broth with addition of 15 g/L bacterial agar. Autoclave and allow the medium to cool, add the appropriate concentration of antibiotics for transformant selection (in this case 100 μg/mL ampicillin), and pour ~25 mL into each plate.
3. **Lysis buffer**: 50 mM Tris-HCl, pH 7.5, 300 mM NaCl, 100 μg/mL Lysozyme, 0.01% (v/v) Tween 20, and 100 μg/mL PMSF.
4. **Equilibration/wash buffer 1:** 50 mM Tris-HCl, 300 mM NaCl, 0.01% (v/v) Tween 20, and 10 mM imidazole.
5. **Wash buffer 2:** 50 mM Tris-HCl, 300 mM NaCl, 0.01% (v/v) Tween 20, and 25 mM imidazole.
6. **Buffer stock solutions**: For 500 mL of 2 M Tris-HCl, dissolve 121.14 g of Tris-base (MW: 121.14 g/mol) in 350 mL of double-distilled water, and adjust the pH to 7.4. Adjust the volume to 500 mL and store at 4 °C. To prepare 500 mL of 2 M NaCl, dissolve 58.44 g NaCl (MW:58.44 g/mol) in 350 mL of double-distilled, adjust the final volume to 500 mL once dissolved, and store it at 4 °C.
7. **MDH aggregation assay buffer** (particularly for diluting SYPRO Orange and MDH): Mix 750 μL of 2 M NaCl with 500 μL of 2 M Tris-HCl (pH 7.5), and add 8.75 mL of double-distilled water.
8. **SYPRO Orange**: Prepare a stock of 50X SYPRO Orange from the commercial stock (5000X for Cat. S5692, Sigma-Aldrich) by diluting the stock 1:100 in assay buffer.
9. **MDH**: The stock concentration of MDH varies with the manufacturer. For monomeric porcine MDH, the molecular weight is 36 kDa, and the molar concentration is 139 μM. Prepare a working MDH stock of 10 μM by mixing the commercial stock and MDH aggregation assay buffer in the ratio 72:928 (retail stock: assay buffer).

10. **Test chaperone proteins**: Depending on the chaperone protein concentration, calculate and dilute the stock using the assay buffer to achieve a working stock concentration of 10 μM for absorbance assays and 50 μM for the fluorescence-based assays.

3 Methods

3.1 His-EcDnaK Protein Purification

1. The purification method for His-EcDnaK is modified from that previously used for trypanosomal heat shock proteins [12].
2. Prepare bacterial cultivation media 2XYT broth and 2XYT agar.
3. To transform the bacterial expression host with the pQE80L-EcDnaK plasmid vector, using commercial or lab-made competent cells. For our protocol, we used calcium-competent BL21-CodonPlus (+) cells (Agilent Technologies) prepared following the published protocol [13]. A typical transformation protocol [14] is described below.
4. For transformation, mix 50 μL competent bacterial cells with 50–100 ng of pQE80L-EcDnaK plasmid DNA, and incubate on ice for 30 min. Include an experimental control of competent cells without plasmid DNA to confirm the successful selection.
5. Heat shock the mixture at 42 °C for 60 s, and immediately incubate on ice for 5 min.
6. Add 450 μL 2XYT broth without antibiotics to the transformation samples and grow at 37 °C in a shaking incubator at 160 rpm for 60 min.
7. Plate 100 μL of transformed and control cells on 2XYT agar plates with 100 μg/mL ampicillin. To select positive clones, incubate the plates inverted at 37 °C for 16–18 h.
8. Inoculate a single colony into 25 mL of 2XYT broth with 100 μg/mL ampicillin, and incubate at 37 °C overnight with shaking at 160 rpm.
9. Dilute the 25 mL overnight culture to 225 mL 2XYT broth supplemented with 100 μg/mL ampicillin, and grow the cells at 37 °C in a shaking incubator at 160 rpm. Monitor the growth by measuring optical density at 600 nm until it reaches between 0.6 and 0.8.
10. Induce His-EcDnaK protein expression with isopropyl β- d-1-thiogalactopyranoside (IPTG) to a final concentration of 0.5 mM, and incubate at 25 °C with shaking at 180 rpm for 3–6 h. It may be necessary to perform optimization experiments to establish the induction time for good quality and yield of your specific protein.

11. Transfer the cultures to centrifuge tubes and collect cells by centrifugation at 6000 x*g* for 15 min.
12. Discard the supernatant, resuspend the pellet in 5 mL lysis buffer, and store at −80 °C overnight or longer prior to subsequent purification.
13. Retrieve the lysate from −80 °C and thaw on ice for 1 h. Sonicate on ice with 6 times 30 s pulses at 60 Hz, with 30 s recovery between pulses. The lysate and all processes downstream should be carried out at 4 °C (*see* **Note 1**).
14. Centrifuge the lysate at 10000 x*g* for 30 min at 4 °C. Transfer the supernatant to a sterile 15 mL falcon tube, and resuspend the pellets in 5 mL lysis buffer.
15. Prepare for Ni-affinity chromatography by loading 1.5 mL HisPur™ Ni-NTA resin into columns. Wash the resin with 30 mL distilled water, and then equilibrate with 30 mL of ice-cold His-tag wash buffer 1/equilibration buffer. Always confirm the pH of buffers (*see* **Note 2**).
16. Load the supernatant on equilibrated resin and incubate at 25 °C for 1 h. Collect the flow through and perform five-column gradient washes with ice-cold wash buffer 1. Perform subsequent washes with wash buffer 2 until the O.D. at 280 nm reaches a baseline level.
17. Place the column resins on ice for 5 min, and elute three subsequent fractions with elution buffer [20 mM Tris-HCl, 150 mM NaCl, 10% glycerol, and 25 mM imidazole]. Confirm the purity of the elutions by SDS-PAGE analysis using standard approaches.
18. Quantify the His-EcDnaK protein in the elution fractions using the bicinchoninic acid assay (BCA) or Bradford technique. Store the purified His-EcDnaK at −80 °C in aliquots of not more than 100 μL until needed. For additional guidelines prior to storage and use of recombinant proteins, (*see* **Notes 3–6**).

3.2 Kinetic and Endpoint Analysis of Aggregation Suppression Analysis Using Absorbance at 360 nm

1. Set the spectrophotometer compartment temperature to 25 °C.
2. Prepare 1 mL reaction components comprising 0.72 μM MDH and five different concentrations of the His-EcDnaK from equimolar to MDH to sub-molar concentrations using four-fold dilutions as shown in Table 1.
3. Control reactions should include the *E. coli* DnaK/holdase protein alone at each titration point, MDH alone (which serves as 100% aggregation), and buffer alone (150 mM NaCl, 100 mM Tris-HCl, pH 7.5).

Table 1
Reaction components for microscale MDH aggregation suppression assay

Reaction component	Reaction master mixes (1 mL reaction for technical triplicates in a 300 μL reaction volume)			
	Holdase test	Buffer control	No-MDH control	MDH alone
2 M NaCl	75 μL	75 μL	75 μL	75 μL
2 M Tris-HCl (pH 7.5)	50 μL	50 μL	50 μL	50 μL
MDH (10 μM)	100 μL	0	0 μL	100 μL
His-EcDnaK or other putative holdase (10 μM)	4.5–72 μL (amount depends on titration, e.g., 72 μL for the equimolar concentration of MDH)	0	4.5–72 μL (equivalent to the amount used in the holdase test reaction)	0
Deionized water	Make up to 1000 μL	875 μL	Make up to 1000 μL	803 μL

4. Pipette 300 μL from each reaction master mix tube to three consecutive wells in a UV-compatible, flat-bottomed 96-well plate (*see* **Notes 7** and **8**). Take the absorbance reading at 340 nm before the start of the experiment.
5. **Endpoint readings:** For endpoint readings, incubate the plate at 48 °C either in an incubator or inside the spectrophotometer with 10 min intervals and linear shaking for 10 s. After 1 h, read absorbance at 340 nm.
6. **Kinetic readings:** Kinetic readings are possible in a spectrophotometer with temperature control. Prepare a running protocol including (a) linear shaking for 10 s, (b) temperature set at 48 °C, and (c) capture absorbance readings at 340 nm for 1 h at intervals of 1 min (Fig. 3).

3.3 Kinetic and Endpoint Analysis of Aggregation Suppression Analysis Fluorescence-Based Detection

1. The fluorescence-based aggregation detection method relies on the SYPRO Orange dye that specifically binds to exposed hydrophobic residues upon the temperature-induced unfolding of the protein. The fluorescent dye method has improved resolution as the dye binds specifically to hydrophobic patches of amino acid residues (Fig. 3).
2. Dilute the SYPRO orange dye to a working stock concentration of 50X in the assay buffer.
3. Prepare test and control reactions as in Table 2.
4. Spin down the master mix reaction and aliquot 25 μL of the reaction mixture into white-walled clear-lidded qPCR tubes in

Table 2
Reaction components for fluorescence-based aggregation suppression assay

Component stock	Reaction master mix (100 μL reaction for technical triplicates in a 25 μL reaction volume)			
	Holdase test reaction	Holdase-only control	MDH-only control	Buffer-only control
SYPRO Orange (50X)	10	10	10	10
Tris-HCl (2 M)	5	5	5	5
NaCl (2 M)	7.5	7.5	7.5	7.5
MDH (100 μM)	5	5	5	5
E. coli DnaK/ holdase (50 μM)	0.5–10 μL (amount depends on titration, e.g., 10 μL for the equimolar concentration of MDH)	0.5–10 μL (equivalent to the amount used in the holdase test reaction)	0	0
Deionized water	Make up to 100 μL	Make up to 100 μL	Make up to 100 μL	Make up to 100 μL

See **Note 10** if the method is adopted for screening inhibitors of the holdase activity

triplicate. Spin down the aliquots briefly to collect the components in the bottom of the qPCR tube.

5. Prepare a running protocol on the CFX96 thermocycler comprising an initial hold at 25 °C for 3 min, followed by 80 cycles of 48 °C for 20 s with a plate read at every cycle using the FRET color module. Other color modules at Sypro Orange's wavelength can be used (*see* **Note 9**).
6. Transfer the RFU quantitative data to any statistical data analysis tool or Excel for analysis.

3.4 Data Analysis

1. **For endpoint analysis,** extract the data from the 96-well plate to an Excel spreadsheet, and calculate the absorbance for MDH alone (absorbance of MDH alone – absorbance of the buffer). This value represents a 100% aggregation. For the mixture of MDH and His-EcDnaK (or other holdase), subtract the absorbance of the His-EcDnaK reaction at the corresponding concentration from the absorbance of the MDH reaction with the His-EcDnaK (MDH + His-EcDnaK – His-EcDnaK at each corresponding concentration). To determine the percentage aggregation suppression, use the following formula:

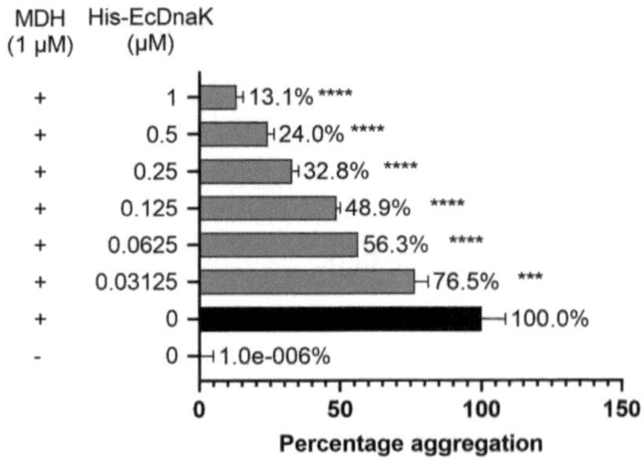

Fig. 2 Endpoint aggregation suppression of MDH by His-EcDnaK. Percentage aggregation of MDH in the presence or absence of His-EcDnaK was calculated from the absorbance at 340 nm after 1 h incubation at 48 °C. Individual groups were compared to MDH alone using a one-way analysis of variance (ANOVA) test. ****, $P \leq 0.0001$; ***, $P \leq 0.001$

$$[A_{340}(\text{MDH} + \text{His-EcDnaK}) - A_{340}(\text{His-EcDnaK})] \times 100$$
$$\text{A}_{340}(\text{MDH})$$

The resulting percentages can be compared on a bar graph (Fig. 2). Depending on the experimental design, statistical analysis can be conducted for the comparison of different His-EcDnaK/ holdase combinations or concentrations against MDH alone, using analysis of variance (ANOVA), or t-test if comparing a single-point concentration of holdase to the MDH aggregation alone.

2. For **kinetic/fluorescence data analysis**, subtract the buffer readings from each MDH alone data point over time. Equally, perform this buffer subtraction for the EcDnaK data points at each concentration. Plot the resultant values on an x–y graph with the x values representing the time (min) against the absorbance at 340 nm or the RFU. Percentage aggregation can then be calculated based on either the linear slope of the resultant curve or the area under the curve (AUC) (Fig. 3).

4 Notes

1. Protein purification steps must be carried out at 4 °C. Most proteins are sensitive to changes in temperature.
2. The pH of Tris-HCl can fluctuate depending on storage temperature. Confirm the pH prior to use even if this had been previously determined.
3. Prior to storage, ensure the purified protein concentrations are below 10 mg/mL. This prevents aggregation.

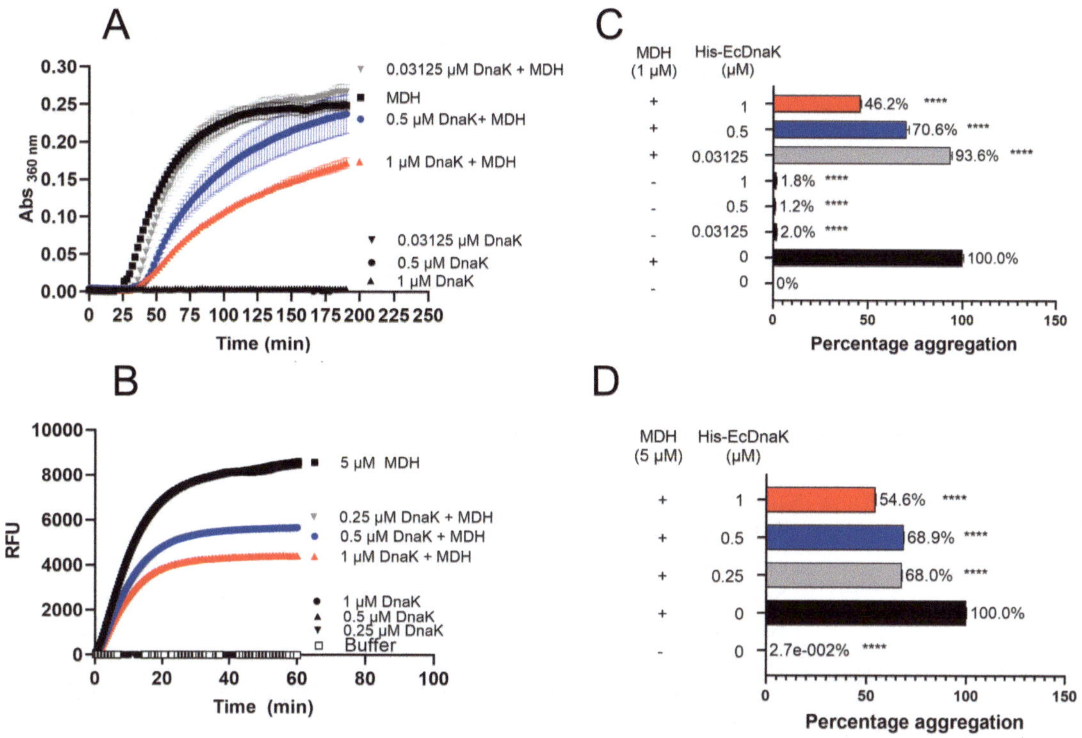

Fig. 3 Kinetic analysis of aggregation suppression by His-EcDnaK. MDH was heated at 48 °C for 1 h, and aggregation was monitored over time by (**a**) absorbance at 340 nm or (**b**) fluorescence by the inclusion of SYPRO Orange dye which binds hydrophobic regions in unfolded proteins. Quantitative data for (**c**) absorbance over time or (**d**) fluorescence over time was calculated by area under the curve, and statistical analysis was done for all groups compared to MDH alone using a one-way analysis of variance (ANOVA) test. ****, $P \leq 0.0001$; ***, $P \leq 0.001$; **, $P \leq 0.01$; *, $P \leq 0.05$

4. If the protein is prone to aggregation, consider increasing the percentage of glycerol. Alternatively, a sodium phosphate buffer system (50 mM) can replace the Tris-HCl buffer.
5. If feasible, the folded status of the proteins (both the His-EcDnaK/holdase and MDH) should be confirmed prior to the assay. This can be done using assays such as the thermal shift assay (TSA) using SYPRO Orange. A clear transition from native conformation (low starting fluorescence) to unfolded form (increasing fluorescence to a maximum plateau) reflects properly folded proteins.
6. Protein preparations should be centrifuged prior to use to remove aggregates.
7. The absorbance readings at 360 nm require a UV-compatible 96-well plate.

8. Mix the reaction components thoroughly. This is to ensure a homogenous combination of the DnaK/holdase and the model substrates.
9. In our experience, the three-color CFX modules FRET, SYBR, and FAM produce comparable results for SYPRO Orange fluorescence and hence can be used interchangeably depending on your system configuration.
10. The assay can be adapted to screen the effects of inhibitors and genetic mutations on the holdase activity. In the case of inhibitors, include controls for the vehicle solvent and the effect of the inhibitors on MDH aggregation alone.

Acknowledgments

This research was supported by grants to A.L.E from the South African Research Chairs Initiative of the Department of Science and Technology (DST) and the NRF (Grant No. 98566), Rhodes University and the Grand Challenges Africa Drug Discovery Programme (which is a partnership between the African Academy of Sciences [AAS], the Bill & Melinda Gates Foundation, Medicines for Malaria Venture [MMV], and the University of Cape Town Drug Discovery and Development Centre [H3D]) (Grant No. GCA/DD/rnd3/043). R.T. and R.O.J. were supported by postgraduate in-region fellowships from the German Academic Exchange Service (DAAD).

References

1. Adio WS, Obunadike CV (2022) Mechanism of Action of Chaperones in Protein Function. World News Nat Sci 44:333–354
2. Moayed F et al (2020) The Anti-Aggregation Holdase Hsp33 Promotes the Formation of Folded Protein Structures. Biophys J 118(1): 85–95. https://doi.org/10.1016/j.bpj.2019.10.040
3. Acebrón SP, Fernández-Sáiz V, Taneva SG, Moro F, Muga A (2008) DnaJ recruits DnaK to protein aggregates. J Biol Chem 283(3): 1381–1390. https://doi.org/10.1074/jbc.M706189200
4. Kang J et al (2022) Novel peptide-based vaccine targeting heat shock protein 90 induces effective antitumor immunity in a HER2+ breast cancer murine model. J Immunother Cancer 10(9):1–14. https://doi.org/10.1136/jitc-2022-004702
5. Xu Y et al (2022) Novel matrinic acid derivatives bearing 2-anilinothiazole structure for non-small cell lung cancer treatment with improved Hsp90 targeting effect. Drug Dev Res 83(6):1434–1454. https://doi.org/10.1002/ddr.21974
6. Pahwa R et al (2022) Inhibition of HSP 90 is associated with potent anti-tumor activity in Papillary Renal Cell Carcinoma. J Exp Clin Cancer Res 41(1):1–17. https://doi.org/10.1186/s13046-022-02416-z
7. Hosfelt J et al (2022) An allosteric inhibitor of bacterial Hsp70 chaperone potentiates antibiotics and mitigates resistance. *Cell Chem Biol* 29(5):854–869.e9. https://doi.org/10.1016/j.chembiol.2021.11.004
8. Fay A, Glickman MS (2014) An Essential Nonredundant Role for Mycobacterial DnaK in Native Protein Folding. *PLoS Genet* 10(7): e1004516. https://doi.org/10.1371/journal.pgen.1004516
9. Cockburn IL et al (2011) Screening for small molecule modulators of Hsp70 chaperone activity using protein aggregation suppression assays: Inhibition of the plasmodial chaperone

PfHsp70-1. Biol Chem 392(5):431–438. https://doi.org/10.1515/BC.2011.040

10. Bentley SJ, Boshoff A (2019) Trypanosoma brucei J-protein 2 functionally co-operates with the cytosolic hsp70 and Hsp70.4 proteins. Int J Mol Sci 20(23):5843. https://doi.org/10.3390/ijms20235843
11. Mabate B et al (2018) Structural and biochemical characterization of Plasmodium falciparum Hsp70-x reveals functional versatility of its C-terminal EEVN motif. Proteins Struct Funct Bioinf 86(11):1189–1201. https://doi.org/10.1002/prot.25600
12. Edkins AL, Ludewig MH, Blatch GL (2004) A Trypanosoma cruzi heat shock protein 40 is able to stimulate the adenosine triphosphate hydrolysis activity of heat shock protein 70 and can substitute for a yeast heat shock protein 40. Int J Biochem Cell Biol 36(8):1585–1598. https://doi.org/10.1016/j.biocel.2004.01.016
13. Chang AY, Chau VWY, Landas JA, Pang Y (2017) Preparation of calcium competent Escherichia coli and heat-shock transformation. J Exp Microbiol Immunol 1:22–25
14. Hanahan D, Jessee J, Bloom FR (1991) Plasmid transformation of Escherichia coli and other bacteria. Methods Enzymol 204(C):63–113. https://doi.org/10.1016/0076-6879(91)04006-A

Chapter 11

Detecting Posttranslational Modifications of Hsp90 Isoforms

Rebecca A. Sager, Sarah J. Backe, Len Neckers, Mark R. Woodford, and Mehdi Mollapour

Abstract

The molecular chaperone heat shock protein 90 (Hsp90) is essential in eukaryotes. Hsp90 chaperones proteins that are important determinants of multistep carcinogenesis. There are multiple Hsp90 isoforms including the cytosolic Hsp90α and Hsp90β as well as GRP94 located in the endoplasmic reticulum and TRAP1 in the mitochondria. The chaperone function of Hsp90 is linked to its ability to bind and hydrolyze ATP. Co-chaperones and posttranslational modifications (such as phosphorylation, SUMOylation, and ubiquitination) are important for Hsp90 stability and regulation of its ATPase activity. Both mammalian and yeast cells can be used to express and purify Hsp90 and TRAP1 and also detect post-translational modifications by immunoblotting.

Key words Heat shock protein 90 (Hsp90), TRAP1, Molecular chaperones, Posttranslational modification, Phosphorylation, SUMOylation, Ubiquitination, *O*-GlcNAcylation

1 Introduction

Heat shock protein 90 (Hsp90) is an essential molecular chaperone in eukaryotes [1, 2] Its cellular functions have been most clearly identified in mammalian cells, *Drosophila*, and baker's yeast *Saccharomyces cerevisiae* [3] There are multiple Hsp90 isoforms including cytosolic Hsp90α and Hsp90β as well as endoplasmic reticulum resident glucose-related protein 94 (GRP94) and the mitochondrial chaperone tumor necrosis factor receptor-associated protein 1 (TRAP1) [4] Hsp90 isoforms create and maintain the functional conformation of a subset of proteins [5–7] These targets (or "clients") are key components of signal transduction pathways and numerous transcription factors, among others. Hsp90 and a discrete set of co-chaperone proteins "hold" these clients in a state from which they can respond to activating signals [8]

Stuart K. Calderwood and Thomas L. Prince (eds.), *Chaperones: Methods and Protocols*, Methods in Molecular Biology, vol. 2693,
https://doi.org/10.1007/978-1-0716-3342-7_11,

Hsp90 chaperone activity depends on ATP binding and hydrolysis [9–11] This is coupled to a conformational cycle involving the opening and closing of a dimeric "molecular clamp" via transient association of Hsp90's N-terminal domain [12, 13] The N-domain also binds the antitumor antibiotics geldanamycin and radicicol, both Hsp90 inhibitors and their derivatives [14–18] In fact, there is now an Hsp90 inhibitor, TAS-116 or pimitespib, approved for clinical use in advanced gastrointestinal stromal tumors (GIST) in Japan as of 2022 [19]

Hsp90 ATPase activity is also regulated by co-chaperones. For example, Hop^{Sti1} [20–22] $p50^{Cdc37}$ [23–25] and $p23^{Sba1}$ [26, 27] have an inhibitory effect on the ATPase cycle of Hsp90, while Aha1 [28–32] and Cpr6 [33, 34] have an activating effect.

Hsp90 is posttranslationally modified (PTMs) by phosphorylation, acetylation, *S*-nitrosylation, methylation, thiocarbamylation, nitration, ubiquitination, *O*-GlcNAcylation, and SUMOylation [35–40]. These PTMs, in concert with co-chaperones, fine-tune Hsp90 chaperone function, which ultimately leads to controlled chaperoning of kinase and non-kinase client proteins of Hsp90 [41, 42] This landscape of Hsp90 PTMs, also referred to as the chaperone code, has several effects on Hsp90 function and further affects Hsp90 inhibitor sensitivity and selectivity and sensitivity to other therapeutics [39, 43]

The most extensively studied Hsp90 PTM is phosphorylation [44–54] Early work has shown that cells treated with the serine/threonine phosphatase inhibitor, okadaic acid, demonstrate Hsp90 hyperphosphorylation and decreased association of the chaperone with $pp60^{v\text{-}Src}$, suggesting a link between Hsp90 phosphorylation and chaperoning of its target proteins [55, 56]. A subsequent study has shown that c-Src directly phosphorylates Tyr300 of Hsp90 under basal conditions, reducing its ability to chaperone client proteins [53]. Phosphorylation events on Hsp90 can promote significant conformational dynamics that assist in co-chaperone recruitment as elegantly demonstrated for Tyr313 phosphorylation and Aha1 recruitment [57]. Hsp90 phosphorylation plays a role in not only Hsp90 chaperone cycle regulation generally but has had distinct roles identified in DNA repair and apoptosis, transcription factor regulation, and cancer development and progression [39].

Hsp90 is also subject to SUMOylation, which is an addition of a small ubiquitin-like modifier to a lysine residue. This modification affects cellular localization or function of a protein rather than a signal for its degradation like ubiquitination. SUMOylation of Lys191 in human Hsp90α (Lys178 in yeast) promotes its binding to the co-chaperone Aha1 and also increases cells' sensitivity to Hsp90 inhibitors [35].

Additionally, non-cytosolic isoforms of Hsp90 also exhibit PTM such as TRAP1 [58–61]. TRAP1 is subject to *O*-GlcNAcylation, which serves to alter its ATPase activity and ultimately affects

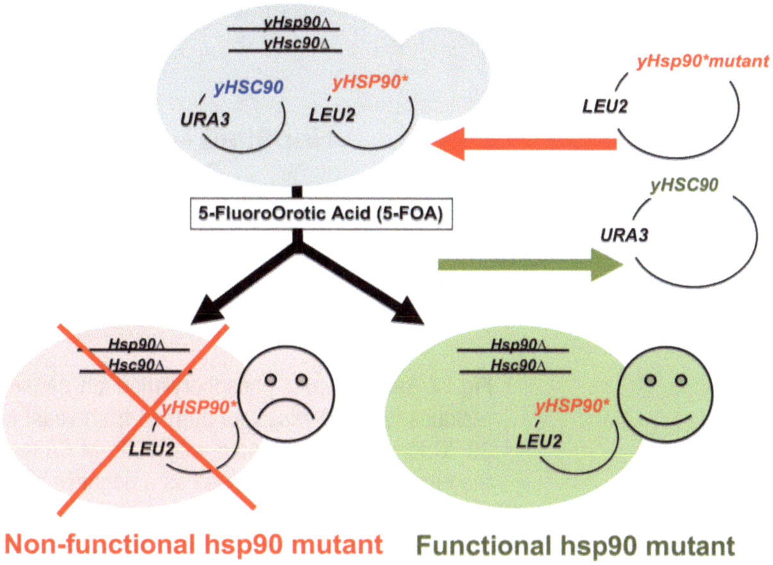

Fig. 1 With plasmid shuffling, a mutant *hsp90* gene can be made to provide all the Hsp90 of the yeast cell (yHsp90 = Hsp82 and yHsc90 = Hsc82). This involves introducing the mutation into yHsp90 on the Leu2 plasmid and then introducing it into haploid yeast cells (yHsp90Δ, yHsp90Δ). The growth of these cells on 5-fluoroorotic acid (5-FOA) will "cure" the yeast cells of the wild-type yHsc90 therefore creating an *hsp90* mutant

mitochondrial metabolism [62] Loss of TRAP1-*O*-GlcNAcylation decreased the binding of TRAP1 to ATP as well as to its client succinate dehydrogenase [62].

The lack of PTM-specific antibodies has made it difficult to study PTMs of Hsp90. There are currently very few phospho-specific Hsp90 antibodies available. Additionally, total *HSP90* gene knockouts are lethal in mammalian systems therefore any PTM Hsp90 mutant must be investigated in a background of highly expressed native mammalian Hsp90 proteins. Simple baker's yeast, *Saccharomyces cerevisiae*, is a well-established and valuable tool for studying various aspects of conserved protein chaperone machinery. The yeast system has provided us with a powerful tool to study Hsp90 phosphorylation, since it readily allows plasmid exchange whereby any introduced Hsp90 gene—provided it is partially functional—can provide 100% of the Hsp90 of the cell (Fig. 1). Such genetic modifications are simply not achievable in cultured mammalian cells. This plasmid exchange (Fig. 1) was used to isolate temperature-sensitive (*ts*) Hsp90 mutants.

This chapter describes multiple techniques for the examination of Hsp90 PTM. First, the isolation and identification of yeast Hsp90 phosphorylation using immunoblotting procedures. Using the yeast system, it is possible to show that Hsp90 is constitutively phosphorylated on serine and threonine residues. However, Hsp90

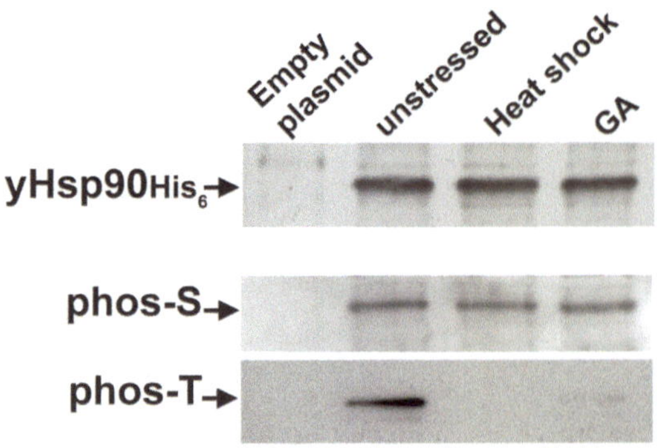

Fig. 2 Yeast Hsp90 phosphorylation on serine (phos-S) and threonine (p-T) residues. yHsp90-His$_6$ was purified from yeast cells that were heat shocked at 39 °C for 40 min or treated with 100 μM GA for 1 h. Wild-type cells containing the empty plasmid were used as a negative control

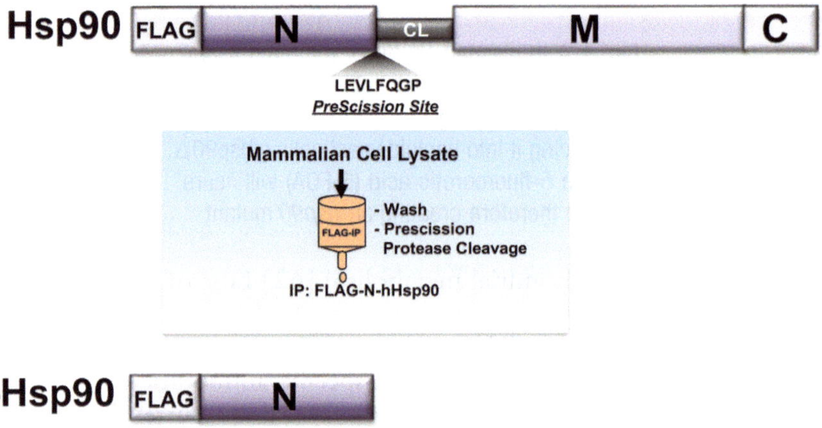

Fig. 3 Schematic representation of hHsp90-FLAG showing the amino (N-), charged linker (CL), middle (M-), and carboxy (C-) domains. Mammalian lysate with the hHsp90-FLAG will be attached to anti-FLAG agarose and hHsp90 N-domain can be isolated by PreScission protease digestion

threonine phosphorylation is lost upon either heat shock stress or treatment with the Hsp90 inhibitor GA (Fig. 2). This chapter also describes the isolation and analysis of the human (h)Hsp90-N-domain from mammalian cells. This is achieved by introducing a PreScission protease cleavage site between the N-domain and adjacent charged linker region of hHsp90α, allowing isolation of the N-domain. Separation of the N-domain containing either wild-type or non-SUMOylated hHsp90α-K191R mutant from the full-length Hsp90 protein allows for better detection of SUMOylated Hsp90 by immunoblotting (Fig. 3). Lastly, analysis of *O*-GlcNAcylation of TRAP1 isolated from mammalian cell mitochondria is described. Initial isolation of crude mitochondrial extract followed by immunoprecipitation of TRAP1 facilitates characterization of

TRAP1 modifications specifically relevant to its function in the mitochondrial sub-compartment. Collectively, these methods provide powerful tools to examine the important signaling mechanisms that help regulate Hsp90 isoform activity.

2 Materials

1. YPD (2% (wt/vol) Bacto peptone, 1% (wt/vol) yeast extract, 2% (wt/vol) glucose, 20 mg/liter adenine).
2. Yeast protein extraction buffer (yEB): 50 mM Tris-HCl, pH 6.8, 100 mM NaCl, 50 mM $MgCl_2$. One tablet of complete EDTA-free protease inhibitor cocktail (Roche) and one tablet of PhosphoSTOP (Roche) are added to 50 mL mEB.
3. 425–600 μm glass beads (acid washed) (Sigma).
4. Dulbecco's Modified Eagle's Medium–high glucose (DMEM; Sigma) supplemented with 10% fetal bovine serum (FBS).
5. Mammalian protein extraction buffer (mEB): 0.1% NP-40, 20 mM Tris-HCl, pH 7.4, 100 mM NaCl, 1 mM $MgCl_2$, 20 mM Na_2MoO_4. One tablet of complete EDTA-free protease inhibitor cocktail (Roche) and one tablet of PhosphoSTOP (Roche) are added to 50 mL mEB. (For detection of SUMO, mEB should also contain 20 mM *N*-ethylmaleimide (NEM), *see* **Note 8**).
6. Mitochondrial isolation buffer (isoB): 250 mM Sucrose, 10 mM Tris-HCl, pH 7.4. One tablet of complete EDTA-free protease inhibitor cocktail (Roche) and one tablet of PhosphoSTOP (Roche) are added to 50 mL isoB.
7. *TransIT®-2020* transfection reagent (Mirus).
8. Bio-Rad Protein Assay Dye solution (Bio-Rad).
9. Ni-NTA agarose (Qiagen).
10. Imidazole (Sigma).
11. Anti-FLAG M2 Affinity Gel agarose (Sigma).
12. PreScission protease (GE Healthcare).
13. PreScission protease cleavage buffer: 50 mM Tris-HCl, 150 mM NaCl, 1 mM EDTA, 1 mM DTT (pH 7.0).
14. SDS-PAGE sample buffer (2×): 125 mM Tris-HCl pH 6.8, 20% glycerol, 2% SDS, 10% 2-mercaptoethanol, 0.01% bromophenol blue, stable at −20 °C. Aliquot and avoid freeze-thaw cycles.
15. Protran BA85, 0.45 μm nitrocellulose membrane (Whatman).
16. Ponceau S solution (Sigma).

17. Tris-buffered saline (TBS): 150 mM NaCl, 25 mM Tris-base. Adjust pH to 7.4 using HCl. Sterile filter and incubate at 4 °C.
18. Albumin, bovine serum (minimum purity 98%).
19. Dried skimmed milk.
20. Phospho-serine antibody (Sigma).
21. Phospho-threonine antibody (Sigma).
22. Phospho-tyrosine antibody (4G10; Millipore).
23. Acetylated lysine antibody (Cell Signaling).
24. Ubiquitin antibody (Santa Cruz Biotechnology).
25. SUMO-1 or SUMO-2/3 antibody (Cell Signaling).
26. 6X-His antibody (Invitrogen).
27. FLAG epitope antibody (Thermo Scientific).
28. GlcNAc antibody (CTD110.6; Cell Signaling).
29. Anti-secondary mouse and/or rabbit antibody (Cell Signaling).
30. ECL plus Western Blotting Detection System (GE Healthcare).
31. X-ray film, X-ray cassette, and X-ray film developing machines.

3 Methods

3.1 Extraction of Total Yeast Protein

1. Grow PP30 cells[9] expressing His_6 linked at the N-domain of Hsp82 (yHsp90) on 150 ml YPD overnight a 28 °C.
2. Harvest and wash cells 2–3 times in ice-cold deionized water (dH_2O).
3. Transfer the cell pellet into a screw cap 2 ml tube.
4. Pellet the cells and remove the supernatant (*see* **Note 1**).
5. Add an equal volume of cell pellets and ice-cold glass beads.
6. Add half the volume of pellet/glass beads, yEB.
7. Bead beat the cells using the Mini-BeadBeater (BioSpec Products, Inc.) for 30 s.
8. Incubate the cells on ice for 30 s.
9. Repeat steps 7 and 8, 10–12 times.
10. Centrifuge the tubes at (10,000 *g*) for 10 min at 4 °C (*see* **Note 2**).
11. Transfer the supernatants into fresh 1.5 ml micro-centrifuge tubes.
12. Centrifuge the tubes at (10,000 *g*) for 10 min at 4 °C (*see* **Note 2**).

13. Transfer the supernatants (soluble protein) into fresh 1.5 ml micro-centrifuge tubes.
14. Determine protein concentrations using Bio-Rad Protein Assay solution (Bio-Rad).
15. Transfer 40 μl of Ni-NTA agarose slurry into a 1.5 ml micro-centrifuge tube. (*see* **Note 3**).
16. Add 1.0 ml of yEB to the Ni-NTA agarose and spin at 10,000 *g* for 1 min at 4 °C.
17. Remove the supernatant and add 1.0 ml of yEB to the Ni-NTA Agarose (*see* **Note 4**).
18. Repeat steps 16–17 four times.
19. Resuspend the Ni-NTA agarose in 30 μM imidazole in yEB and incubate at 4 °C for 30 min (*see* **Note 5**).
20. Repeat steps 16–17 twice and remove the supernatant.
21. Add 1 mg of total protein to the Ni-NTA agarose in a total volume of 600 μl.
22. Incubate the total proteins/Ni-NTA agarose at 4 °C for 2 h (*see* **Note 6**).
23. Centrifuge the tubes at (1000 *g*) for 1 min at 4 °C.
24. Gently remove the supernatant (*see* **Note 7**).
25. Add 1 ml of yEB to the Ni-NTA Agarose.
26. Repeat steps 23–25 five times.
27. Wash the Ni-NTA agarose with 30 μM imidazole in yEB.
28. Wash the Ni-NTA agarose with yEB once.
29. Centrifuge the micro-centrifuge tube at 15,000 *g* for 1 min at 4 °C.
30. Remove as much supernatant as possible.
31. Add 40 μl of the protein sample buffer.
32. Boil the samples for 3–5 min.
33. Proceed to Subheading 3.3 for Western blotting and PTM detection.

3.2 Extraction of Total Protein from HEK293 Cells and Immunoprecipitation (IP) of hHsp90

1. Transfect HEK293 cells (~40% confluent in a 10 cm dish; growing in DMEM +10% FBS) with 2 μg hHSP90-FLAG using *Trans*IT-2020 reagent (Mirus), and incubate overnight at 37 °C, 5%CO_2.
2. Place plates on ice and aspirate media. Wash 2× with cold PBS. Remove all remaining PBS from the plate.
3. Add 200 μL cold mEB to the plate. Scrape cells and transfer them to a 1.5 mL micro-centrifuge tube on ice (*see* **Note 8**).
4. Sonicate the lysate and return to ice.

5. Centrifuge the tubes at (10,000× *g*) for 8 min at 4 °C (*see* **Note 2**).
6. Transfer supernatants (soluble protein) into fresh 1.5 mL micro-centrifuge tubes.
7. Determine protein concentrations using Bio-Rad Protein Assay solution (Bio-Rad).
8. Transfer 50 μL anti-FLAG M2 Affinity Gel agarose (Sigma) into a 1.5 mL micro-centrifuge tube.
9. Add 500 μL mEB to the anti-FLAG agarose and spin at 10,000× *g* for 1 min. Remove supernatant (*see* **Note 4**).
10. Repeat step 9 four times.
11. Add 1 mg of total protein to the anti-FLAG agarose in a total volume of 500 μL.
12. Incubate the total protein/anti-FLAG agarose at 4 °C for 2 h on a rotator (*see* **Note 6**).
13. Centrifuge the tubes at 1000× *g* for 1 min.
14. Gently remove the supernatant (*see* **Note** 7).
15. Add 500 μL mEB to the anti-FLAG agarose.
16. Repeat steps 13–15 five times.
17. Add 500 μL mEB to the anti-FLAG agarose.
18. Centrifuge at 15,000× *g* for 1 min.
19. Remove as much supernatant as possible (*see* **Note 9** for optional PreScission protease cleavage, Fig. 3).
20. Add 40 μL of the protein sample buffer.
21. Boil the samples for 3-5 min.
22. Proceed to Subheading 3.4 for Western blotting and PTM detection.

3.3 Mitochondrial Isolation from HEK293 Cells, Disruption, and IP of TRAP1

1. Transfect HEK293 cells (~40% confluent in a 10 cm dish; growing in DMEM +10% FBS) with 2 μg TRAP1-FLAG using TransIT®-2020 reagent (Mirus), and incubate overnight at 37 °C, 5%CO_2 (*see* **Note 10**).
2. Place plates on ice and aspirate media. Wash 2× with cold PBS. Remove all remaining PBS from the plate.
3. Add 100 μL cold isoB to the plate. Scrape cells and transfer them to a 1 mL Dounce tissue homogenizer on ice.
4. Homogenize cell suspension for ten strokes with a tight-fitting pestle.
5. Centrifuge homogenate at (600× *g*) for 15 min at 4 °C.

6. Transfer supernatant to a new tube (discarding the residual pellet of cell debris) and centrifuge at (10,000× *g*) for 15 min at 4 °C.
7. Remove supernatant containing cytosolic fraction.
8. Wash the remaining mitochondrial pellet 1× with isoB and centrifuge at (10,000× *g*) for 1 min at 4 °C.
9. Remove supernatant and resuspend the mitochondrial pellet in 200 μL mEB.
10. Sonicate the mitochondria for 10 s, amplitude 60, and return to ice.
11. Determine protein concentration using Bio-Rad Protein Assay solution (Bio-Rad).
12. Transfer 25 μL anti-FLAG M2 Affinity Gel agarose (Sigma) into a 1.5 mL micro-centrifuge tube.
13. 9. Add 500 μL mEB to the anti-FLAG agarose and spin at 10,000× *g* for 1 min. Remove supernatant (*see* **Note 4**).
14. 10. Repeat step 9 four times.
15. 11. Add 250 μg of mitochondrial lysate to the anti-FLAG agarose in a total volume of 500 μL.
16. Incubate the total protein/anti-FLAG agarose at 4 °C for 2 h on a rotator (*see* **Note 6**).
17. Centrifuge the tubes at 1000× *g* for 1 min.
18. Gently remove the supernatant (*see* **Note** 7).
19. Add 500 μL mEB to the anti-FLAG agarose.
20. Repeat steps 13–15 five times.
21. Add 500 μL mEB to the anti-FLAG agarose.
22. Centrifuge at 15,000× *g* for 1 min.
23. Remove as much supernatant as possible.
24. Add 50 μL of the protein sample buffer.
25. Boil the samples for 3-5 min.
26. Proceed to Subheading 3.4 for Western blotting and PTM detection.

3.4 Western Blotting and Detection of Hsp90 PTMs

1. Centrifuge the samples at 10,000 *g* and load the supernatant onto a 7.5% SDS-PAGE Tris-HCl gel (*see* **Note 11**).
2. Transfer the proteins from SDS-PAGE gel onto a 0.45 μm nitrocellulose membrane (Bio-Rad) (*see* **Note 12**).
3. Examine the quality and efficiency of the transfer by staining the membrane with Ponceau S solution (Sigma) for 1 min (see **Note 13**).
4. Wash the membrane with dH_2O.

Table 1
Primary antibodies for Hsp90 posttranslational modifications detection by Western blot

PTM	Antibody	Manufacturer	Dilution	Diluent	Time and temperature	Species
Phosphorylation	Phospho-serine (PSR-45)	Sigma (cat no. P5747)	1:500–1:1000	1% BSA in TBS-T	o/n 4 °C	Mouse
	Phospho-threonine (PTR-8)	Sigma (cat no. P6623)	1:500–1:1000	1% BSA in TBS-T	o/n 4 °C	Mouse
	Phospho-tyrosine (4G10)	Millipore (cat no. 05–321)				Mouse
Acetylation	Acetylated lysine	Cell signaling (#9441)				Rabbit
Ubiquitination	Ubiquitin (P4D1)	Santa Cruz (sc-8017)	1:500–1:2000	5% milk in TBS-T	o/n 4 °C	Mouse
SUMOylation	SUMO-1 (2A12)	Cell signaling (#5718)				Mouse
	SUMO-2/3 (18H8)	Cell signaling (#4971)				Rabbit
O-GlcNAcylation	GlcNAc (CTD110.6)	Cell signaling (#9875)	1:1000	5% milk in TBS-T	o/n 4 °C	Mouse

5. Incubate the membrane in 5% milk in TBS-T for 15–20 min at room temperature.
6. Wash the membrane with 1× TBS-T for 5 min at room temperature.
7. Repeat step 6 three times.
8. Incubate the membrane with the primary antibody (*see* Table 1).
9. Wash the membrane three times with 1× TBS-T for 5 min at room temperature.
10. Incubate the membrane with 1:2000 dilution of secondary anti-mouse or anti-rabbit antibody in 5% milk-TBS-T for 1 h at room temperature.
11. Wash the membrane three times with 1× TBS-T for 5 min at room temperature.
12. Remove 1XTBS-T and then apply ECL plus (GE Healthcare) to nitrocellulose membrane for 2-3 min.

13. Drain nitrocellulose membrane of excess developing solution (do not let dry).
14. Wrap the blot in Saran wrap.
15. Place the blot in the X-ray film cassette (*see* **Note 14**).
16. Expose the blots to X-ray films by placing X-ray film directly against the western blot at different lengths of time.

4 Notes

1. The cell pellet must be kept on ice.
2. At this stage, a Bio-Rad Protein Assay solution (Bio-Rad) should be prepared.
3. Ni-NTA agarose is precharged with Ni2+ ions and appears blue. It is provided as a 50% slurry in 30% ethanol.
4. Do not disturb the Ni-NTA or anti-FLAG agarose pellet.
5. Imidazole at low concentrations is commonly used in the binding and wash buffer to minimize the binding of unwanted host cell proteins.
6. Use Eppendorf Thermomixer R to gently mix total proteins/Ni-NTA agarose solution.
7. Avoid disturbing the Ni-NTA or anti-FLAG agarose.
8. For the detection of SUMO, mEB should always contain 20 mM *N*-ethylmaleimide (NEM).
9. PreScission protease cleavage: incubate hHsp90-FLAG bound to anti-FLAG agarose with two units of PreScission protease in 50 mM Tris-HCl, 150 mM NaCl, 1 mM EDTA, and 1 mM DTT (pH 7.0) at 10 °C for 16 h.
10. Use a minimum of three plates for a minimally sufficient yield of mitochondria, and scale as appropriate.
11. Criterion precast gels from Bio-Rad are suitable for this purpose.
12. The high MW setting on the Bio-Rad Trans-Blot Turbo transfer system is suitable for this purpose.
13. Prepare 5% dry milk (LabScientific Inc.) in 1× TBS-T (O.1% Tween 20, Sigma) buffer before examining the membrane.
14. This procedure must be performed in the dark.

References

1. Prodromou C (2016) Mechanisms of Hsp90 regulation. Biochem J 473(16):2439–2452. https://doi.org/10.1042/BCJ20160005
2. Rohl A, Rohrberg J, Buchner J (2013) The chaperone Hsp90: changing partners for demanding clients. Trends Biochem Sci 38(5):253–262. https://doi.org/10.1016/j.tibs.2013.02.003
3. Taipale M, Jarosz DF, Lindquist S (2010) HSP90 at the hub of protein homeostasis: emerging mechanistic insights. Nat Rev Mol Cell Biol 11(7):515–528. https://doi.org/10.1038/nrm2918
4. Schopf FH, Biebl MM, Buchner J (2017) The HSP90 chaperone machinery. Nat Rev Mol Cell Biol 18(6):345–360. https://doi.org/10.1038/nrm.2017.20
5. Karagoz GE, Rudiger SG (2015) Hsp90 interaction with clients. Trends Biochem Sci 40(2): 117–125. https://doi.org/10.1016/j.tibs.2014.12.002
6. Oroz J, Kim JH, Chang BJ, Zweckstetter M (2017) Mechanistic basis for the recognition of a misfolded protein by the molecular chaperone Hsp90. Nat Struct Mol Biol 24:407. https://doi.org/10.1038/nsmb.3380
7. Zierer BK, Rubbelke M, Tippel F et al (2016) Importance of cycle timing for the function of the molecular chaperone Hsp90. Nat Struct Mol Biol 23(11):1020–1028. https://doi.org/10.1038/nsmb.3305
8. Verba KA, Wang RY, Arakawa A et al (2016) Atomic structure of Hsp90-Cdc37-Cdk4 reveals that Hsp90 traps and stabilizes an unfolded kinase. Science 352(6293): 1542–1547. https://doi.org/10.1126/science.aaf5023
9. Panaretou B, Prodromou C, Roe SM et al (1998) ATP binding and hydrolysis are essential to the function of the Hsp90 molecular chaperone in vivo. EMBO J 17(16): 4829–4836. (In eng). https://doi.org/10.1093/emboj/17.16.4829
10. Obermann WM, Sondermann H, Russo AA, Pavletich NP, Hartl FU (1998) In vivo function of Hsp90 is dependent on ATP binding and ATP hydrolysis. J Cell Biol 143(4): 901–910. (In eng) (http://www.ncbi.nlm.nih.gov/entrez/query.fcgi?cmd=Retrieve&db=PubMed&dopt=Citation&list_uids=9817749)
11. Grenert JP, Johnson BD, Toft DO (1999) The importance of ATP binding and hydrolysis by hsp90 in formation and function of protein heterocomplexes. J Biol Chem 274(25): 17525–17533. (In eng) (http://www.ncbi.nlm.nih.gov/entrez/query.fcgi?cmd=Retrieve&db=PubMed&dopt=Citation&list_uids=10364185)
12. Prodromou C, Roe SM, O'Brien R, Ladbury JE, Piper PW, Pearl LH (1997) Identification and structural characterization of the ATP/ADP-binding site in the Hsp90 molecular chaperone. Cell 90(1):65–75. (In eng). S0092-8674(00)80314-1 [pii]
13. Grenert JP, Sullivan WP, Fadden P et al (1997) The amino-terminal domain of heat shock protein 90 (hsp90) that binds geldanamycin is an ATP/ADP switch domain that regulates hsp90 conformation. J Biol Chem 272(38): 23843–23850. (In eng) (http://www.ncbi.nlm.nih.gov/entrez/query.fcgi?cmd=Retrieve&db=PubMed&dopt=Citation&list_uids=9295332)
14. Stebbins CE, Russo AA, Schneider C, Rosen N, Hartl FU, Pavletich NP (1997) Crystal structure of an Hsp90-geldanamycin complex: targeting of a protein chaperone by an antitumor agent. Cell 89(2):239–250. (In eng). S0092-8674(00)80203-2 [pii]
15. Roe SM, Prodromou C, O'Brien R, Ladbury JE, Piper PW, Pearl LH (1999) Structural basis for inhibition of the Hsp90 molecular chaperone by the antitumor antibiotics radicicol and geldanamycin. J Med Chem 42(2):260–266. (In eng). https://doi.org/10.1021/jm980403y
16. Workman P, Burrows F, Neckers L, Rosen N (2007) Drugging the cancer chaperone HSP90: combinatorial therapeutic exploitation of oncogene addiction and tumor stress. Ann N Y Acad Sci 1113:202–216. (In eng). annals.1391.012 [pii] 1196/annals.1391.012
17. Neckers L, Trepel JB (2014) Stressing the development of small molecules targeting HSP90. Clin Cancer Res 20(2):275–277. https://doi.org/10.1158/1078-0432.CCR-13-2571
18. Neckers L, Workman P (2012) Hsp90 molecular chaperone inhibitors: are we there yet? Clin Cancer Res 18(1):64–76. (In eng). 18/1/64 [pii] 1158/1078-0432.CCR-11-1000
19. Kurokawa Y, Honma Y, Sawaki A et al (2022) Pimitespib in patients with advanced gastrointestinal stromal tumor (CHAPTER-GIST-301): a randomized, double-blind, placebo-controlled phase III trial. Ann Oncol 33(9): 959–967. https://doi.org/10.1016/j.annonc.2022.05.518

20. Schmid AB, Lagleder S, Grawert MA et al (2012) The architecture of functional modules in the Hsp90 co-chaperone Sti1/Hop. EMBO J 31(6):1506–1517. https://doi.org/10.1038/emboj.2011.472
21. Richter K, Muschler P, Hainzl O, Reinstein J, Buchner J (2003) Sti1 is a non-competitive inhibitor of the Hsp90 ATPase. Binding prevents the N-terminal dimerization reaction during the ATPase cycle. J Biol Chem 278(12):10328–10333. (In eng). https://doi.org/10.1074/jbc.M213094200
22. Song Y, Masison DC (2005) Independent regulation of Hsp70 and Hsp90 chaperones by Hsp70/Hsp90-organizing protein Sti1 (Hop1). J Biol Chem 280(40):34178–34185. (In eng). M505420200 [pii]. https://doi.org/10.1074/jbc.M505420200
23. Lee P, Shabbir A, Cardozo C, Caplan AJ (2004) Sti1 and Cdc37 can stabilize Hsp90 in chaperone complexes with a protein kinase. Mol Biol Cell 15(4):1785–1792. (In eng) E03-07-0480 [pii]. https://doi.org/10.1091/mbc.E03-07-0480
24. MacLean M, Picard D (2003) Cdc37 goes beyond Hsp90 and kinases. Cell Stress Chaperones 8(2):114–119. (In eng) (http://www.ncbi.nlm.nih.gov/entrez/query.fcgi?cmd=Retrieve&db=PubMed&dopt=Citation&list_uids=14627196)
25. Vaughan CK, Mollapour M, Smith JR et al (2008) Hsp90-dependent activation of protein kinases is regulated by chaperone-targeted dephosphorylation of Cdc37. Mol Cell 31(6):886–895. (In eng). S1097-2765(08)00541-8 [pii]. https://doi.org/10.1016/j.molcel.2008.07.021
26. McLaughlin SH, Sobott F, Yao ZP et al (2006) The co-chaperone p23 arrests the Hsp90 ATPase cycle to trap client proteins. J Mol Biol 356(3):746–758. (In eng). S0022-2836(05)01533-0 [pii]. https://doi.org/10.1016/j.jmb.2005.11.085
27. Picard D (2006) Intracellular dynamics of the Hsp90 co-chaperone p23 is dictated by Hsp90. Exp Cell Res 312(2):198–204. (In eng). S0014-4827(05)00479-9 [pii]. https://doi.org/10.1016/j.yexcr.2005.36.009
28. Lotz GP, Lin H, Harst A, Obermann WM (2003) Aha1 binds to the middle domain of Hsp90, contributes to client protein activation, and stimulates the ATPase activity of the molecular chaperone. J Biol Chem 278(19):17228–17235. (In eng). M212761200 [pii]. https://doi.org/10.1074/jbc.M212761200
29. Meyer P, Prodromou C, Liao C et al (2004) Structural basis for recruitment of the ATPase activator Aha1 to the Hsp90 chaperone machinery. EMBO J 23(3):511–519. (In eng) [pii]. https://doi.org/10.1038/sj.emboj.7600060
30. Panaretou B, Siligardi G, Meyer P et al (2002) Activation of the ATPase activity of hsp90 by the stress-regulated cochaperone aha1. Mol Cell 10(6):1307–1318. (In eng). S1097276502007852 [pii]
31. Wolmarans A, Lee B, Spyracopoulos L, LaPointe P (2016) The mechanism of Hsp90 ATPase stimulation by Aha1. Sci Rep 6:33179. https://doi.org/10.1038/srep33179
32. Retzlaff M, Hagn F, Mitschke L et al (2010) Asymmetric activation of the hsp90 dimer by its cochaperone aha1. Mol Cell 37(3):344–354. https://doi.org/10.1016/j.molcel.2010.01.006
33. Johnson JL, Halas A, Flom G (2007) Nucleotide-dependent interaction of Saccharomyces cerevisiae Hsp90 with the cochaperone proteins Sti1, Cpr6, and Sba1. Mol Cell Biol 27(2):768–776. (In eng). MCB.01034-06 [pii]. https://doi.org/10.1128/MCB.01034-06
34. Mayr C, Richter K, Lilie H, Buchner J (2000) Cpr6 and Cpr7, two closely related Hsp90-associated immunophilins from Saccharomyces cerevisiae, differ in their functional properties. J Biol Chem 275(44):34140–34146. (In eng) M005251200 [pii]. https://doi.org/10.1074/jbc.M005251200
35. Mollapour M, Bourboulia D, Beebe K et al (2014) Asymmetric Hsp90 N domain SUMOylation recruits Aha1 and ATP-competitive inhibitors. Mol Cell 53(2):317–329. https://doi.org/10.1016/j.molcel.2013.12.007
36. Woodford MR, Dunn D, Miller JB, Jamal S, Neckers L, Mollapour M (2016) Impact of posttranslational modifications on the anticancer activity of Hsp90 inhibitors. Adv Cancer Res 129:31–50. https://doi.org/10.1016/bs.acr.2015.09.002
37. Mollapour M, Neckers L (2012) Post-translational modifications of Hsp90 and their contributions to chaperone regulation. Biochim Biophys Acta 1823(3):648–655. (In eng). S0167-4889(11)00217-5 [pii]. https://doi.org/10.1016/j.bbamcr.2011.07.018
38. Preuss KD, Pfreundschuh M, Weigert M, Fadle N, Regitz E, Kubuschok B (2015) Sumoylated HSP90 is a dominantly inherited plasma cell dyscrasias risk factor. J Clin Invest 125(1):316–323. https://doi.org/10.1172/JCI76802

39. Backe SJ, Sager RA, Woodford MR, Makedon AM, Mollapour M (2020) Post-translational modifications of Hsp90 and translating the chaperone code. J Biol Chem 295(32): 11099–11117. https://doi.org/10.1074/jbc.REV120.011833
40. Woodford MR, Hughes M, Sager RA et al (2019) Mutation of the co-chaperone Tsc1 in bladder cancer diminishes Hsp90 acetylation and reduces drug sensitivity and selectivity. Oncotarget 10(56):5824–5834. https://doi.org/10.18632/oncotarget.27217
41. Soroka J, Wandinger SK, Mausbacher N et al (2012) Conformational switching of the molecular chaperone Hsp90 via regulated phosphorylation. Mol Cell 45(4):517–528. (In eng). S1097-2765(12)00040-8 [pii]. https://doi.org/10.1016/j.molcel.2011.12.031
42. Xu W, Mollapour M, Prodromou C et al (2012) Dynamic tyrosine phosphorylation modulates cycling of the HSP90-P50 (CDC37)-AHA1 chaperone machine. Mol Cell 47(3):434–443. https://doi.org/10.1016/j.molcel.2012.05.015
43. Woodford MR, Backe SJ, Wengert LA, Dunn DM, Bourboulia D, Mollapour M (2021) Hsp90 chaperone code and the tumor suppressor VHL cooperatively regulate the mitotic checkpoint. Cell Stress Chaperones 26:965. https://doi.org/10.1007/s12192-021-01240-2
44. Nguyen MT, Kniess RA, Daturpalli S et al (2017) Isoform-specific phosphorylation in human Hsp90beta affects interaction with clients and the Cochaperone Cdc37. J Mol Biol 429(5):732–752. https://doi.org/10.1016/j.jmb.2017.01.011
45. Woodford MR, Truman AW, Dunn DM et al (2016) Mps1 mediated phosphorylation of Hsp90 confers renal cell carcinoma sensitivity and selectivity to Hsp90 inhibitors. Cell Rep 14(4):872–884. https://doi.org/10.1016/j.celrep.2015.12.084
46. Muller P, Ruckova E, Halada P et al (2012. (In Eng)) C-terminal phosphorylation of Hsp70 and Hsp90 regulates alternate binding to co-chaperones CHIP and HOP to determine cellular protein folding/degradation balances. Oncogene 32:3101. https://doi.org/10.1038/onc.2012.314
47. Scroggins BT, Neckers L (2007) Post-translational modification of heat shock protein 90: impact on chaperone function. Expert Opin Drug Discov 2(10):1403–1414
48. Garnier C, Lafitte D, Jorgensen TJ, Jensen ON, Briand C, Peyrot V (2001) Phosphorylation and oligomerization states of native pig brain HSP90 studied by mass spectrometry. Eur J Biochem 268(8):2402–2407. (In eng). ejb2121 [pii]
49. Zhao YG, Gilmore R, Leone G, Coffey MC, Weber B, Lee PW (2001) Hsp90 phosphorylation is linked to its chaperoning function. Assembly of the reovirus cell attachment protein. J Biol Chem 276(35):32822–32827. (In eng) M105562200 [pii]. https://doi.org/10.1074/jbc.M105562200
50. Fontana J, Fulton D, Chen Y et al (2002) Domain mapping studies reveal that the M domain of hsp90 serves as a molecular scaffold to regulate Akt-dependent phosphorylation of endothelial nitric oxide synthase and NO release. Circ Res 90(8):866–873. (In eng) (http://www.ncbi.nlm.nih.gov/entrez/query.fcgi?cmd=Retrieve&db=PubMed&dopt=Citation&list_uids=11988487)
51. Adinolfi E, Kim M, Young MT, Di Virgilio F, Surprenant A (2003) Tyrosine phosphorylation of HSP90 within the P2X7 receptor complex negatively regulates P2X7 receptors. J Biol Chem 278(39):37344–37351. (In eng) M301508200 [pii]. https://doi.org/10.1074/jbc.M301508200
52. Wang C, Chen J (2003) Phosphorylation and hsp90 binding mediate heat shock stabilization of p53. J Biol Chem 278(3):2066–2071. (In eng) M206697200 [pii]. https://doi.org/10.1074/jbc.M206697200
53. Duval M, Le Boeuf F, Huot J, Gratton JP (2007) Src-mediated phosphorylation of Hsp90 in response to vascular endothelial growth factor (VEGF) is required for VEGF receptor-2 signaling to endothelial NO synthase. Mol Biol Cell 18(11):4659–4668. (In eng). E07-05-0467 [pii]. https://doi.org/10.1091/mbc.E07-05-0467
54. Miyata Y, Yahara I (1992) The 90-kDa heat shock protein, HSP90, binds and protects casein kinase II from self-aggregation and enhances its kinase activity. J Biol Chem 267(10):7042–7047. (In eng) (http://www.ncbi.nlm.nih.gov/entrez/query.fcgi?cmd=Retrieve&db=PubMed&dopt=Citation&list_uids=1551911)
55. Mimnaugh EG, Worland PJ, Whitesell L, Neckers LM (1995) Possible role for serine/threonine phosphorylation in the regulation of the heteroprotein complex between the hsp90 stress protein and the pp60v-src tyrosine kinase. J Biol Chem 270(48):28654–28659. (In eng) (http://www.ncbi.nlm.nih.gov/entrez/query.fcgi?cmd=Retrieve&db=PubMed&dopt=Citation&list_uids=7499384)

56. Ogiso H, Kagi N, Matsumoto E et al (2004) Phosphorylation analysis of 90 kDa heat shock protein within the cytosolic aryl hydrocarbon receptor complex. Biochemistry 43(49): 15510–15519. (In eng). https://doi.org/10.1021/bi048736m
57. Xu W, Beebe K, Chavez JD et al (2019) Hsp90 middle domain phosphorylation initiates a complex conformational program to recruit the ATPase-stimulating cochaperone Aha1. Nat Commun 10(1):2574. https://doi.org/10.1038/s41467-019-10463-y
58. Wengert LA, Backe SJ, Bourboulia D, Mollapour M, Woodford MR (2022) TRAP1 chaperones the metabolic switch in cancer. Biomol Ther 12(6):786. https://doi.org/10.3390/biom12060786
59. Masgras I, Ciscato F, Brunati AM et al (2017) Absence of Neurofibromin induces an oncogenic metabolic switch via mitochondrial ERK-mediated phosphorylation of the chaperone TRAP1. Cell Rep 18(3):659–672. https://doi.org/10.1016/j.celrep.2016.12.056
60. Rizza S, Montagna C, Cardaci S et al (2016) S-nitrosylation of the mitochondrial chaperone TRAP1 sensitizes hepatocellular carcinoma cells to inhibitors of succinate dehydrogenase. Cancer Res 76(14):4170–4182. https://doi.org/10.1158/0008-5472.CAN-15-2637
61. Park HK, Hong JH, Oh YT et al (2019) Interplay between TRAP1 and Sirtuin-3 modulates mitochondrial respiration and oxidative stress to maintain Stemness of glioma stem cells. Cancer Res 79(7):1369–1382. https://doi.org/10.1158/0008-5472.CAN-18-2558
62. Kim S, Backe SJ, Wengert LA et al (2022) O-GlcNAcylation suppresses TRAP1 activity and promotes mitochondrial respiration. Cell Stress Chaperones 27(5):573–585. https://doi.org/10.1007/s12192-022-01293-x

Chapter 12

Multiple Targeting of HSP Isoforms to Challenge Isoform Specificity and Compensatory Expression

Kisho Ono and Takanori Eguchi

Abstract

Heat shock proteins (HSPs) are molecular chaperones that assist in protein folding, trafficking, and metabolism. Intracellular chaperone functions of HSPs had been well-investigated, but extracellular and exosomal HSPs have been recently found. Exosomal HSPs are intercellularly transferred, while extracellular HSPs play cytokine-like roles called chaperokines. We have shown that exosomal HSPs play key roles in intercellular communication between tongue carcinoma and tumor-associated macrophages in the tumor microenvironment. Notably, HSP90 isoforms consist of HSP90alpha, HSP90beta, mitochondrial TRAP1, and GRP94 in the endoplasmic reticulum. Moreover, many pseudogenes of HSP90 can be transcribed into RNA. Besides, the function of HSP90 is defined by their cochaperones, such as CDC37 or AHA1. Therefore, isoform-specific small interfering RNA (siRNA) is necessary for precisely targeting each HSP90 isoform and cochaperone. Nevertheless, we often encountered compensatory expression of HSP90 isoforms in the knockdown studies. Here, we provide dual and triple knockdown methods to target multiple RNA for challenging isoform-specific roles and compensatory expression of intracellular, extracellular, and exosomal HSPs.

Key words Heat shock proteins, HSP90, CDC37, Cochaperone, Exosomes, Extracellular vesicles, siRNA, Chaperone knockdown

1 Introduction

Heat shock protein 90 (HSP90) is an ATP-dependent molecular chaperone that supports protein homeostasis. More than several hundred client proteins whose conformation, stability, and activity are regulated by HSP90-binding are known to be factored in cancer cell survival and proliferation [1, 2]. In other words, inhibition of HSP90 induces the blockade of signaling pathways in cancer cells by making the client protein structurally unstable and degrading it, leading to growth inhibition and apoptosis of cancer cells. Against these findings, several HSP90 inhibitors are in clinical development as anti-cancer targets. However, the structural origin of the inhibitors and ontarget side effects have limited the dosage and

Stuart K. Calderwood and Thomas L. Prince (eds.), *Chaperones: Methods and Protocols*, Methods in Molecular Biology, vol. 2693, https://doi.org/10.1007/978-1-0716-3342-7_12,

Table 1
Isoforms of HSP90

	Localization	Functions/properties
HSP90α	Cytoplasm, nucleus	Molecular chaperone, induced by stress, secreted extracellularly
HSP90β	Cytoplasm, nucleus	Molecular chaperone, constitutively expressed
Grp94	Endoplasmic reticulum	Transport of cancer peptides, Maturation and secretion of insulin-like growth factor
Trap1	Mitochondria	Oxidative stress protection

administration schedule, and they have yet to demonstrate sufficient therapeutic efficacy [3, 4]. There is also a need for research on the deactivation of HSP90, which is involved in the malignant mechanism of cancer, by means other than inhibitors.

Four members are known as the HSP90 family (Table 1) [3, 5, 6]. First, the cytoplasmic and nuclear localization of HSP90 is mediated by the proteotoxic stress-induced HSP90α encoded by *HSP90AA1* and the constitutively expressed HSP90β encoded by *HSP90AB1*. HSP90α is induced by heat shock and other stresses. HSP90β, on the other hand, is constantly expressed and accounts for 1–2% of total intracellular proteins. In addition, glucose-regulated protein 94 (GRP94) is localized in the endoplasmic reticulum and tumor necrosis factor receptor-associated protein 1 (TRAP1) in mitochondria. All four members have ATPase activity. In particular, HSP90α is often overexpressed in cancer cells and is secreted extracellularly as a soluble protein or as a cargo protein of exosome/extracellular vesicles (EVs) and has been reported to promote cancer cell invasion and metastasis by activating matrix metalloproteinases (MMPs) localized in the extracellular matrix (ECM) [5, 7–9]. Our recent EV-proteome analysis of oral cancer cells revealed that highly metastatic oral cancer cells secrete large amounts of HSP90-rich exosomes/EVs [5, 8].

HSP90 functions in a chaperone complex with auxiliary factors, which are called cochaperones, of which more than 20 are known (Table 2) [10–12]. The roles of these cochaperones include regulation of HSP90 ATPase activity, control of the chaperone cycle, and mobilization and degradation of specific client proteins into the chaperone complex. Among them, CDC37 is a cochaperone that is an important partner of HSP90 and assists in molecular chaperone activity, particularly for protein kinase regulation [13]. The HSP90/CDC37 complex regulates the folding of many protein kinases and plays a central role in intracellular signaling networks, making CDC37 a potential intermediate in carcinogenesis [13–18].

Our previous data were the first to demonstrate the efficacy of triple depletion of the interacting CDC37/HSP90α/HSP90β

Table 2
Cochaperones of HSP90

Cochaperone		Main function
AHA1	Activator of HSP90 ATPase homolog 1	Modulates HSP90 ATPase activity
Cdc37/p50	Cell division cycle 37 homolog	Recruit protein kinases
CyP40	Cyclophilin 40	Modulates HSP90 ATPase activity
CHIP	Carboxy-terminus of HSP70-interacting protein	Client protein degradation
HIP	HSP70-interacting protein	HSP70/HSP90 adaptor protein
HOP	HSP70/HSP90-organizing protein	HSP70/HSP90 adaptor protein
Hsp40	Heat shock protein 40	Modulates HSP90 ATPase activity
Hsp70	Heat shock protein 70	Molecular chaperone
FKBP51, FKBP52	51, 52 kDa FK506 binding protein that binds the immunosuppressive drug FK506 without initiating immunosuppression	Maturation and activation of steroid hormone receptors
p23	Prostaglandin E synthase 3	Modulates HSP90 ATPase activity
SGT1	Suppressor of G2 allele of SKP1	Kinetochore complex formation
TAH1	Tpr-containing protein associated with HSP90	Modulates ribonucleoproteins and chromatin remodeling
UNC45	Unc-45 myosin chaperone B	Myosin folding and assembly

complex (Fig. 1) [5, 8]. Notably, depletion of either HSP90α, HSP90β, or CDC37 was shown to cause an opposing increase in other chaperones [5, 8]. The compensatory response observed in this HSP90-CDC37 system supported our original idea of depleting the triple chaperone. The results showed that triple depletion of CDC37/HSP90α/β by siRNA significantly reduced the survival activity of oral cancer cells [5, 8]. Furthermore, triple depletion of CDC37/HSP90α/β by siRNA resulted in HSP90 knockdown in cancer cells and in secreted exosomes/EVs (Fig. 2) [8].

Exosomes/EVs, nano-sized vesicles secreted by cells, contain internalized proteins and nucleic acids and act as cell-to-cell communication tools. In addition, exosomes/EVs have been shown to potentially deliver molecules to specific cells in remote organs and tissues. HSP90 is one of the major proteins secreted extracellularly through exosomes/EVs and has been shown to increase HSP90

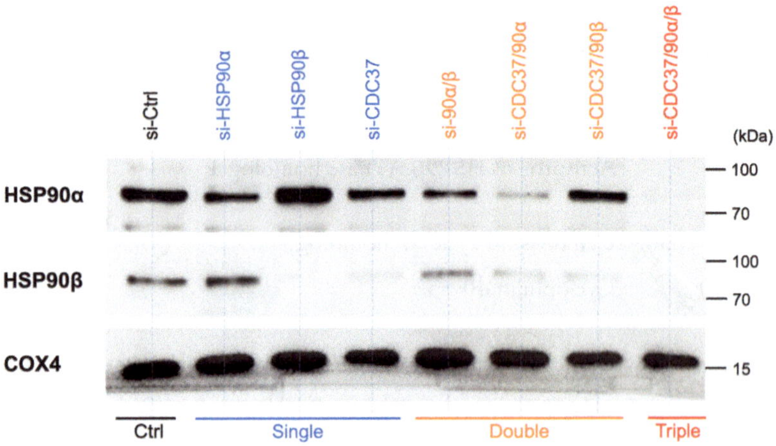

Fig. 1 Single or multiple chaperone knockdown using siRNA targeting HSP90 in cancer cells. Raw data of western blotting showing HSP90α and HSP90β were presented. Oral carcinoma cells were transfected with siRNAs targeting CDC37, HSP90α, and/or HSP90β or nontargeting siRNA (si-Ctrl) for 48 h. COX4, loading control. (The completed data was published in ref. [8])

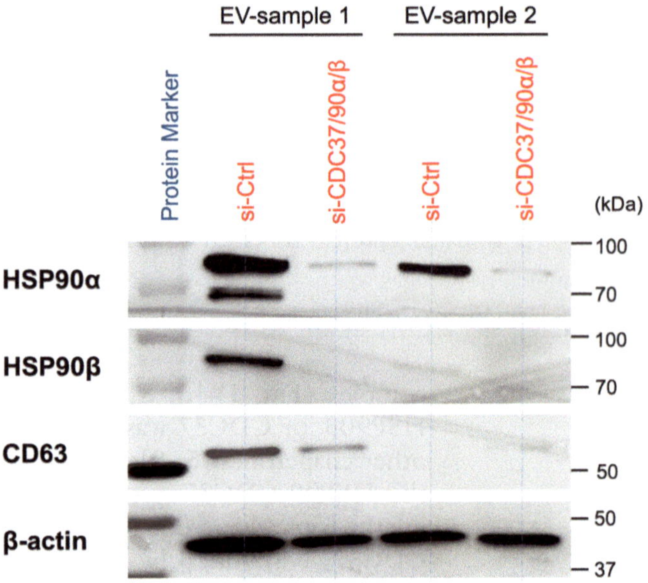

Fig. 2 Depletion of HSP90 in EVs by triple chaperone knockdown. Raw data of western blotting showing HSP90α, HSP90β, and CD63 (EV marker) were presented. Oral cancer cells were transfected with siRNAs targeting all of CDC37, HSP90α, and HSP90β or nontargeting siRNA (si-Ctrl) for 48 h, and EVs were collected. EV samples were purified from two independent experiments. β-actin, loading control. (The completed data was published in ref. [8])

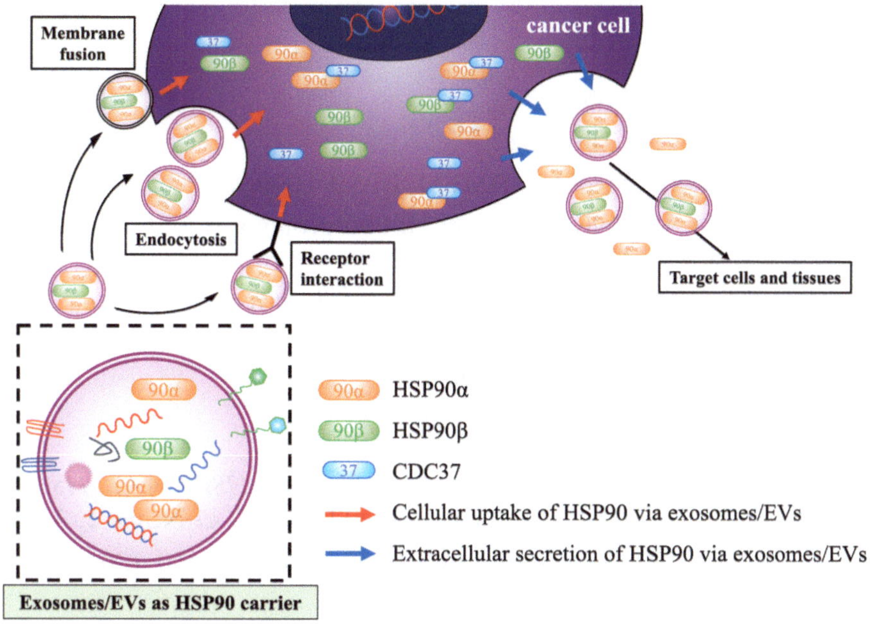

Fig. 3 Schematic diagram of intracellular uptake and extracellular secretion of HSP90 via exosomes/EVs. HSP90α is predominantly expressed in the extracellular space and exosomes/EVs compared to HSP90β. HSP90β is expressed permanently intracellularly, but some are expressed within exosomes/EVs. CDC37 is expressed only intracellularly and forms a complex with HSP90. (This schema was considered from data in ref. [8])

content in exosomes/EVs in response to cellular stress (hypoxia, acidosis, metabolic deficits, nutrient deprivation, etc.) and to play an important role in cell-to-cell communication as described above (Fig. 3) [15]. It has also been reported that HSP90 mediates the fusion of multivesicular bodies (MVBs) with the plasma membrane and induces exosome/EV secretion and that exosomes/EVs with HSP90α on their surface secreted from cancer cells induce migration of cancer cells and stromal cells as autocrine or paracrine factors [19, 20]. In our EV-HSP90 depletion experiments, malignant transformation-promoting activity was significantly reduced [8]. However, there are many functions and recipient molecules of chaperones, including HSP90 in exosomes/EVs, that have not yet been elucidated, and novel experimental systems are needed to solve these problems in the future.

This chapter outlines an experimental approach for elucidating the molecular mechanism of exosome/EV chaperones by applying the RNAi method. We provide protocols of chaperone knockdown using siRNA transfection (Subheading 3.1), exosome/EV sampling using the polymer-based precipitation (PBP) method (Subheading 3.2), the harvest of cellular proteins (Subheading 3.3), protein assay (Subheading 3.4), and sample preparation to western blotting (Subheading 3.5, 3.6, 3.7, 3.8, and 3.9).

2 Materials

2.1 Transfection

2.1.1 Cell Culture for Transfection

1. PBS
2. Centrifuge
3. Medium (DMEM, RPMI 1640, etc.)
4. Cells producing HSP90 proteins at high levels, e.g., HSC-3-M3 cells (*see* **Note 1**)
5. Serum for maintaining cells, if required
6. Cell culture dishes (15 cm, 10 cm, and 6 cm) or plates (6-well, 24-well, and 96-well)
7. Trypsin–EDTA (0.25%)

2.1.2 Preparation of siRNAs for Transfections

1. HSP90 targeting siRNA, e.g., *HSP90AA1*, *HSP90AB1*, and *CDC37* (10 μM stock) (Table 3) (*see* **Notes 2**, **3**, and **4**)
2. The control nontargeting siRNA (10 μM stock) (Nippon Gene)
3. Transfection reagent, e.g., Lipofectamine® RNAiMAX Reagent (Invitrogen)
4. Opti-MEM® reduced-serum medium
5. Eppendorf tubes

2.2 Simple Isolation of Exosomes/EVs

2.2.1 Cell Culture for Exosome/EV Isolation

1. PBS
2. Centrifuge
3. Medium (DMEM, RPMI 1640, etc.)
4. Cells transfected with target siRNA (Subheading 2.1) (*see* **Note 1**)
5. Serum for maintaining cells, if required

Table 3
List of sequences of siRNA

Name of siRNA	Sequence (5′ to 3′)
hCDC37.NM7065–433 sense	gcaagaaggagaagagcauTT
hCDC37.NM7065–433 antisense	augcucuucuccuucuugcTT
hCDC37.NM7065–584 sense	gaaacagaucaagcacuuuTT
hCDC37.NM7065–584 antisense	aaagugcuugaucuguuucTT
hHSP90AA1.NM5348–415 sense	gcugcauauuaaccuuauaTT
hHSP90AA1.NM5348–415 antisense	uauaagguuaauaugcagcTT
hHSP90AA1.NM5348–2010 sense	caaacauggagagaaucauTT

6. EV/exosome-free medium or EV/exosome-free serum (*see* **Note 5**)
7. Cell culture dishes (15 cm, 10 cm, and 6 cm)
8. Trypsin–EDTA (0.25%)

2.2.2 Isolation of Exosomes/EVs Using the PBP Method

The materials listed below are the minimum requirements for the PBP method with ultrafiltration. See another chapter [21] for other methods.

1. (Optional) 0.2 μm pore filter, 0.8 μm pore filter, and syringes (*see* **Notes 6** and 7)
2. Amicon Ultra-15 Centrifugal Filter Devices for M.W. 100 kD
3. Polyethylene glycol (PEG) polymer, e.g., Total Exosome Isolation Reagent (Thermo Fisher)
4. Centrifuge
5. PBS
6. Vesicle/particle analyzer (*see* **Note 8**) and cuvettes
7. Transmission electron microscopy (TEM) (e.g., H-7650, Hitachi)

2.3 Buffers to Lyse Membranes of Cells and Vesicles

To determine the subcellular and subvesicular localization of chaperone proteins, choose a buffer from the following three: lysis buffer, RIPA buffer, or sample buffer. Lysis buffer is a standard for western blotting, as this buffer contains SDS, which denatures proteins and detergents such as Triton X-100 and NP-40 to dissolve lipid membranes. RIPA buffer is milder than the lysis buffer. Protein 3D structures and multi-protein complexes can be kept in the RIPA buffer. Therefore, RIPA buffer is useful for analyzing protein–protein interaction by immunoprecipitation (*see* **Note 9**) and enzyme activities (*see* **Note 10**). SDS sample buffer contains SDS at a high concentration and is used for sample preparation for western blotting. Therefore, adding the sample buffer directly to cells is time-saving, although protein samples in the SDS sample buffer are not applicable for protein assay. Therefore, the number of cells should be equalized among the samples for equal sample loading in western blotting.

1. Cell lysis buffer: 150 mM NaCl, 1% NP-40, 1% Na-deoxycholate, 0.1% SDS, 50 mM Tris–HCl, pH 7.5, 1 mM PMSF, 25 mM NaF, 2% Triton X-100 (*see* **Notes 11 and 12**). Add protease inhibitors and phosphatase inhibitors right before use.
2. 10 × RIPA buffer (also called RIPA lysis buffer): 0.5 M Tris–HCl pH 7.6, 1.5 M NaCl, 5% sodium deoxycholate, 10% NP-40, 1% SDS (*see* **Notes 12 and 13**). Add protease inhibitors and phosphatase inhibitors right before use.

3. 100 × protease inhibitor cocktail (*see* **Note 14**): PMSF (or AEBSF), aprotinin, leupeptin, pepstatin A, bestatin, E-64, and EDTA in DMSO.
4. Phosphatase inhibitor cocktail: sodium fluoride (NaF), sodium orthovanadate (Na_3VO_4), sodium pyrophosphate decahydrate, and β-glycerophosphate.
5. 6 × SDS sample buffer: 0.375 M Tris–HCl pH 6.8, 12% SDS, 60% glycerol, 0.6 M DTT, 0.06% bromophenol blue (*see* **Note 15**).
6. Beta-mercaptoethanol (β-ME) (*see* **Notes 16**, **17**, and **18**).
7. Cell fractionation kit (*see* **Note 19**).
8. Ice bag and ice bucket (or cold room) (*see* **Note 20**).
9. Cell scraper.
10. 25G needle and syringe.
11. Liquid nitrogen.

2.4 Protein Assay

1. Micro BCA protein assay reagent (Thermo Fisher) (*see* **Notes 21** and **22**)
2. Microplate reader

2.5 Sample Preparation for Western Blotting

1. 6 × SDS sample buffer
2. β-ME (*see* **Notes 16**, **17**, and **18**)
3. Protein marker
4. Peltier heating block incubator
5. Parafilm or plastic material to seal the lids of tubes
6. Low binding/adhesion microcentrifuge tubes (*see* **Note 23**)
7. Gel loading tips
8. Mini centrifuge

2.6 SDS-Page

1. Polyacrylamide gel (PAG): 10%, 12%, or gradient (*see* **Note 24**)
2. Electrophoresis tank
3. Power source
4. 10 × SDS running buffer: Tris-base 30.3 g; glycine 144.4 g; SDS 10 g. Dissolve in 1 L of Milli-Q-filtered H_2O

2.7 Protein Transfer

1. Transfer buffer: 25 mM Tris–HCl, 192 mM glycine, 0.05% SDS, 10% methanol
2. Protein transfer apparatus (*see* **Note 25**)
3. Power source
4. 100% methanol

5. Polyvinylidene fluoride (PVDF) membrane
6. Whatman filter papers
7. Plastic bat
8. Ice bag

2.8 Immunoblotting

1. TBS-T: Tris-buffered saline with 0.1% Tween 20, pH 7.6
2. Blocking agent, e.g., skim milk
3. Antibodies for detection of chaperone proteins (*see* **Note 26**), e.g., anti-HSP90α (GTX109753, GeneTex), HSP90β (GTX101448, GeneTex), CDC37 (D11A3, Cell Signaling Technology) antibody, exosome markers (*see* **Note 27**), and loading control (GAPDH or β-actin)
4. Horseradish peroxidase (HRP)-conjugated secondary antibodies
5. Stripping reagent, e.g., Western Blot Stripping Buffer (TaKaRa)

2.9 Imaging and Data Analysis

1. An imaging system, e.g., ChemiDoc MP Imaging System (Bio-Rad)
2. ECL detection substrate
3. ImageJ software

3 Methods

3.1 Transfection

3.1.1 Cell Culture and Transfection

1. Seed cells to be 60–80% confluent at transfection, and incubate overnight at 37 °C (Day 0). For example, seeding of 1–4 x 10^4 cells in 96-well plates, 0.5–2 x 10^5 cells in 24-well plates, and 0.25–1 x 10^6 cells in 6-well plates is reasonable.
2. Ensure that cultured cells are 60–80% confluent (Day 1).
3. Dilute Lipofectamine® RNAiMAX Reagent in Opti-MEM® (*see* **Note 28**). For example, the appropriate liquid volume of Opti-MEM® and Lipofectamine® RNAiMAX Reagent is 25 μl and 1.5 μl for 96-well plates, 50 μl and 3 μl for 24-well plates, and 150 μl and 9 μl for 6-well plates, respectively.
4. Dilute siRNA in Opti-MEM® (*see* **Note 4**). For example, the appropriate liquid volume of Opti-MEM® and siRNA (10 μM) is 25 μl and 0.5 μl (5 pmol) for 96-well plates, 50 μl and 1 μl (10 pmol) for 24-well plates, and 150 μl and 3 μl (30 pmol) for 6-well plates, respectively.
5. Add all diluted siRNA to all diluted Lipofectamine® RNAi-MAX Reagent (1:1 ratio), and incubate for 5 min at room temperature.

6. Add siRNA-lipid complex to cells and incubate for 1–3 days at 37 °C (*see* **Note 29**). For example, the appropriate liquid volume of siRNA-lipid complex per well is 10 μl (1 pmol) for 96-well plates, 50 μl (5 pmol) for 24-well plates, and 250 μl (25 pmol) for 6-well plates, respectively.
7. (Optional) Visualize and analyze transfected cells (*see* **Note 30**).
8. Go to Subheading 3.2 to prepare EVs/exosomes from a 10 cm dish.

3.2 Cell Culture and Isolation of EVs/Exosomes Using the PBP Method

We provide a simple protocol for isolating EVs/exosomes from 2D cultured cells using the PBP often combined with the UF method for detecting HSP90-EVs. Only a few 10 cm dishes are required to isolate EVs/exosomes from the cell culture supernatant and prepare WCL. See another chapter for the comprehensive and detailed methodology for detecting chaperone protein factors in EVs and 3D cultured tissues and cells [21].

3.2.1 2D Cell Culture for Preparation of sEV/Exosomes and WCL

1. Seed cells in 10 ml medium in a few 10 cm dishes (*see* **Note 31**). For example, seed 1 × 10^6 cells per 10 cm dish.
2. Culture cells to be 60–80% confluency. For a medium change, aspirate the culture medium carefully. Do not aspirate cells.
3. Wash cells with PBS carefully.
4. Culture the cells in 4–10 ml EV/exosome-free or serum-free medium for 1–2 days. Reducing the volume of the medium improves the concentration of sEVs secreted (*see* **Note 1**).
5. Transfer the culture supernatant to 50 mL tubes.
6. Go to Subheading 3.3 for the preparation of WCL from a 10 cm dish.

3.2.2 (Optional) Removal of Large EVs

1. Centrifuge the culture supernatant at 3000 × *g* for 15 min at 4 °C to remove cell debris.
2. Filter the supernatant with a 0.2 μm syringe filter (*see* **Notes 6** and 7).

3.2.3 Concentrating the EV Fraction

1. Apply the pass-through to an Amicon Ultra-15 Centrifugal Filter Devices for M.W. 100 kD.
2. Centrifuge at 5000 × *g* at 4 °C for concentrating the sample to less than 1 mL (*see* **Note 32**). The vesicle-free HSP90 proteins (less than 100 kD) will pass through the filter.

3.2.4 Isolation of EVs

1. Transfer the concentrate above the filter into a low-adhesive microtube.
2. Suspend the sample in the same volume of PEG polymer, and incubate overnight at 4 °C.

3. Centrifuge the sample at 10,000 × g for 60 min at 4 °C.
4. Aspirate the supernatant and leave an EV/exosome pellet.
5. Suspend EVs/exosomes in 100 μL PBS. For western blotting, suspend EVs/exosomes in RIPA buffer and incubate for 30 min or overnight on ice (*see* **Note 33**).
6. See another chapter for visualizing EVs/exosomes using TEM, counting EVs/exosomes, and analyzing vesicle/particle size [21].
7. Store the isolated EVs/exosomes at 4 °C for up to 1 week or at ≤ −20 °C for long-term storage.

3.2.5 Non-vesicular Fraction (Including Vesicle-Free HSP90 Proteins)

1. Prepare the pass-through from Subheading 3.2.3, **step 2**. This fraction contains vesicle-free HSP90 proteins.
2. Apply the pass-through to an Amicon Ultra-15 Centrifugal Filter Devices for M.W. 10 kD.
3. Centrifuge at 5000 × *g* at 4 °C for concentrating the non-EV sample.
4. Use the concentrate (remaining above the filter) as a non-EV/exosome fraction containing HSP90 proteins.

3.3 Harvest of Cellular Proteins (WCL)

One of the 10 cm dishes is enough for the preparation of WCL. Handling protein samples on ice or in a cold room is recommended as proteins are often stable at lower temperatures. RT to 37 °C often promotes enzymatic reactions, such as proteolysis. Four methods are available: (Subheading 3.3.2) cell lysis buffer protocol, (Subheading 3.3.3) RIPA buffer and homogenization protocol, (Subheading 3.3.4) trypsin and RIPA buffer protocol (*see* **Note 34**), and (Subheading 3.3.5) sample buffer protocol and cell fractionation protocol (*see* **Note 19**). Trypsin digests proteins on the surface of cells, ECM, and EVs bound with cells or ECM, such as chaperone proteins, metalloproteinases, and tetraspanins. Therefore, the cell lysis buffer protocol or the RIPA + homogenization protocol is recommended for detecting chaperone proteins.

3.3.1 Common Steps

1. Prepare an ice bag and ice bucket (*see* **Note 20**).
2. Prepare 1× lysis buffer with protease inhibitor cocktail, 1× RIPA buffer with protease inhibitor cocktail, or 2 × SDS sample buffer with β-ME (*see* **Notes 16**, **17**, and **18**).
3. Collect cell culture supernatant.
4. Wash the cells with PBS in an ice bag. Proteins are more stable at 4 °C than RT. Collect protein samples in an ice bag or in a cold room.
5. Tilt the dish diagonally. Accumulate the remained medium and buffer and aspirate it.

3.3.2 Cell Lysis Buffer Protocol

1. Add 1× cell lysis buffer to cover the cells, e.g., 2 mL for a 10 cm dish.
2. Gently shake or swirl the dish for 15 min.
3. Scrape the samples on the dish, collect, and transfer them to 1.5 mL or 2.0 mL microcentrifuge tubes.
4. Centrifuge at 15,000 × g at 4 °C for 20 min.
5. Transfer the supernatant (as a protein sample) to another tube (*see* **Note 35**).
6. Snap freeze in liquid nitrogen and store at −80 °C.

3.3.3 RIPA Buffer and Homogenization Protocol

1. Add 1× RIPA buffer to cover the cells, e.g., 2 mL for a 10 cm dish.
2. Scrape and collect the cells using a cell scraper.
3. Homogenize samples with a 25G needle syringe for ten strokes.
4. Incubate samples for 30 min or overnight (*see* **Note 33**) on ice.
5. Centrifuge at 15,000 × g at 4 °C for 20 min to remove cell debris.
6. Transfer the supernatant (as a protein sample) to another tube (*see* **Note 35**).
7. Snap freeze in liquid nitrogen and store at −80 °C.

3.3.4 Trypsin and RIPA Buffer Protocol

1. Collect the cells using a cell scraper or trypsin–EDTA solution from dushes to a new centrifuge tube. In the case using trypsin–EDTA, add a medium containing serum after cell collection.
2. Centrifuge cell suspension at 1000 rpm (200 × g) for 5 min at −80 °C.
3. Remove the supernatant.
4. Suspend cell pellet with 1 mL PBS.
5. Transfer all the suspension into new 1.5 mL tubes.
6. Repeat **steps 2** and **3**.
7. Suspend cell pellet with 1 mL PBS.
8. Repeat **steps 2** and **3**.
9. Add 1 mL RIPA buffer and suspend cells by pipetting.
10. Incubate on ice for 5–30 min or overnight (*see* **Note 33**).
11. Centrifuge the cell lysate at 14,500 rpm (20,000 × g) for 10–30 min at 4 °C.
12. Transfer the supernatant into a new 1.5 mL tube (*see* **Note 35**).
13. Snap freeze in liquid nitrogen and store at −80 °C.

3.3.5 Sample Buffer Protocol

1. Add 2× SDS sample buffer to cover the cells, e.g., 2 mL for a 10 cm dish.
2. Scrape the samples on the dish, collect, and transfer them to 1.5 mL or 2.0 mL microcentrifuge tubes.
3. Boil the sample at 95 °C for 5 min within a block incubator.
4. Store the samples at 4 °C or − 20 °C.

3.4 Protein Assay

EVs/exosomes are surrounded by lipid membranes, and proteins exist inside and on the surface of EVs/exosomes. Therefore, dissolve the membrane using lysis buffer or RIPA buffer before protein assay and western blotting. A highly sensitive protein assay kit, e.g., micro BCA kit, is necessary for exosomes/EVs, while a regular BCA kit is enough for WCL.

1. Mix a part of the isolated EV/exosome sample within the lysis buffer or RIPA buffer. Pipette a few times for mixing.
2. Incubate for 30 min or overnight on ice to dissolve the EV/exosome membrane.
3. Dilute the WCL 1:10–1:100 with PBS (*see* **Note 36**).
4. Perform protein assay using a micro BCA protein assay kit, according to the manufacturer's protocol.

3.5 Sample Preparation for Western Blotting

1. Make a plan for the SDS-PAGE and western blotting, e.g., amounts/volumes of proteins for loading, reducing or nonreducing, lane ordering, blank lanes, etc.
2. Mix a part of the isolated EV/exosome sample with lysis buffer or RIPA buffer (*see* **Note 33**). Pipette a few times for mixing.
3. Incubate for 30 min on ice to dissolve the EV/exosome membrane.
4. Prepare protein samples (EV/exosome, non-EV, and/or WCL) from an equal number of cells, equal protein amounts, or an equal number of exosome/EV samples (*see* **Notes 37** and **38**) in low-binding tubes.
5. Adjust the volumes by adding PBS.
6. Prepare 6 × SDS sample buffer. If required, add β-ME to the sample buffer. The final concentration of β-ME should be 5%.
7. Mix the 6 × SDS sample buffer, protein samples, and water.
8. Seal the lids of tubes with plastic materials or parafilm (to avoid the lids opening by heating).
9. If you added β-ME, boil the samples at 95 °C for 5 min. If you did not add β-ME, boil the samples at 70 °C for 10 min.
10. Spin down.

3.6 SDS-Page

1. Set a PAG in an electrophoresis tank.
2. Pore the running buffer in the tank.
3. Load equal volumes of samples to each well. A blank lane may be important.
4. Run at 100 V for an initial 10–20 min and then 200 V for the next 30–60 min (*see* **Note 39**).
5. Stop the running before the leading dye reaches the bottom of the gel.

3.7 Wet Transfer

1. Soak a PVDF membrane in 100% methanol for 10–30 s for hydrophilizing.
2. Soak the membrane in the transfer buffer, and gently shake for more than 10 min for equilibrating.
3. "Sandwich" wet system sequentially using filter paper and foam.
4. Transfer at 90 V for 90 min with cooling.

3.8 Immunoblotting

1. (Optional) Cut the membrane for detecting multiple protein types (*see* **Note 40**).
2. Soak the membrane in 10% skim milk in TBS-T for blocking.
3. Gently swirl for 30–60 min at RT.
4. Add anti-HSP90 primary antibody to 10% skim milk in TBS-T and mix well. For more abundant proteins, 5% skim milk is enough for blocking.
5. Soak the membrane in the anti-HSP90 antibody solution, and gently swirl for 2 days at 4 °C (*see* **Note 41**). For abundant proteins, incubation overnight is enough.
6. Wash the membrane with TBS-T for 5 min three times and 10 min three times.
7. Prepare HRP-conjugated secondary antibody in TBS-T with 5–10% skim milk and mix well.
8. Soak the membrane with the HRP-conjugated secondary antibody and gently swirl at RT for 30–60 min.
9. Wash the membrane with TBS-T for 5 min three times and 10 min three times.
10. Add the ECL substrate to the membrane and make a reaction.
11. Visualize the immunoreactive bands by the imaging system (*see* **Note 42**).
12. Quantify the densitometric analysis using Image J (*see* **Note 43**).

3.9 Stripping

1. Wash the membrane subjected to chemiluminescence detection with TBS-T for 10 min.
2. Immerse the membrane in a stripping agent and shake gently for 5–10 min at RT to detach the antibody.
3. After stripping, wash the membrane with TBS-T for 10 min three times.
4. Perform immunoblotting protocol with another antibody type.

4 Notes

1. Cells producing the high-level proteins or HSP90s of interest have an advantage for detection. Besides, stimulating cells with cell stress, such as heat shock stress (HSS), is also effective for detecting HSPs efficiently [15].
2. If multiple subsequent experiments are planned, siRNAs can be stored at 4 °C for up to 1 month with no observable degradation. The stock concentration should be no less than 10 μM.
3. It is recommended that a mixture of two types of siRNA duplex be used for targeting each mRNA.
4. When multiple knockdowns (double or triple) are performed by mixing multiple siRNAs, the total siRNA concentration used should be the same as for single knockdowns, e.g., for comparison with a single knockdown using 30 pmol concentrations of siRNA, adjust 15 pmol each of two siRNAs for double knockdown and 10 pmol each of three siRNAs for the triple knockdown.
5. The serum contains EV/exosomes. Therefore, culture cells in serum-free or EV/exosome-free medium before EV/exosome isolation. EV/exosome-free serum is preparable by the exosome/EV isolation methods of this chapter. Use serum as a starting material in this case. Alternatively, commercially available serum-free medium such as mTeSR1 is useful for serum-free, EV-free culture [22, 23].
6. The 0.2 μm filter syringe is useful for separating large EVs and the selective purification of exosomes/sEVs (50–200 nm). Do not use a 0.2 μm filter syringe if you intend to analyze large EVs, such as microvesicles, stressome, and/or apoptotic bodies [15, 22].
7. The high pressure in the filtration step can destruct the structure of EVs/exosomes [24, 25].
8. ZetaSizer (Malvern) is useful for vesicle/particle distribution analysis. For counting vesicles/particles and their distribution

analysis, use Video Drop (Meiwa Fosis), qNano (Meiwa Fosis), or NanoSight (Malvern).

9. Practical examples of IP are available in ref. [26–28].
10. Practical examples of zymography detecting metalloproteinase activities are available in ref. [29].
11. Practical examples using lysis buffer are available in refs. [27, 30, 31].
12. Detergents such as NP-40 or Triton X-100 are needed to dissolve the EV membrane for protein assay and western blotting. Otherwise, proteins will be predominantly detected at very high molecular weight bands. See failure examples in supplemental figures in the ref. [32].
13. Practical examples using RIPA buffer are available in refs. [33–35].
14. Proteases in the EVs/exosomes can degrade EV/exosome proteins. Such proteolysis affects downstream applications such as western blotting. Add a protease inhibitor cocktail to the EV/exosome samples to inhibit protease-dependent protein degradation. EDTA is a metalloproteinase inhibitor, and EDTA(+) and EDTA-free versions are commercially available.
15. A reducing agent DTT in this SDS sample buffer and boiling at 70 °C for 10 min is often enough for reducing proteins. For stronger reduction, use β-ME.
16. β-ME is a powerful reductant that cuts bisulfate bonds between cysteines in proteins. The reduction of proteins is needed for many antibody types to detect the proteins. However, another antibody type favors 3D structures of the protein, for which β-ME is useless. See user manuals of antibodies carefully to know if the β-ME is needed or unnecessary.
17. Confirm a strong odor of β-ME right before use. If it does not smell strong, it is expired.
18. Nonreducing preparation of protein samples without β-ME was needed for anti-CD9 (D252–3, MBL), anti-CD63 (EXOAB-CD63A-1, System Biosciences), and anti-EpCAM (VU1D9, CST) [5, 8, 23]. We used β-ME to reduce proteins for other antibodies' reactions.
19. Cell fractionation kits are available, e.g., NE-PER (to fractionate nucleus and cytoplasm), Mem-PER (to fractionate integral and membrane-associated proteins), Cell Surface Protein kit, and subcellular protein fractionation (to fractionate nucleus, cytoplasm, membrane, cytoskeleton, and chromatin-bound). See the practical example in ref. [26, 27, 33].

20. Handling protein samples on ice or in a cold room is recommended as proteins are often stable at lower temperatures. RT to 37 °C often promotes enzymatic reactions, such as proteolysis.
21. Exosome/EV protein concentration is often very low and under the detection limit of the regular BCA assay kit. Therefore, a micro BCA kit is necessary.
22. Body fluids such as serum and urine are rich in albumin and globulin bound with EVs/exosomes. In such cases, protein assay is useless.
23. Low binding/adhesion microcentrifuge tubes are useful to avoid sample loss.
24. Make or buy gels with an appropriate polyacrylamide concentration for your proteins of interest.
25. Electroblotting (also called semidry blotting) is often enough for proteins smaller than 50 kD and fast (in 30 min). Double membranes can catch smaller proteins. The wet transfer method is efficient for proteins larger than 50 kD and takes several hours to overnight.
26. See refs. [5, 8, 15] for practical examples of western blotting of chaperone proteins.
27. There are several candidates for exosomal markers, including membrane proteins of the tetraspanin family such as CD9 [5, 36, 37], CD63, [5, 8, 22, 32], and CD81, as well as EpCAM/CD326 [23], Tsg101, Alix, and HSPs [38, 39], including HSP70 and HSP90 [35, 40]. However, these markers may not be included in the EVs/exosomes depending on the cell type from which they are derived.
28. RNA-Lipofectamine® RNAiMAX complexes must be made in a serum-free medium such as Opti-MEM® reduced serum medium and can be added directly to cells in the culture medium, in the presence or absence of serum/antibiotic.
29. It is not necessary to remove complexes or change/add medium after transfection.
30. At this stage, the experimenter may consider proceeding to WCL conditioning and western blotting to confirm knockdown efficiency (Subheadings 3.3–3.9).
31. For the PBP method, 2–4 10 cm dishes are required. For the SEC method, 5–20 15 cm dishes are required.
32. Use 12 ml medium for a 15 cm dish and 4 ml medium for a 10 cm dish for concentrating EVs/exosomes in the culture supernatant.
33. Incubating cells in RIPA buffer overnight may increase cell lysis and protein harvest efficiencies.

34. Trypsin digests proteins on the surface of cells, ECM, and EVs bound with cells or ECM, such as chaperone proteins, EpCAM, and tetraspanins.
35. The pellet can be analyzed as an insoluble fraction that contains chromatin [41].
36. Protein concentrations in WCL can be quantified using the standard BCA protein assay kit, while EVs/exosomes need a micro BCA kit. Dilute the WCL for the micro BCA kit.
37. Load equal numbers of vesicles per lane or equal protein amount per lane [8, 36, 37]. Otherwise, loading the EV/exosome samples prepared from equal numbers of cells is also applicable [22]. In the case of the SEC method, load equal liquid volumes from each fraction [32].
38. Growth factors [35], HSPs [40, 42, 43], cytokines, chemokines [44], metalloproteinases [22], and CCN2/CTGF [34] are found in both EV and non-EV fractions.
39. These sequential steps are effective in preventing too much heating that triggers proteolysis.
40. Cutting the membrane enables us to detect multiple proteins in different molecular weights in the same sample/lane.
41. Blocking membranes in 5% skim milk is a general protocol. When evaluating protein phosphorylation, consider blocking with 5% BSA. In addition, if needed, using 10% skim milk can reduce nonspecific reactions. Also, blocking the membrane overnight at 4 °C and reacting with primary antibodies for 2 days at 4 °C can increase the specificity.
42. Take photographs with short, medium, and long exposure periods [5, 23, 35, 45].
43. Some reviewers or journals request relative quantification of band intensities, reproducibility, and statistics in western blotting [8].

Acknowledgments

K.O. was supported by JSPS Kakenhi (grant numbers 19 K24072 and 21 K17115), Wesco Scientific Promotion Foundation, and the KAWASAKI Foundation for Medical Science and Medical Welfare. T.E. was supported by JSPS Kakenhi, grant numbers 17 K11642-TE, 20 K09904-CS, 19H03817-MT, 20H03888-HN, 20 K20611-MT, 20H03888-HN, 21H03119-TY, and 21 K08902-HY, and Wesco Scientific Promotion Foundation. The authors thank Chiharu Sogawa, Manh Tien Tran, Kuniaki Okamoto, and Stuart K. Calderwood for useful information, discussion, materials, and experimentation.

References

1. Trepel J, Mollapour M, Giaccone G, Neckers L (2010) Targeting the dynamic HSP90 complex in cancer. Nat Rev Cancer 10(8):537–549. https://doi.org/10.1038/nrc2887
2. Whitesell L, Lindquist SL (2005) HSP90 and the chaperoning of cancer. Nat Rev Cancer 5(10):761–772. https://doi.org/10.1038/nrc1716
3. Neckers L, Workman P (2012) Hsp90 molecular chaperone inhibitors: are we there yet? Clin Cancer Res 18(1):64–76. https://doi.org/10.1158/1078-0432.Ccr-11-1000
4. Scaltriti M, Dawood S, Cortes J (2012) Molecular pathways: targeting hsp90--who benefits and who does not. Clin Cancer Res 18(17): 4508–4513. https://doi.org/10.1158/1078-0432.Ccr-11-2138
5. Ono K, Eguchi T, Sogawa C, Calderwood SK, Futagawa J, Kasai T et al (2018) HSP-enriched properties of extracellular vesicles involve survival of metastatic oral cancer cells. J Cell Biochem 119(9):7350–7362. https://doi.org/10.1002/jcb.27039
6. Taha EA, Ono K, Eguchi T (2019) Roles of Extracellular HSPs as Biomarkers in Immune Surveillance and Immune Evasion. Int J Mol Sci 20(18):4588. https://doi.org/10.3390/ijms20184588
7. Wong DS, Jay DG (2016) Emerging roles of extracellular Hsp90 in cancer. Adv Cancer Res 129:141–163. https://doi.org/10.1016/bs.acr.2016.01.001
8. Ono K, Sogawa C, Kawai H, Tran MT, Taha EA, Lu Y et al (2020) Triple knockdown of CDC37, HSP90-alpha and HSP90-beta diminishes extracellular vesicles-driven malignancy events and macrophage M2 polarization in oral cancer. J Extracell Vesicles 9(1): 1769373. https://doi.org/10.1080/20013078.2020.1769373
9. Eustace BK, Sakurai T, Stewart JK, Yimlamai D, Unger C, Zehetmeier C et al (2004) Functional proteomic screens reveal an essential extracellular role for hsp90 alpha in cancer cell invasiveness. Nat Cell Biol 6(6): 507–514. https://doi.org/10.1038/ncb1131
10. Wandinger SK, Richter K, Buchner J (2008) The Hsp90 chaperone machinery. J Biol Chem 283(27):18473–18477. https://doi.org/10.1074/jbc.R800007200
11. Rong B, Yang S (2018) Molecular mechanism and targeted therapy of Hsp90 involved in lung cancer: new discoveries and developments (review). Int J Oncol 52(2):321–336. https://doi.org/10.3892/ijo.2017.4214
12. Bohush A, Bieganowski P, Filipek A (2019) Hsp90 and Its Cochaperones in Neurodegenerative Diseases. Int J Mol Sci 20(20):4976. https://doi.org/10.3390/ijms20204976
13. Calderwood SK (2015) Cdc37 as a cochaperone to Hsp90. Subcell Biochem 78:103–112. https://doi.org/10.1007/978-3-319-11731-7_5
14. Calderwood SK (2013) Molecular cochaperones: tumor growth and cancer treatment. Scientifica (Cairo) 2013:217513. https://doi.org/10.1155/2013/217513
15. Eguchi T, Sogawa C, Ono K, Matsumoto M, Tran MT, Okusha Y et al (2020) Cell stress induced stressome release including damaged membrane vesicles and extracellular HSP90 by prostate cancer cells. Cells 9(3):755. https://doi.org/10.3390/cells9030755
16. Gray PJ Jr, Prince T, Cheng J, Stevenson MA, Calderwood SK (2008) Targeting the oncogene and kinome chaperone CDC37. Nat Rev Cancer 8(7):491–495. https://doi.org/10.1038/nrc2420
17. Gray PJ Jr, Stevenson MA, Calderwood SK (2007) Targeting Cdc37 inhibits multiple signaling pathways and induces growth arrest in prostate cancer cells. Cancer Res 67(24): 11942–11950. https://doi.org/10.1158/0008-5472.Can-07-3162
18. Eguchi T, Prince TL, Tran MT, Sogawa C, Lang BJ, Calderwood SK (2019) MZF1 and SCAND1 Reciprocally Regulate CDC37 Gene Expression in Prostate Cancer. Cancers (Basel) 11(6):792. https://doi.org/10.3390/cancers11060792
19. Lauwers E, Wang YC, Gallardo R, Van der Kant R, Michiels E, Swerts J et al (2018) Hsp90 Mediates Membrane Deformation and Exosome Release. Mol Cell 71(5):689–702.e9. https://doi.org/10.1016/j.molcel.2018.07.016
20. Tang X, Chang C, Guo J, Lincoln V, Liang C, Chen M et al (2019) Tumour-secreted Hsp90α on external surface of exosomes mediates tumour - stromal cell communication via autocrine and paracrine mechanisms. Sci Rep 9(1): 15108. https://doi.org/10.1038/s41598-019-51704-w
21. Eguchi T, Okusha Y, Lu Y, Ono K, Taha EA, Fukuoka S (2022) Comprehensive method for exosome isolation and proteome analysis for detection of CCN factors in/on exosomes. In: Takigawa M (ed) CCN proteins methods and protocols, 2nd edn. Humana Press, New York

22. Taha EA, Sogawa C, Okusha Y, Kawai H, Oo MW, Elseoudi A et al (2020) Knockout of MMP3 Weakens Solid Tumor Organoids and Cancer Extracellular Vesicles. Cancers (Basel) 12(5):1260. https://doi.org/10.3390/cancers12051260
23. Eguchi T, Sogawa C, Okusha Y, Uchibe K, Iinuma R, Ono K et al (2018) Organoids with cancer stem cell-like properties secrete exosomes and HSP90 in a 3D NanoEnvironment. PLoS One 13(2):e0191109. https://doi.org/10.1371/journal.pone.0191109
24. Thery C, Witwer KW, Aikawa E, Alcaraz MJ, Anderson JD, Andriantsitohaina R et al (2018) Minimal information for studies of extracellular vesicles 2018 (MISEV2018): a position statement of the International Society for Extracellular Vesicles and update of the MISEV2014 guidelines. J Extracell Vesicles 7(1):1535750. https://doi.org/10.1080/20013078.2018.1535750
25. Yang D, Zhang W, Zhang H, Zhang F, Chen L, Ma L et al (2020) Progress, opportunity, and perspective on exosome isolation - efforts for efficient exosome-based theranostics. Theranostics 10(8):3684–3707. https://doi.org/10.7150/thno.41580
26. Tran MT, Okusha Y, Feng Y, Morimatsu M, Wei P, Sogawa C et al (2020) The inhibitory role of Rab11b in Osteoclastogenesis through triggering lysosome-induced degradation of c-Fms and RANK surface receptors. Int J Mol Sci 21(24):9352
27. Eguchi T, Kubota S, Kawata K, Mukudai Y, Uehara J, Ohgawara T et al (2008) Novel transcription-factor-like function of human matrix metalloproteinase 3 regulating the CTGF/CCN2 gene. Mol Cell Biol 28(7):2391–2413. https://doi.org/10.1128/MCB.01288-07
28. Eguchi T, Kubota S, Kondo S, Shimo T, Hattori T, Nakanishi T et al (2001) Regulatory mechanism of human connective tissue growth factor (CTGF/Hcs24) gene expression in a human chondrocytic cell line, HCS-2/8. J Biochem 130(1):79–87. https://doi.org/10.1093/oxfordjournals.jbchem.a002965
29. Okusha Y, Eguchi T, Sogawa C, Okui T, Nakano K, Okamoto K et al (2018) The intranuclear PEX domain of MMP involves proliferation, migration, and metastasis of aggressive adenocarcinoma cells. J Cell Biochem 119(9):7363–7376. https://doi.org/10.1002/jcb.27040
30. Arai K, Eguchi T, Rahman MM, Sakamoto R, Masuda N, Nakatsura T et al (2016) A novel high-throughput 3D screening system for EMT inhibitors: a pilot screening discovered the EMT inhibitory activity of CDK2 inhibitor SU9516. PLoS One 11(9):e0162394. https://doi.org/10.1371/journal.pone.0162394
31. Eguchi T, Prince TL, Tran MT, Sogawa C, Lang BJ, Calderwood SK (2019) MZF1 and SCAND1 reciprocally regulate CDC37 gene expression in prostate cancer. Cancers (Basel) 11(6):1–15. https://doi.org/10.3390/cancers11060792
32. Lu Y, Eguchi T, Sogawa C, Taha EA, Tran MT, Nara T et al (2021) Exosome-based molecular transfer activity of macrophage-like cells involves viability of oral carcinoma cells: size exclusion chromatography and concentration filter method. Cells 10(6):1328. https://doi.org/10.3390/cells10061328
33. Tran MT, Okusha Y, Feng Y, Sogawa C, Eguchi T, Kadowaki T et al (1868) A novel role of HSP90 in regulating osteoclastogenesis by abrogating Rab11b-driven transport. Biochim Biophys Acta, Mol Cell Res 2021(10):119096. https://doi.org/10.1016/j.bbamcr.2021.119096
34. Okusha Y, Eguchi T, Tran MT, Sogawa C, Yoshida K, Itagaki M et al (2020) Extracellular vesicles enriched with moonlighting metalloproteinase are highly transmissive, pro-tumorigenic, and trans-activates cellular communication network factor (CCN2/CTGF): CRISPR against cancer. Cancers (Basel) 12(4):881. https://doi.org/10.3390/cancers12040881
35. Eguchi T, Sogawa C, Ono K, Matsumoto M, Tran MT, Okusha Y et al (2020) Cell stress induced Stressome release including damaged membrane vesicles and extracellular HSP90 by prostate cancer cells. Cell 9(3):1–24. https://doi.org/10.3390/cells9030755
36. Fujiwara T, Eguchi T, Sogawa C, Ono K, Murakami J, Ibaragi S et al (2018) Carcinogenic epithelial-mesenchymal transition initiated by oral cancer exosomes is inhibited by anti-EGFR antibody cetuximab. Oral Oncol 86:251–257. https://doi.org/10.1016/j.oraloncology.2018.09.030
37. Fujiwara T, Eguchi T, Sogawa C, Ono K, Murakami J, Ibaragi S et al (2018) Anti-EGFR antibody cetuximab is secreted by oral squamous cell carcinoma and alters EGF-driven mesenchymal transition. Biochem Biophys Res Commun 503(3):1267–1272
38. Taha EA, Ono K, Eguchi T (2019) Roles of extracellular HSPs as biomarkers in immune surveillance and immune evasion. Int J Mol Sci 20(18):1–32. https://doi.org/10.3390/ijms20184588
39. Yamamoto T, Eguchi T (2021) Heat shock proteins and periodontitis — crossreaction

between bacterial and human HSP in periodontal infection linking with cardiovascular diseases. In: AAA A, Kaur P (eds) Heat shock proteins in inflammatory diseases, Heat Shock Proteins. Springer Nature, Cham, pp 19–32

40. Eguchi T, Ono K, Kawata K, Okamoto K, Calderwood SK (2019) Regulatory roles of HSP90-Rich extracellular vesicles. In: AAA A, Kaur P (eds) Heat shock protein 90 in human diseases and disorders, Heat Shock Proteins. Springer Nature, Cham, pp 3–17
41. Eguchi T, Calderwood SK, Takigawa M, Kubota S, Kozaki K (2017) Intracellular MMP3 promotes HSP gene expression in collaboration with Chromobox proteins. J Cell Biochem 118(1):43–51. https://doi.org/10.1002/jcb.25607
42. Eguchi T, Taha EA (2021) Extracellular Vesicle-associated Moonlighting Proteins: Heat Shock Proteins and Metalloproteinases. In: AAA A, Kaur P (eds) Heat Shock Proteins, Heat Shock Proteins. Springer Nature, Cham, pp 1–18
43. Eguchi T, Lang BJ, Murshid A, Prince T, Gong J, Calderwood SK. Regulatory roles for Hsp70 in cancer incidence and tumor progression. In: Galigniana MD, editor. Frontiers in Structural Biology. Bentham Science; 2018. p. 1–22
44. Lima LG, Ham S, Shin H, Chai EPZ, Lek ESH, Lobb RJ et al (2021) Tumor microenvironmental cytokines bound to cancer exosomes determine uptake by cytokine receptor-expressing cells and biodistribution. Nat Commun 12(1):3543. https://doi.org/10.1038/s41467-021-23946-8
45. Sogawa C, Eguchi T, Tran MT, Ishige M, Trin K, Okusha Y et al (2020) Antiparkinson drug Benztropine suppresses tumor growth, circulating tumor cells, and metastasis by acting on SLC6A3/DAT and reducing STAT3. Cancers (Basel) 12(2):1–22

Chapter 13

Using a Modified Proximity Ligation Protocol to Study the Interaction Between Chaperones and Associated Proteins

Simone Baldan, Anatoli B. Meriin, and Michael Y. Sherman

Abstract

Molecular chaperones can interact with multiple proteins to form large networks. Understanding these interactions may shed light on the complexity of the chaperone functions. Here we developed a protocol for a modified proximity ligation-based methodology (PLA) for the detection of protein–protein interactions in order to understand how the Hsp70-Bag3 complex interacts with components of the Hippo signaling pathway. These experiments helped to elucidate the mechanisms of transmission of the proteotoxic stress signal to the Hippo pathway. The modified PLA technology has many advantages compared to co-immunoprecipitation protocols. It has higher sensitivity, is quantitative, and can be done in a 96-well format.

Key words Hsp70, Bag3, LATS1/2, Yap, Hippo pathway, Protein complex, Proximity ligation assay

1 Introduction

Molecular chaperones often cooperate in complex protein folding and degradation processes [1]. For example, in the folding of newly synthesized proteins, there is a cascade of chaperones that transfer a newly synthesized polypeptide between various types of chaperones in a chain of reactions [1–3]. Withholding chaperones like Hsp70 family members, there are multiple interactions with (a) proteins that regulate the activity of Hsp70 (e.g., J-domain proteins) [4, 5], (b) proteins that are recruited to Hsp70 in order to modify the bound polypeptide (e.g., ubiquitin ligase CHIP) [4], and proteins that target Hsp70, often in a complex with a bound unfolded polypeptide, to different structures (e.g., Bag1 or Bag3 that target to proteasome or autophagic vacuoles, as well as to components of

Authors Simone Baldan and Anatoli B. Meriin have equally contributed to this chapter.

Stuart K. Calderwood and Thomas L. Prince (eds.), *Chaperones: Methods and Protocols*, Methods in Molecular Biology, vol. 2693, https://doi.org/10.1007/978-1-0716-3342-7_13,

various signaling pathways) [4–6]. Overall, chaperones and co-chaperones form very complex and dynamic interaction networks that are also tissue-specific [7]. Through these interactions, chaperones control numerous intracellular processes in a precise manner, both spatially and temporally. Understanding these networks requires accurate measurements of interactions of their components, including various chaperones, co-chaperones, and other interacting proteins.

A most commonly used approach toward assessing protein interactions in general and in application to the chaperone networks is co-immunoprecipitation, where protein complexes are precipitated with an antibody against one of the components of the complex, and then other components are detected either via immunoblotting with other antibodies or via mass spectroscopy. This method has a number of drawbacks, e.g., losing unstable interactions, which diminishes the ability to quantify the interaction; detecting false interactions between proteins normally existing in different intracellular compartments; and requirement of large amounts of lysates or overexpression of proteins due to the low efficiency of Co-IP and limited sensitivity of immunoblotting.

Recently, a proximity ligation approach has been developed [8, 9], which is a modification of the immuno-PCR method [10, 11]. The idea of in situ proximity ligation between two proteins of interest involves two corresponding antibodies, each covalently attached to a unique oligonucleotide. If the proteins interact in a cell, the antibodies would bind in close proximity to each other, allowing for the interaction of the attached oligonucleotides and their ligation. The ligated DNA product can be detected by various methods. Being much more quantitative and much more sensitive than Co-IP, the proximity ligation approach is currently used mostly in its commercial version called DuoLink® Proximity Ligation Assay (Sigma). For the detection of the ligated DNA product, DuoLink uses a fluorescence-based assay that allows not only detecting protein–protein interactions but also visualizing them at specific cellular locations. However, it has some downsides, since fluorescence-based measurements of interactions between proteins are difficult to quantify, and the DuoLink® kit is expensive, which prohibits a large number of measurements.

The proximity ligation assay can be adapted to do multiplex measurements of proteins, their modifications, or interactions in complex mixtures, e.g., cells and tissues homogenates. We adapted the PLA technology to monitor protein interactions in fixed cells on a 96-well plate. This approach proved to be inexpensive, highly sensitive, and highly quantitative.

Over several years, we have been studying how the Hsp70-Bag3 module regulates various signaling pathways [12]. Bag3 has protein–protein interaction motifs, including WW-domain and PXXP-motif that interacts with SH3-domain proteins

[13]. Through these and other surfaces, Bag3 interacts with components of multiple signaling pathways, including Src, myc, NF-kB, Hifl, and others [8, 14–16]. Therefore, Bag3 serves as a scaffold that links Hsp70, either alone or in complex with bound unfolded polypeptides, with the signaling pathways. Via these interactions, modulation of activity or levels of Hsp70 can regulate multiple signaling pathways and affect various stages of cancer development [6, 17–19].

Previously, we demonstrated that the natural function of the Hsp70-Bag3 (HB complex) is to monitor abnormal polypeptides in the cytosol upon proteotoxic stress in order to regulate the activities of the signaling pathways [8]. One such signaling pathway regulated by the HB complex is the Hippo pathway, which plays an important role in integrating cell density or mechanical stress cues. LATS1, the major effector kinase of the Hippo pathway, was shown previously to directly interact with Bag3 in the HB complex [20]. This interaction is mediated by the WW-domain of Bag3 [20] and depends on the association of the latter with Hsp70 [8].

Major targets of LATS1 include the transcriptional effectors YAP and Taz [21]. Their phosphorylation stabilizes the association of these factors with 14-3-3 proteins that anchor YAP in the cytoplasm and prevent its nuclear translocation [21]. Accordingly, reduced LATS1 activity results in YAP nuclear translocation and activation of transcription of the target genes [22]. LATS1-dependent YAP phosphorylation is mediated by the scaffold protein AmotL2 [23, 24]. In our recent works, we successfully used the modified PLA to monitor changes in protein–protein interaction in order to understand how Hsp70-Bag detects cellular proteotoxicity and regulates Hippo and other signaling pathways [25].

1.1 Results

1.1.1 Using the Modified PLA Method for Detection of Complexes of Hsp70-Bag3 with Components of the Hippo Pathway

Our goal was to study how the Hsp70-Bag3 module is affected by the buildup of abnormal polypeptides and how it impacts LATS1-mediated phosphorylation and nuclear translocation of YAP. In this work, we utilize the modified PLA assay (Fig. 1).

Based on indirect observations, we hypothesized that stress-induced misfolding/damage of newly synthesized proteins specifically is monitored by the HB complex. Accordingly, we generated a model polypeptide that is degraded right after the release from ribosomes, GFP-NS, and studied its interaction with Bag3. We discovered that upon mild proteasome inhibition, this polypeptide interacts with Bag3 [8]. Moreover, this interaction depended on its ubiquitination by LTN1, a component of the ribosome-associated quality control system, and on the ribosome-associated chaperone NAC [8]. Noteworthy, in this work we assessed the protein–protein interactions using the DuoLink® kit and our modified PLA assay in parallel and obtained similar results [8].

The modified PLA assay detected a significant association between Amotl2 and Bag3 in naïve cells. This interaction was

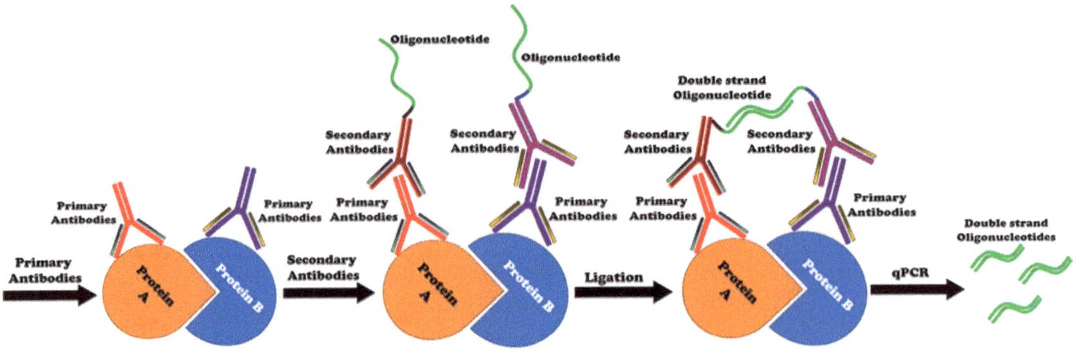

Fig. 1 Schematic representation of the modified proximity ligation assay

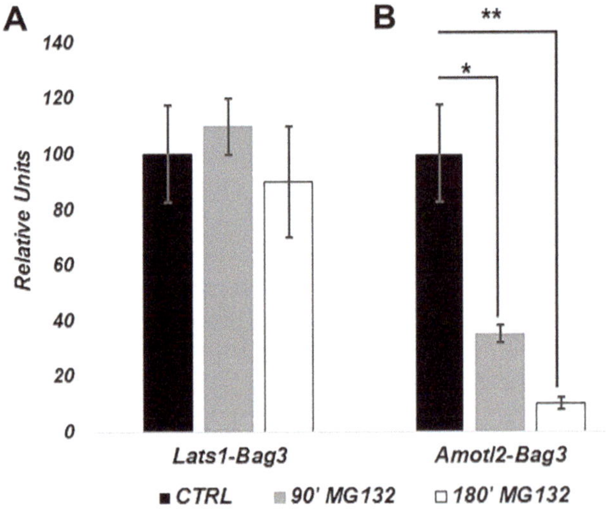

Fig. 2 Effects of proteasome inhibition on complexes of LATS1 and Amotl2 with Bag3 measured by IPAD assay. a. Effects of proteasome inhibition on the association between LATS1 and Bag3. b. Effects of proteasome inhibition on the association between Amotl2 and Bag3. MCF10A cells were treated with 5 μM MG132 for the indicated time periods, cells were fixed, and the interactions were measured. The statistical values were calculated using the unpaired two-tailed t-test (*—$p \leq 0.05$; **—$p \leq 0.01$)

strongly reduced in response to proteasome inhibitor MG132, while LAST1 remained associated with Bag3 (Fig. 2a, b). Under these conditions, the association of Amotl2 with YAP measured by the assay was also dramatically reduced (Fig. 3a). Therefore, our observations indicated that in naïve cells the LATS-Amotl2-YAP complex associates with the HB module and that Amotl2 dissociates from this complex upon proteotoxic stress [9].

We previously reported that Bag3 is required for the downregulation of YAP phosphorylation in response to proteotoxic stress [8], raising the possibility that Bag3 is required for the

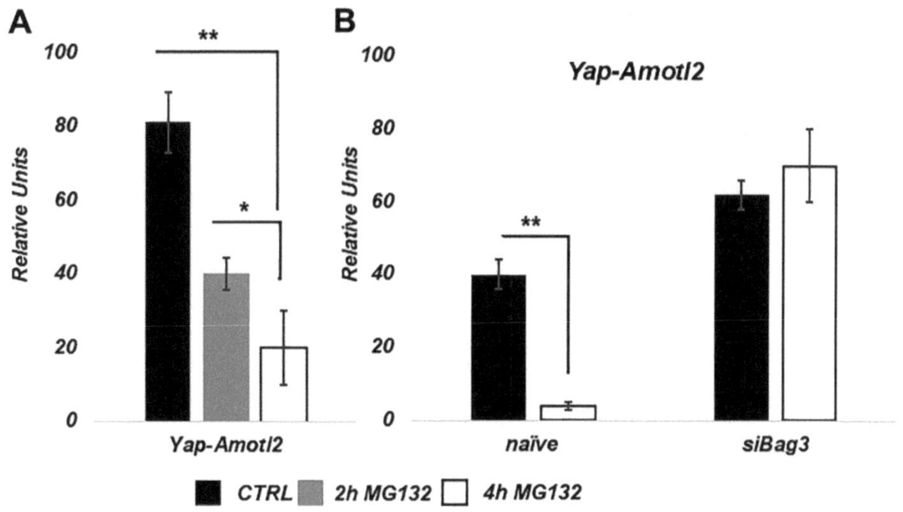

Fig. 3 Effects of Bag3 on Amotl2-YAP complexes. a. Effects of proteasome inhibition on the association between AmotL2 and YAP. b. In Bag3-depleted cells, AmotL2 remains in the complex with YAP even upon proteasome inhibition. Conditions are the same as in Fig. 2

dissociation of Amotl2 from LATS1-AmotL2-YAP complexes. To test this possibility, we measured the interaction between Amotl2 and YAP in the presence and absence of Bag3. Figure 3b shows that the association between Amotl2 and YAP was higher in Bag3-depleted cells. Importantly, upon proteasome inhibition, Amotl2-YAP interaction was reduced in control cells, while in Bag3-depleted cells, there was no dissociation of Amotl2 from YAP (Fig. 3b). These data indicate that upon proteotoxic stress Bag3 facilitates dissociation of Amotl2 from YAP, which prevents phosphorylation of YAP by LATS1 (Fig. 4), leading to the overall activation of the Hippo pathway. This could be a common example of the mechanism of transmitting the proteotoxic stress signal to downstream signaling pathways.

1.2 Modified Proximity Ligation IPAD Technology

The basis for the approach is the idea that if two proteins are present in a complex with fixed permeabilized cells, the corresponding antibodies bind these proteins in proximity to each other. Such a proximity localization could be quantitatively detected, since each antibody has a unique double-stranded relatively short (20 bp) oligos covalently attached. The sticky ends of the two oligos attached to a pair of antibodies against the interacting proteins are complementary to each other and thus can be ligated. As noted above, a standard way of detecting the ligation products is by a fluorescence-based DuoLink kit. However, since it has certain drawbacks, here we detected the ligated oligonucleotides covalently linked to the antibodies by quantitative PCR of the ligated product.

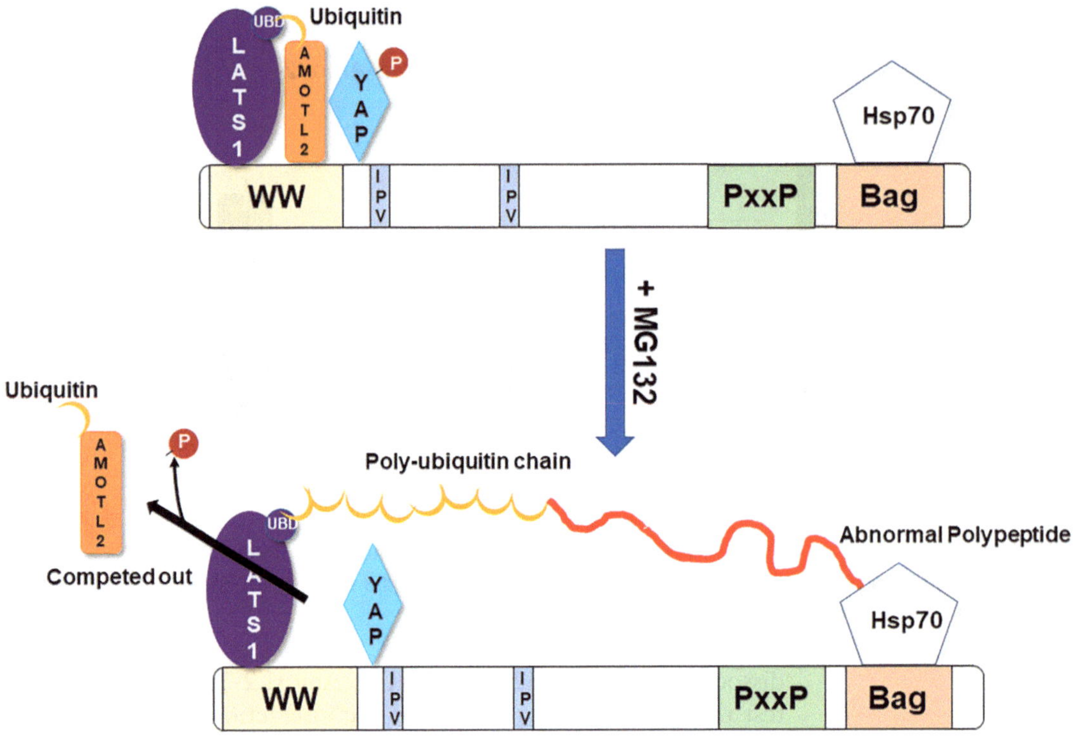

Fig. 4 A proposed mechanism of AmotL2 dissociation from HB complex upon proteotoxic stress. In naïve cells, LATS1, Amotl2, and YAP associate with Bag3. Amotl2 is mono-ubiquitinated and directly interacts with LATS1 via its UBD domain. Upon proteasome inhibition, recruitment of poly-ubiquitylated proteins to Hsp70-Bag3 competitively inhibits association of monoubiquitin on AmotL2 with the UBA domain of LATS. This leads to dissociation of AmotL2, disruption of interaction between LATS and YAP, and dephosphorylation and activation of YAP

To avoid attaching oligos to every primary antibody, we used secondary antibodies conjugated to oligos. They obviously can be used with any pair of regular (unmodified) primary antibodies. However, using such universalization requires that two primary antibodies should be raised in distinct hosts, so each secondary antibody would bind only to one of the primary ones. The protein–protein interaction measurements by modified PLA need to be done in biological triplicates or even quadruplicates. Briefly, treated and control cells are fixed and permeabilized in the wells of the 96-well plates, and the wells are incubated with a blocking solution. Of note, negative controls for the specificity are required, including omitting primary antibodies to either of the proteins. More strict negative controls would require omission of one of the interacting proteins from the cells. It could be easily done if either of these proteins is exogenously expressed in the cells. If both proteins of interest are endogenous, shRNA depletion of one of them could be done. Therefore, for each experimental point, there needs to be a sample with the complete set of antibodies and at least two negative controls, each in triplicate—12 wells total. Pipetting needs to be done with high precision, and thus a robotic liquid handler is desirable, but not essential. To compensate for a possible loss of cells under certain treatments, a signal from each experimental condition can be normalized by the amount of an endogenous protein, whose levels are not affected by experimental conditions.

Following incubation with primary antibodies, the plate is washed, and samples are incubated with the secondary oligo-conjugated antibodies. After the washout of unbound secondary antibodies, the ligase and ligase buffer are added. The ligated antibodies are separated from immobilized antigens by heating the plate, and the solution with the DNA product is transferred to the PCR plates. The relative amounts of the DNA product in different wells are quantified by qPCR.

2 Materials

All solutions are prepared using sterile double-distilled water and stored at room temperature unless otherwise specified. Most buffers are commercially available.

1. PBS: phosphate buffered saline, pH 7.4. The solution contains 137 mM of NaCl, 2.7 mM of KCl, 8.0 mM of Na2HPO4, and 2.0 mM of KH2PO4. Mix the solution and adjust pH to 7.4, and top up to the desired volume with double-distilled water.
2. PBST: phosphate buffered saline with 0.05% Tween® 20, pH 7.4. The solution contains 137 mM of NaCl, 2.7 mM of KCl, 8.0 mM of Na2HPO4, and 2.0 mM of KH2PO4. Mix the solution and adjust pH to 7.4, and top up to the desired volume with double-distilled water. Add 0.5475 g of Tween® 20 for 1 L of solution.
3. BCS: blocking column solution, 200 μg/mL BSA, 10 mM HEPES, 150 mM NaCl. In 5 mL of PBS, in a suitable container, add 2 mL of a solution of 1 mg/mL of BSA in PBS, 0.1 mL of HEPES 1 M, and 1.5 mL of NaCl 1 M. Add double-distilled water until the volume is 10 mL.
4. 50 mM Tris-HCl: 50 mM Tris-HCl, pH 7.4 in PBS. In 80 mL of distilled or double-distilled water, in a suitable container, add 5.0 mL of Tris-HCl 1 M, and adjust the pH to 7.4 using HCl if required. Add double-distilled water until the volume is 100 L.
5. TE: 10 mM Tris-HCl, 1 mM EDTA saline solution, pH 8.0. In 80 mL of distilled or double-distilled water, in a suitable container, add 1.0 mL of 1 M Tris-HCl (pH 8.0) and 0.2 mL of 0.5 M EDTA (pH 8.0). Add double-distilled water until the volume is 100 mL.

6. Blocking buffer: 3% BSA, 300 nM Donkey Gamma Globulin (Jackson ImmunoResearch Ca. N# 017–000002), 0.1 mg/mL ssDNA (Sigma-Aldrich Cat. N# D9156), 5 mM EDTA in PBST. In 8 mL of PBST, add 0.1 mL of 0.5 M EDTA (pH 8.0), 0.1 mL of 10 mg/mL ssDNA, 44.78 μL of 67 μM Donkey Gamma Globulin, and 0.3 g of BSA. Add PBST until the volume is 10 mL. Mix until the BSA is dissolved. Store at 4 °C.
7. 10× ligation buffer: 400 mM Tris-HCl (pH 8.0), 100 mM MgCl2. In 4 mL of distilled or double-distilled water, in a suitable container, add 4 mL of Tris-HCl 1 M (pH 8.0) and 1 mL of 1 M MgCl2. Add double-distilled water until the volume is 10 mL.
8. 4% formaldehyde: In 4 mL of PBS, in a suitable container, add 1 mL of 16% methanol-free formaldehyde solution from a freshly opened ampule (*see* **Note 1**).
9. 0.2% Triton X-100: 0.2% Triton X-100 in PBS. In 45 mL of PBS, in a suitable container add 0.107 g of Triton X-100. Add PBS until the volume is 50 mL.
10. Primary antibodies: Select the primary antibodies against the protein of your interest (*see* **Note 2**).
11. Secondary antibodies: Select the secondary antibodies against the two sources of the primary antibodies. Mouse or rabbit: Jackson ImmunoResearch—AffiniPure Donkey Anti-Rabbit IgG (H + L), Code 711–005-152, and AffiniPure Donkey Anti-Mouse IgG (H + L), Code 715–005-150 (*see* **Note 2**).

3 Methods

3.1 Conjugation Between Antibodies and Oligonucleotides

The oligos were SH-modified on the 5′-end and conjugated with SMCC to the NH2-groups on the IgGs according to the SMCC conjugation protocol: https://www.thermofisher.com/document-connect/document-connect.html?url=https://assets.thermofisher.com/TFS-Assets%2FLSG%2Fmanuals%2FMAN0011295_SMCC_SulfoSMCC_UG.pdf

3.2 Antibody Purification

The resulting conjugates need to be purified from the unbound oligonucleotides. The latter can significantly increase the nonspecific background in PCR reactions.

1. Block the column (Vivaspin 4 Turbo 100 kDa MWCO Pes, Sartorius Stedim). Fill the column with the BCS for 10 min at 4 °C.

2. Spin down at 13000 rpm for 2 to 5 min (*see* **Note 3**). If there is some BSC remaining in the upper volume, remove it.
3. Add the conjugated antibodies to the column. Spin it down for 4 min at 13,000 rpm to get rid of the free oligonucleotides.
4. Add 1 mg total of ssDNA to the upper volume, and bring the volume to 100 μL with PBS. Incubate it for 2 h at 4 °C (*see* **Note 4**).
5. After the incubation time, spin it down for 2 min at 13,000 rpm.
6. Add an extra 60 μL of PBS. Spin it down for 2 min at 13,00 rpm.
7. Repeat step 6.
8. Collect the upper volume that contains the conjugated antibodies purified.
9. You can check the quality of the purification by running an agarose gel. The samples should be all the washes, starting conjugated antibody solution, and purified conjugated antibody solution. You should not see any band corresponding to free oligos after purification (*see* **Note 5**).

3.3 IPAD Protocol

Carry out all procedures at room temperature unless otherwise specified. Use appropriate volume to prevent wells from drying out, and **be careful to avoid cross-contamination between wells**.

1. Plate cells on a 96-well plastic flat bottom plate (*see* **Note 6**), to have it at the desired confluency (40–60% in our experiments), on the day of your experiment.
2. Fix cells after treatment with 4% formaldehyde for 10 min.
3. Rinse 1× with 50 mM Tris-HCl (pH 7.4) in PBS.
4. Wash 1× with 50 mM Tris-HCl (pH 7.4) in PBS, for 5 min.
5. Permeabilize with 0.2% Triton X-100 in PBS for 5 min. The cells can be left in a cold room before or after permeabilization, if need to interrupt.
6. Leave the wells with blocking buffer o.n. at 4 °C.
7. Incubate the selected wells with both primary antibodies against the two interacting proteins of interest. Prepare the antibodies in the blocking buffer at a concentration used for immunofluorescence. Incubate for 2 h.
8. Wash 3× with PBST, 5 min each time.
9. Incubate the wells with one of the two secondary antibodies in the blocking buffer with a dilution of 1:2000. Incubate for 1 h.
10. Wash 3× with PBST, 5 min each time.

11. Repeat step [10] with the second secondary antibody.
12. Wash 3× with PBST, 5 min each time.
13. Wash 2× with TE, 10 min each time (*see* **Note 7**).
14. Dilute on ice T4 DNA ligase (Thermo Fisher Cat. N# 15-224-041) at the final concentration of 5 U/mL in 1× Ligation Buffer, and supplement the solution with 1 mM ATP. Add 40 μL per selected well and leave the solution for 30 min.
15. Wash 1× with PBST for 5 min.
16. Wash 2× with TE, 5 min each time.
17. Add 50 μL of TE to the wells, carefully seal the plate with autoclave tape, wrap it with aluminum foil, and then leave the plate on the bottom of a heating block at 95–100 °C with metal holders on top of it, for 25–30 min.
18. Take out the plate, and let it cool down with the metal holders sitting on top of it (*see* **Note 8**).
19. Spin the condensate into the wells.
20. Run qPCR using the ligation products in the well from the previous step as a template (*see* **Note 9**). Follow the manufacturer's protocol for the qPCR (*see* **Note 10**).
21. Place a qPCR 96-well plate on ice, and dispense 20 μL per well of reaction mix and mix gently. Add 2 μL per well of your template. qPCR thermal profile from the manufacturer's protocol of the qPCR mix.

4 Notes

1. Use alcohol-free formaldehyde.
2. Be careful that you are using primary antibodies from different sources to detect the proteins in the complex. Antibodies1 vs Protein1 must have a different source from the Antibodies2 vs Protein2. The same approach is valid per the secondary antibodies. Secondary AB1 vs primary AB1 and secondary AB2 vs primary AB2.
3. Be careful to avoid blocking the column with the BSA.
4. The total upper volume depends on how much conjugated antibody solution you want to purify. Adjust the final volume by the column's maximum upper volume.
5. If you are not satisfied with the level of purification, you can try to repeat step 6 one more time.

6. Choose a type of 96-well plate that does not melt at 95–100 °C.
7. Washing is a crucial step; it allows to avoid some residual secondary antibodies that can cause signals during qPCR.
8. At this step, the plate can be stored at −20 °C.
9. There is no restriction for the type of qPCR Mix that can be used, but it should contain ROX. In our experiments, we automatically pipetted 20 μl of full PCR mixture without templates and added 2 μl of a ligation product.
10. Calculate the amount of reaction mix you need according to the number of templates, keep in mind the triplicates, and add a few extra wells in case of mistakes while preparing the qPCR 96-well plate.

References

1. Hartl FU, Bracher A, Hayer-Hartl M (2011) Molecular chaperones in protein folding and proteostasis. Nature 475:324–332
2. Chaperone machines for protein folding, unfolding and disaggregation | Nature Reviews Molecular Cell Biology, https://www.nature.com/articles/nrm3658
3. Morán Luengo T, Mayer MP, Rüdiger SGD (2019) The Hsp70-Hsp90 chaperone cascade in protein folding. Trends Cell Biol 29:164–177
4. Rosenzweig R, Nillegoda NB, Mayer MP et al (2019) The Hsp70 chaperone network. Nat Rev Mol Cell Biol 20:665–680
5. Kampinga HH, Craig EA (2010) The HSP70 chaperone machinery: J proteins as drivers of functional specificity. Nat Rev Mol Cell Biol 11:579–592
6. Albakova Z, Armeev GA, Kanevskiy LM et al (2020) HSP70 multi-functionality in cancer. Cell 9:E587
7. Shemesh N, Jubran J, Dror S et al (2021) The landscape of molecular chaperones across human tissues reveals a layered architecture of core and variable chaperones. Nat Commun 12:2180
8. Meriin AB, Narayanan A, Meng L et al (2018) Hsp70-Bag3 complex is a hub for proteotoxicity-induced signaling that controls protein aggregation. Proc Natl Acad Sci USA 115:E7043–E7052
9. Baldan S, Meriin AB, Yaglom J et al (2021) The Hsp70-Bag3 complex modulates the phosphorylation and nuclear translocation of hippo pathway protein yap. J Cell Sci 134: jcs259107
10. Adler M, Wacker R, Niemeyer CM (2008) Sensitivity by combination: immuno-PCR and related technologies. Analyst 133:702–718
11. Malou N, Raoult D (2011) Immuno-PCR: a promising ultrasensitive diagnostic method to detect antigens and antibodies. Trends Microbiol 19:295–302
12. Sherman MY, Gabai V (2022) The role of Bag3 in cell signaling. J Cell Biochem 123:43–53
13. Colvin TA, Gabai VL, Gong J et al (2014) Hsp70-Bag3 interactions regulate cancer-related signaling networks. Cancer Res 74: 4731–4740
14. Ammirante M, Rosati A, Arra C et al (2010) IKK{gamma} protein is a target of BAG3 regulatory activity in human tumor growth. Proc Natl Acad Sci USA 107:7497–7502
15. Ulbricht A, Höhfeld J (2013) Tension-induced autophagy: may the chaperone be with you. Autophagy 9:920–922
16. Hiebel C, Stürner E, Hoffmeister M et al (2020) BAG3 proteomic signature under Proteostasis stress. Cell 9:E2416
17. Garrido C, Schmitt E, Candé C et al (2003) HSP27 and HSP70: potentially oncogenic apoptosis inhibitors. Cell Cycle 2:578–583
18. Sherman MY, Gabai VL (2015) Hsp70 in cancer: back to the future. Oncogene 34:4153–4161
19. Murphy ME (2013) The HSP70 family and cancer. Carcinogenesis 34:1181–1188
20. Ulbricht A, Eppler FJ, Tapia VE et al (2013) Cellular Mechanotransduction relies on tension-induced and chaperone-assisted autophagy. Curr Biol 23:430–435

21. Zhu C, Li L, Zhao B (2015) The regulation and function of YAP transcription co-activator. Acta Biochim Biophys Sin 47:16–28
22. Meng Z, Moroishi T, Guan K-L (2016) Mechanisms of hippo pathway regulation. Genes Dev 30:1–17
23. Chan EHY, Nousiainen M, Chalamalasetty RB et al (2005) The Ste20-like kinase Mst2 activates the human large tumor suppressor kinase Lats1. Oncogene 24:2076–2086
24. Varelas X (2014) The hippo pathway effectors TAZ and YAP in development, homeostasis and disease. Development 141:1614–1626
25. Yaglom JA, Wang Y, Li A et al (2018) Cancer cell responses to Hsp70 inhibitor JG-98: comparison with Hsp90 inhibitors and finding synergistic drug combinations. Sci Rep 8:3010

Chapter 14

Use of Native-PAGE for the Identification of Epichaperomes in Cell Lines

Tanaya Roychowdhury, Anand R. Santhaseela, Sahil Sharma, Palak Panchal, Anna Rodina, and Gabriela Chiosis

Abstract

Epichaperomes are disease-associated pathologic scaffolds, composed of tightly bound chaperones, co-chaperones, and other factors. They mediate anomalous protein–protein interactions inside cells, which aberrantly affects the function of protein networks, and in turn, cellular phenotypes. Epichaperome study necessitates the implementation of methods that retain these protein complexes in their native cellular states for analysis. Here we describe a protocol for detection and composition analysis of epichaperomes in cell homogenates through native polyacrylamide gel electrophoresis.

Key words Native polyacrylamide gel electrophoresis (native-PAGE), Epichaperomes, Immunoblot, High-order assemblies, Multimolecular protein complexes, Non-denaturing gel, Heat shock protein 90 (HSP90), HSP70, Chaperones, Oligomers

1 Introduction

Polyacrylamide gel electrophoresis (PAGE) in the presence of a detergent-like sodium dodecyl sulfate (SDS) and presence or absence of denaturing and reducing substance is the most used technique for the separation of proteins. The introduction of SDS was the first hallmark in protein electrophoresis [1, 2]. SDS imparts a uniform negative charge to any protein (around 1.4 g/1 g of protein), making the individual charge of the protein irrelevant in terms of its mobility in a polyacrylamide gel [3]. SDS forms micelle around the protein molecules with a net negative charge depending on the size of the protein. Thus, the mobility of the protein is determined by the size of the protein effectively. Although SDS-PAGE is a very powerful technique for resolving a mixture of proteins in a sample, the obvious disadvantage is that the samples are boiled in Laemmli buffer that contains SDS and sometimes reducing agents like β-mercaptoethanol (β-ME) or dithiothreitol

Stuart K. Calderwood and Thomas L. Prince (eds.), *Chaperones: Methods and Protocols*, Methods in Molecular Biology, vol. 2693, https://doi.org/10.1007/978-1-0716-3342-7_14,

(DTT), which disintegrates most of the higher molecular weight complexes. However, electrophoresis can also be performed in the absence of any detergent. Under such conditions, protein complexes may be retained. The separation will be driven by charge and be dependent mainly on the primary sequence and structure of the protein and the pH of the electrophoresis buffer. This method of separation of proteins and complexes is referred to as native-PAGE.

Native-PAGE allows for the separation from cells and tissues of protein complexes in their native states. The method can be used for identification of physiological protein–protein interactions, oligomeric states, and native protein mass. These native complexes can be recovered from gels using diffusion and electroelution and can be subjected to downstream applications, like electron microscopy and 2D crystallization, or for further analysis for activity [4]. Thus, native-PAGE serves as an important tool for the analysis of higher molecular weight complexes isolated from tissues and cells to understand the changes in pattern and activity under different physiological conditions.

Here we demonstrate the use of native-PAGE for the study of epichaperomes. Not to be confused with folding chaperones, which have evolved to be dynamic and short-lived [5–7], epichaperomes are long-lived heterooligomeric assemblies of tightly bound chaperones, co-chaperones, and other factors [8–17]. Functionally, epichaperomes are also distinct from chaperones. They act as pathologic scaffolds remodeling protein–protein interaction networks, rather than serving as folders of proteins in protein synthesis and degradation pathways [8–17]. Epichaperomes are principally localized to diseased cells and tissues and represent a fraction of the total chaperone pools [8–17]. In contrast, the ubiquitous folding chaperones are abundantly expressed in all cells and across normal and disease conditions [18, 19]. Epichaperomes composition is context-dependent, meaning that a complement of epichaperome structures, each with distinct composition, forms in distinct disease conditions. For example, in cancer cells, heat shock protein 90 (HSP90) recruits heat shock cognate 70 (HSC70), HSP70-HSP90 organizing protein (HOP), and HSP110 and other co-chaperones and other factors, to function as a network to provide a survival advantage to cancer cells and tumor-supporting cells in the microenvironment [8, 14, 20]. In Parkinson's disease, within midbrain dopaminergic neurons exposed to toxic stressors (e.g., rotenone), HSP90 recruits HSP60 into epichaperomes, and these epichaperome structures act to aberrantly rewire the interaction of proteins involved in dopamine synthesis pathways [10]. In neurons exposed to genetic stressors (e.g., PARKIN mutation), HSP90 recruits HSC70, HOP, HSP40, and several other co-chaperones to epichaperomes, to rewire the interaction of numerous proteins involved in inflammatory signaling pathways [10]. HSC70 is also an epichaperome constituent along with HSP90 in Alzheimer's disease where epichaperomes negatively impact the interaction of proteins integral for synaptic plasticity [12]. Methods and protocols to study the presence and abundance of these disease-promoting heterooligomers in distinct biological contexts are of high importance.

Here, we describe in detail the step-by-step protocol for identification of HSP90-incorporating epichaperome complexes using native-PAGE from cell lines and evaluation of total chaperone levels using SDS-PAGE (Fig. 1). We discuss the impact of sample storage and handling on epichaperome stability (Fig. 2), the effect of gel gradient on the separation and detection of epichaperomes (Fig. 3), and the influence of gel type on epichaperome separation and detection efficiency (Fig. 4). Finally, we confirm the reproducibility of the protocol in the hands of a second investigator (Fig. 5).

2 Materials

Prepare all solutions using deionized water at 25 °C and analytical-grade reagents. Prepare and store all reagents at room temperature (unless indicated otherwise). Diligently follow all waste disposal regulations when disposing of waste materials. This protocol uses cancer cells. In order to use biologicals requiring biosafety level 2 (BSL2), the project was registered with the Institutional Biosafety Committee (IBC), and the personnel was trained in proper handling and use of hazardous materials. Biohazardous materials were handled and disposed of according to applicable state and federal regulations.

2.1 Cell Culture

1. Cell lines: (a) breast cancer cell lines, MDA-MB-468 (Catalog# HTB-132, RRID: CVCL_0419) and MDA-MB-453 (Catalog# HTB-131, RRID: CVCL_0418), (b); pancreatic cancer cell lines, ASPC1 (Catalog# CRL-1682, RRID: CVCL_0152) and MiaPaCa2 (Catalog# CRL-1420, RRID: CVCL_0428); and (c) normal colon fibroblasts, CCD-18 (Catalog# CRL-1459, RRID: CVCL_2379) were purchased from American Type Culture Collection (ATCC). Vials are stored in liquid N_2 when not in use (*see* **Note 1**).
2. Gibco™ Dulbecco's Modified Eagle's Medium (DMEM) supplemented with L-glutamine, from Fisher Scientific (Catalog #11–965-084), for culturing of cells (stored at 4 °C) (*see* **Note 2**).
3. Gibco™ Fetal Bovine Serum (FBS), from Fisher Scientific (Catalog #10-082-147), for culturing of cells (stored at −20 °C, in 50 mL aliquots, thawed on ice prior to use).

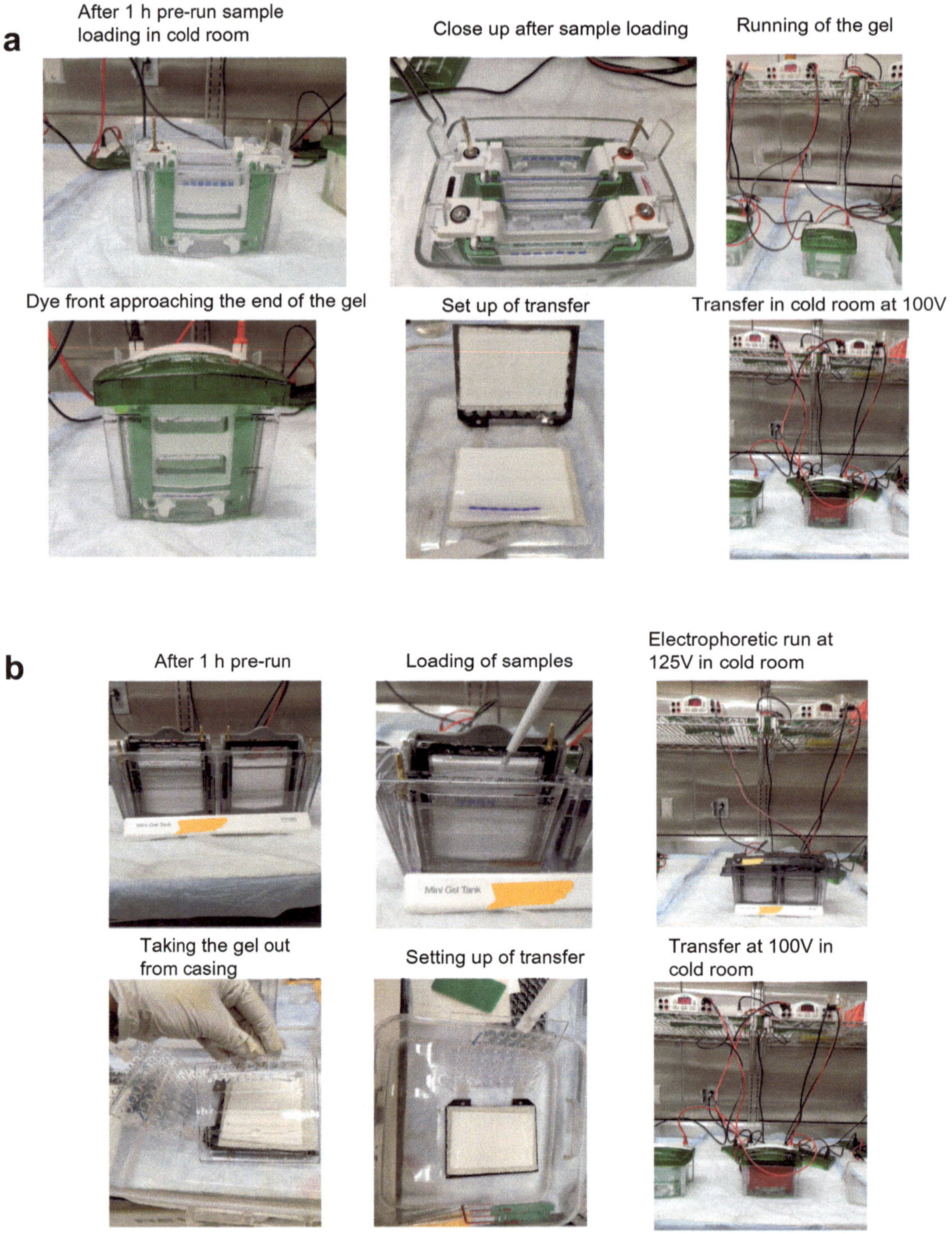

Fig. 1 Electrophoresis system setup. Snapshots of the setup at different stages, as described in the protocol, are provided. a: Cell lysates (100 μg) were loaded onto the gel. Pre-run, for gel equilibration, was done at 100V (cold room). The gel was run at 125V in the cold room. The run time for the 4 to 8 % hand cast gels was 2 h, for the 4 to 10% hand cast gel was 2.5 h, and for the 4 to 20% precast gel (Biorad) was 4 h. Transfer was done with 0.02% SDS in 1× transfer buffer for 2 h (cold room). PVDF membrane was used for the protein transfer process, unless otherwise indicated. b: Setup for Invitrogen 4 to 12% precast gels. The run time for the 4–12% precast gel (Invitrogen) was 3.5 h

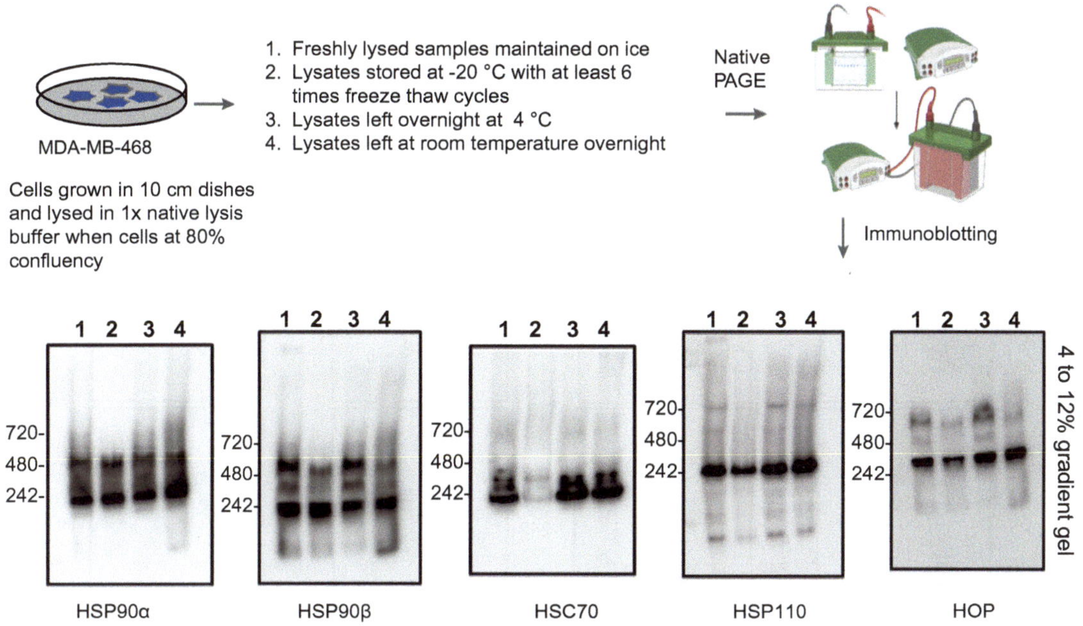

Fig. 2 The impact of sample storage and handling on epichaperome stability. Cell lysates (100 μg) were loaded onto the gel (Invitrogen precast 4–12% gradient gel) freshly prepared (1), or after being stored as indicated. For (2) lysates were stored at -20 °C but thawed at least 6 times before use. Pre-run, for gel equilibration, was done at 100V (cold room). The gel was run at 125V in the cold room for 3.5 h. Transfer was done with 0.02% SDS in 1× transfer buffer for 2 h (cold room). PVDF membrane was used for the protein transfer process

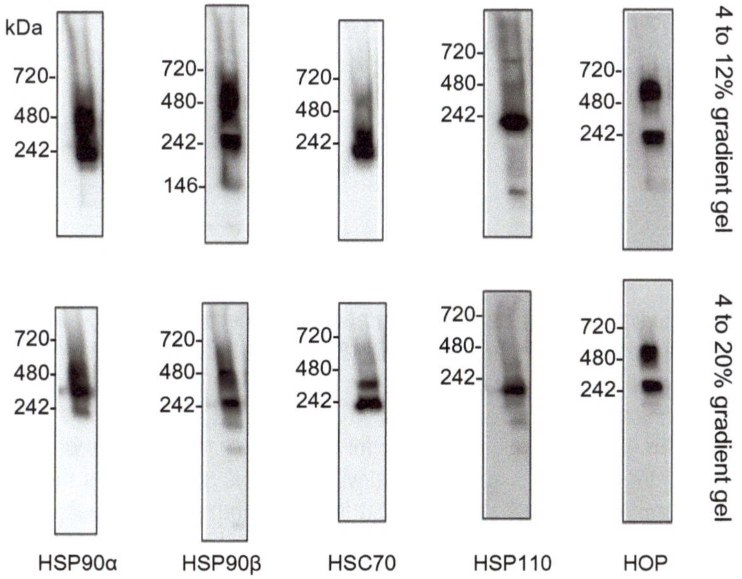

Fig. 3 The effect of gel gradient on the separation and detection of epichaperomes in the MDA-MB-468 cancer cell line. Several chaperone and co-chaperone components of epichaperomes were detected as indicated. Setup same as in Fig. 1 for cell lysates (100 μg) loaded onto the Invitrogen precast 4–12% gradient gel or the Biorad 4 to 20% gradient gel

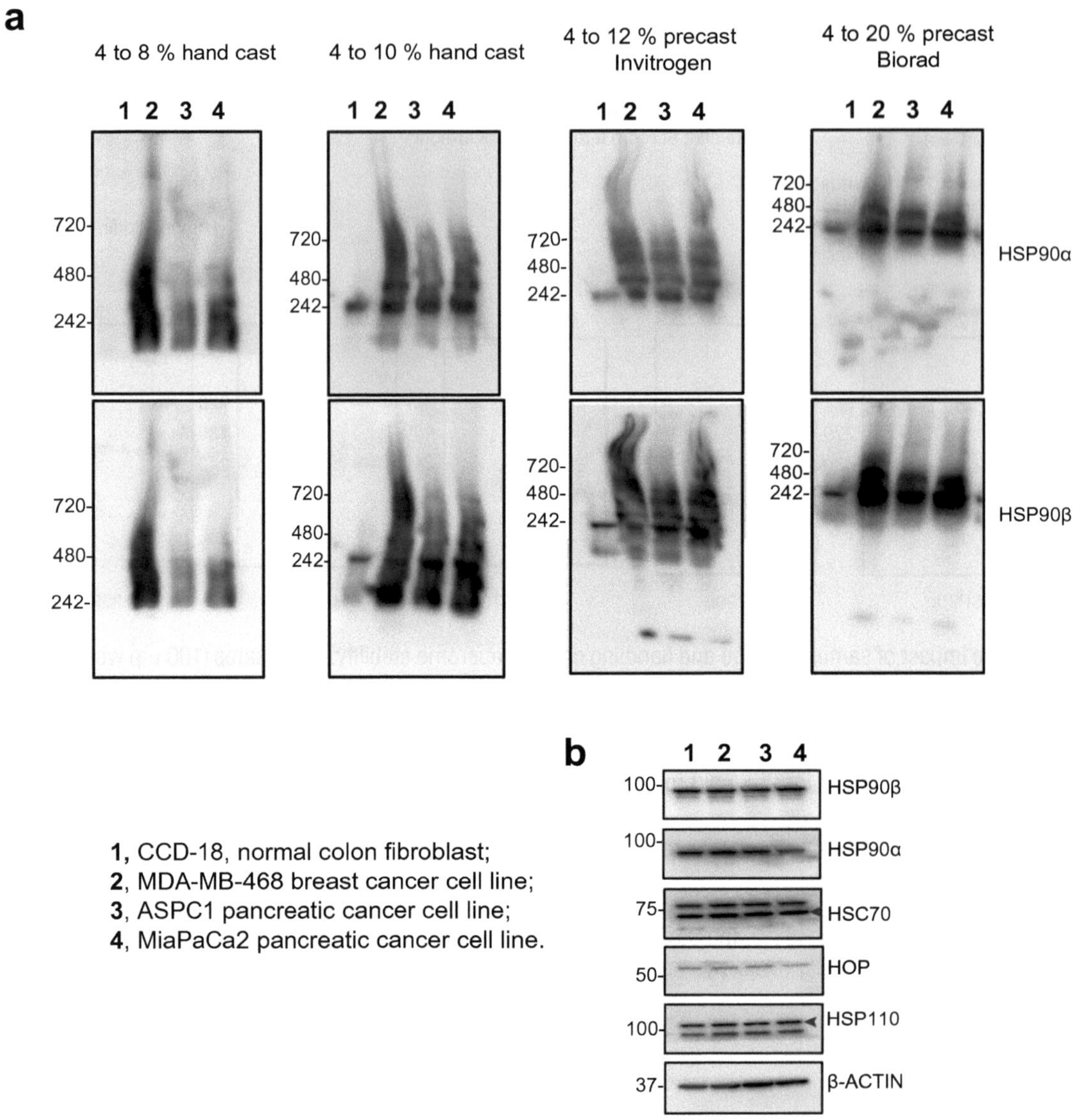

Fig. 4 The influence of gel type on epichaperome separation and detection efficiency. a: Native PAGE profile of epichaperomes in the indicated cancer cell lines. CCD-18 contains little to no epichaperomes and is used as control to show the biochemical signature of chaperones on Native PAGE. Note: the protocol used is same as in Fig. 1, but a roller was used to remove air bubbles, which stretched the gel. Most affected are the 4–12 % precast Invitrogen gels (1mm thickness). The run time for the 4–8 % hand cast gels was 2 h, for the 4 to 10 % hand cast gel was 2.5 h, for the 4–12 % precast gel (Invitrogen) was 3.5 h, and for the 4 to 20% precast gel (Biorad) was 4 h. The hand cast gel was stored at 4 °C for one month prior to use. b: SDS PAGE profile (i.e., total levels) of chaperones in samples from panel (a). Note: additional bands for HSP110 and HSC70 are due to incomplete membrane stripping prior to re-blotting

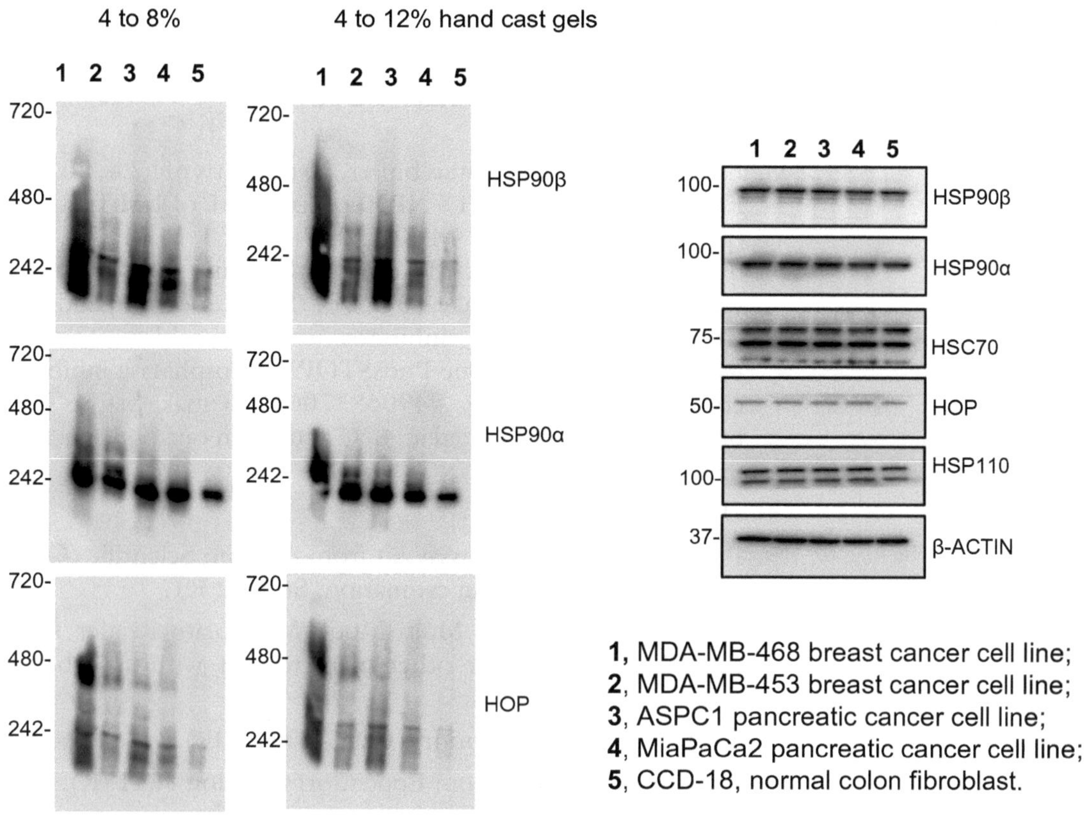

Fig. 5 Reproducibility by a second investigator. Same as in Fig. 4 for samples run by a different investigator using hand cast gels. A nitrocellulose membrane was used for transfer in this case (for the Native gel only)

4. Corning™ Penicillin-Streptomycin Solution (100x), from Fisher Scientific (Catalog #MT30002CI, storage temperature − 20 °C).
5. Gibco™ PBS (pH 7.4), from Fisher Scientific (Catalog #10-010-023). Store at 4 °C, brought to room temperature (RT) before use.
6. Gibco™ Trypsin-EDTA (0.25%), from Fisher Scientific (Catalog #25-200-056), for cell trypsinization (storage temperature, −20 °C, thawed before use).
7. 100 mm cell culture dish from Fisher Scientific (Catalog #08-772-6). Store at RT.
8. 175 cm^2 cell culture flask from Fisher Scientific (Catalog #10-126-13). Store at RT.
9. CO_2 incubator from NuAire (Catalog #NU-8700).

2.2 Cell Lysate Preparation for Native-PAGE

1. Gibco™ PBS (pH 7.4), from Fisher Scientific (Catalog #10-010-023). Store at 4 °C.
2. Cell scraper, from Fisher Scientific (Catalog #08-100-240). Store at RT.
3. Lysis buffer (1 x native lysis buffer, 20 mM Tris pH 7.4, 20 mM KCl, 5 mM $MgCl_2$, 0.01% NP40): Mix 10 mL of 1 M Tris-HCl (pH 7.4), 10 mL of 1 M KCl, 2.5 mL of 1 M $MgCl_2$, and 0.5 mL of 10% NP40. Make up to 500 mL with deionized water. Store at 4 °C. Take 10 mL of this solution and add one Complete™ Protease inhibitor cocktail tablet (Catalog #11697498001) and one PhosSTOP™ phosphatase inhibitor cocktail tablet (Catalog #4906837001) to make native lysis buffer for cell lysis (stored at 4 °C and taken out on ice before use).
4. Dry ice.
5. Pierce™ BCA protein assay kit from Thermo Scientific (Catalog #23225) for protein estimation. Store at RT.
6. SpectraMax Paradigm Multi-Mode Microplate Reader from Molecular Devices (or other plate readers with absorbance measurement capabilities).
7. Centrifuge from Eppendorf (Catalog #5810R).
8. Benchtop centrifuge from Eppendorf (Catalog #5417R).
9. Fisher Scientific Isotemp 220 Digital Water Bath (Product Code: FS-220).

2.3 Native Polyacrylamide Gel

10. Running gel buffer (1 × Tris-glycine native buffer, 25 mM Tris, 192 mM glycine, 0.1% SDS): Dissolve 3.0 g of Tris-base and 14.4 g of glycine in 500 mL of deionized water, and make up to 1000 mL with deionized water. Store at 4 °C (*see* **Note 3**).
11. Acrylamide/bis-acrylamide 37.5:1 (40% solution) from Fisher Scientific (Catalog #BP1410-01). Store at 4 °C.
12. 1 M Tris-HCl (pH 8.0) form TEKnova (Catalog #T1080). Store at room temperature.
13. Transfer buffer (25 mM Tris pH 8.3, 192 mM glycine, 20% methanol): Dissolve 3.0 g of Tris-base and 14.4 g of Glycine in 500 mL of deionized water, and make up to 800 mL with deionized water. Add 200 mL of methanol to make the final volume of 1000 mL. Add 2 mL of 10% SDS solution per 1000 mL of the transfer buffer before the protein transfer procedure. Store at 4 °C.
14. Ammonium persulfate solution (10%): Dissolve 1 g of ammonium persulfate in 5 mL of deionized buffer, and make up to 10 mL with deionized water. Store at 4 °C. It is an oxidizing

agent that is used with TEMED to catalyze the polymerization of acrylamide and bis-acrylamide.

15. TEMED (N,N,N′,N′-tetramethylethylenediamine) from Fisher Scientific (Code# BP150-20). It is a free radical stabilizer and an essential catalyst for polyacrylamide gel polymerization. Store at 4 °C.
16. Loading buffer (6×): Mix 3.75 mL of Tris-HCl 1 M, pH 8, 6 mL glycerol, and 6 mg bromophenol blue, and make up to 10 mL with deionized water. Preferable storage at −20 °C but can also be stored at 4 °C, if in regular use.
17. Precast gels: NuPAGE™ 4 to 12%, Bis-Tris, 1.0–1.5 mm, mini protein gels from Thermo Fisher (Catalog #NP0321BOX), and 4 to 20% Mini-PROTEAN® TGX™ Precast Protein Gels, 10-well, 50 μL from BioRad (Catalog #4561094).
18. Electrophoretic running unit: Mini-PROTEAN Tetra Vertical Electrophoresis Cell, 4-gel (Catalog #1658004EDU).
19. Electrophoretic transfer unit: Mini Trans-Blot Electrophoretic Transfer Cell (Catalog# 1703930).
20. Invitrogen Mini Gel Tank (Catalog #A25977) (suitable for the Invitrogen precast gels only).
21. Gradient Maker from C.B.S. Scientific Co. (Model #GM-40) for the hand cast gels.
22. Bio-Rad PowerPac Universal Power Supply (Catalog #1645070).
23. Membrane for protein transfer: Immobilon®-P PVDF (PolyVinyliDeneFluoride). Membrane (Catalog #IPVH00010) and Amersham Protran 0.2 μm Nitrocellulose Blotting Membrane (Catalog #10600006). Store at RT.
24. Tris buffer saline Tween 20 (TBST): Mix 20 mL of 1 M Tris-HCl (pH 7.4) and 8.78 g of NaCl in 100 mL of deionized water. Make up to 1000 mL of deionized water, and add 1 mL of Tween 20. Store at RT.
25. Blocking solution: Dissolve 0.5 g of bovine serum albumin (BSA) in 5 mL of TBST, and make up to 10 mL with TBST.
26. Molecular weight marker: NativeMark™ Unstained Protein Standard (Catalog #LC0725). Store at−20 °C.
27. Ponceau S solution from MilliporeSigma (Catalog #P7170). Store at RT.
28. Thermo Scientific SuperSignal™ (SuperSignal West Dura Extended Duration Substrate is a luminol-based enhanced chemiluminescence (ECL) horseradish peroxidase (HRP) substrate) detection kit. Store at 4 °C.

2.4 Membrane Stripping

1. Stripping solution: 5 mM EDTA in deionized water. It is a solution for removing primary and secondary antibodies from probed Western blot membranes. Alternatively, a commercially made stripping solution can also be used.

2.5 SDS-Page

1. ProtoGel Resolving Buffer (4×) from National Diagnostics (Order# EC-892).
2. ProtoGel Stacking Buffer (4×) from National Diagnostics (Order# EC-893).
3. SDS-PAGE 1× running buffer: Dilute 100 mL of 10× TGS Buffer from Fisher Scientific (Catalog #BP1341) with deionized water to make the final volume up to 1000 mL.
4. Laemmli Sample Buffer (5×): Mix 1 g of SDS, 5 mL of glycerol, 3.125 mL of Tris-HCl (pH 6.8), 5 mg of bromophenol blue, and 0.775 g of DTT, and make up to 10 mL with deionized water.
5. Acrylamide/bis-acrylamide 37.5:1 (40% solution), ammonium persulfate solution, TEMED, TBST, and Ponceau S solution were prepared or purchased as mentioned above.
6. Transfer buffer (25 mM Tris pH 8.3, 192 mM glycine, 20% methanol): Dissolve 3.0 g of Tris-base and 14.4 g of glycine in 500 mL of deionized water, and make up to 800 mL with deionized water. Add 200 mL of methanol to make the final volume of 1000 mL.
7. Precision Plus Protein™ Kaleidoscope™ Prestained Protein Standards from Bio-Rad (Catalog #1610375). Store at −20 °C.

2.6 Detection

1. A panel of anti-chaperone antibodies has been screened to identify the ones recognizing the target protein in its native form. These native-cognate antibodies were used in the native-PAGE analysis of epichaperome assemblies: HSP90β (SMC-107, RRID:AB_854214, store at −20 °C) and HSP110 (SPC-195, RRID:AB_2119373, store at −20 °C) antibodies from StressMarq; HSC70 (SPA-815, RRID: AB_10617277, store at −20 °C) and HOP (SRA-1500, RRID:AB_10618972, store at −20 °C) from Enzo; and HSP90α (ab2928, RRID:AB_303423, store at −20 °C) from Abcam. All the above antibodies were used at 1:2000 dilution. Anti-β-actin antibody (A1978, RRID: AB_476692, store at −20 °C) was purchased from Sigma-Aldrich and used at 1:5000 dilution.
2. Species-specific HRP-conjugated secondary antibodies were purchased from Southern Biotech (stored at 4 °C), and used at a 1:5000 dilution: anti-mouse (Catalog #1030-05), anti-rat (Catalog #3030–05), and anti-rabbit (Catalog #4010-05).

3. ChemiDoc MP imaging system from Bio-Rad (Catalog #17001402).

3 Methods

All procedures are performed at room temperature unless otherwise specified.

3.1 Cell Culture and Cell Lysate Preparation for Native- and SDS-PAGE

1. Perform the cell culture in sterile conditions by operating under a Class II Biological Safety Cabinet.
2. Maintain the cells in culture at 37 °C in Dulbecco's Modified Eagle's Medium (DMEM) supplemented with 1× penicillin and streptomycin and 10% FBS (Sigma) inside an incubator with 5% CO_2 supply (*see* **Note 4**).
3. For the experiment, seed the cells in 10 cm dishes (in 12 mL medium) and grow till 80% confluency to harvest for cell lysis (*see* **Note 5**).
4. Discard the media and wash the cells with ice-cold 1 × PBS, and then scrape off using cell scrapers. Cell scrapers should be washed with ice-cold 1× PBS before use. Collect the cell suspension in 1 × PBS following by centrifugation at 3500 rpm for 5 min at 4 °C. Discard the supernatant and add the native lysis buffer at around 1.5 times the cell pellet volume.
5. For lysis, incubate the samples on dry ice for 10 min. Remove the sample from dry ice, and incubate it at 37 °C water bath for 2 min. Repeat the procedure three times followed by incubation on ice for 30 min. Centrifuge the lysates at 12,000 rpm for 20 min at 4 °C.
6. Collect the supernatant and determine the protein concentration with the BCA kit.
7. Prepare the samples for loading onto the native gel by mixing 100 μg of lysate with 6× loading dye (*see* **Note 6**).

3.2 Cell Lysate Preparation for SDS-PAGE

1. For sample preparation, heat 20 μg of protein in 5× Laemmli buffer at 95 °C for 10 min.

3.3 Preparation of Precast Gels for Running Native-PAGE

1. Take the NuPAGE™ 4 to 12% gel out of the plastic bag. Remove the white tape at the bottom of the gel. Place the gel into the Invitrogen mini gel tank, and take out the comb. Clear out the individual lanes using a 1 mL syringe and fill the tank with 1× cold native running buffer (approximate volume 500 mL, maintained at 4 °C). If using the 4 to 20% Mini-PROTEAN® TGX™ Precast Protein Gel, place the gel in the Mini-PROTEAN Tetra Vertical Electrophoresis Cell, and take out the comb. Clear the lines using a 1 mL syringe and fill the

Table 1
Preparation of the hand cast gels for Native-PAGE

Components	4%[a]	8% [a]	10% [a]
40% Acrylamide (in mL)	1	1	1.25
Deionized water (in mL)	5.2	2.1	1.875
1 M Tris-HCl, pH 8 (in mL)	3.8	1.9	1.875
10% APS (in μL)	0.080	0.040	0.040
TEMED (in μL)	0.006	0.003	0.003

[a]Polyacrylamide percentage

tank with 1× cold native running buffer (i.e., maintained at 4 °C).

2. Pre-run the gel without the samples for 1 h at 100 V in the cold room (i.e., maintained at 4 °C).

3.4 Preparation of Handmade Continuous Gels for Running Native-PAGE

1. Clean and dry the glass plates by whiping them with a tissue paper. Thoroughly clean the 1.5 mm spacers and the comb. Assemble the glass plates, spacers, and comb as described by the manufacturer.
2. Prepare the high- and low-percentage gel solutions as per Table 1. (*see* **Note 7**).
3. Prepare the gradient generator (*see* **Note 8**). Ensure that the generator is clean and without bubbles inside the tube and in the channel connecting both chambers.
4. A gradient generator consists of two chambers, A and B, connected to each other. Chamber B contains the low-percentage non-denaturing polyacrylamide gel (4%) and chamber A contains the high-percentage non-denaturing polyacrylamide gel (8%, 10%). A plastic tubing, through which the gel gets poured into the gel cassette, is connected to the lower bottom of container A.
5. After pouring the gels into the appropriate chambers, allow approximately 0.3 mL of the gel solution to flow through the tubing to ensure continuous gel flow and removal of air bubbles. Then, open the valve between the chambers allowing for the low-percentage gel to mix with the high-percentage gel. Make sure the flow rate is low. A faster flow will cause turbulence in the gel and may disrupt the gradient.
6. Attach a 200 μL pipette tip to the free end of the tube, and insert it into the assembled glass plates to allow the gradient gel to be poured in. Fill up the cassette with the gel.

7. After the process is complete, wash the gradient generator unit immediately to avoid acrylamide polymerization inside the gradient generator.
8. Place the comb into the gel cassette.
9. Set the gel aside for 20–30 min for the gel to polymerize.
10. Load the samples into the native gel wells. Use one well to add the molecular weight marker (8 μL of unstained protein standard (NativeMark™, Invitrogen)). Start the run at 125 V. Perform this step while in the cold room (i.e., maintained at 4 °C).
11. For 4% to 8% gradient gel, the run is continued until the dye front reaches the end of the gel, which takes approximately 2 h. For 4% to 10% gradient gel, the electrophoresis is continued to allow the dye front to run out of the gel, which takes approximately 2.5 h. For the 4% to 12% Invitrogen precast gel, the dye front is allowed to run out, which takes approximately 3.5 h. For the 4% to 20% Bio-Rad precast gel, the dye front is allowed to run out (approximately 4 h) (*see* **Note 9**).

3.5 SDS-Page

1. Use the 4 to 20% precast gel from Bio-Rad or the 4 to 12% precast gel from Invitrogen for running the SDS-PAGE. Alternatively, prepare a hand cast gel (Table 2).
2. Clean and dry the glass plates, clean the 1.5 mm spacers and the comb thoroughly, and assemble the unit as described by the manufacturer.
3. Using a glass marker, draw a line on the glass plate around 3 cm below the top of the plate. This indicates the level to which the resolving gel will be poured.

Table 2
Preparation of the hand cast gels for SDS-PAGE

Components	4% stacking	10% resolving
Acrylamide/bis-acrylamide 40%	0.8	5.32
Resolving gel buffer (mL)	–	5.32
Stacking gel buffer (mL)	2	
Water (mL)	5.1	10.4
SDS 20% (μL)	40	106.4
APS10% (μL)	40	106.4
TEMED (μL)	8	8

4. Prepare both the stacking and the resolving gel solutions at room temperature by combining all reagents according to Table 2, except for APS and TEMED (*see* **Note 10**).
5. Add APS and TEMED to the resolving gel solution (*see* Table 2 for the specified amount), and mix well by swirling gently. Pour the acrylamide gel mixture into the gel cassette.
6. Overlay the gel with isopropanol.
7. Allow the gel to polymerize for 1 h.
8. Remove the water/isopropanol overlay from the top of the gel, wash with water, and drain the excess water with strips of filter paper.
9. Add APS and TEMED to the stacking gel solution, and then pour the stacking solution gently on top of the resolving gel. This should be done gently to avoid bubbles.
10. Place the comb carefully onto the top of the gel and allow the gel to polymerize.
11. Following polymerization, set the gel in the Bio-Rad apparatus and fill up the tank with the running buffer till the recommended mark.
12. Remove the comb carefully, clear the lanes using a 1 mL syringe, and then load 20 μg of sample into each well. Use precision Plus Protein™ Kaleidoscope™ Prestained Protein Standard as a molecular size marker.
13. Run the gel at 100 V at room temperature till the dye runs out of the gel.

3.6 Protein Transfer and Immunoblotting

1. After the electrophoresis run is over, disassemble the gel unit, and set up the transfer process. For the Invitrogen and Bio-Rad precast gels, carefully remove the plastic casing such that the gel is not torn apart.
2. Activate the PVDF membrane for transfer by incubating it in methanol for 2 min, followed by washing it in distilled water and 1 × cold transfer buffer.
3. Set up the gel for transfer in a way that the gel is on the cathode side (black side of the transfer cassette) and the PVDF membrane is placed on top of the gel facing the anode side (the transparent side of the cassette) (*see* **Note 11**).
4. Use a roller to remove any air bubbles between the gel and PVDF membrane (*see* **Note 12**).
5. Place the cassette into the transfer unit, fill it with 1× cold transfer buffer (kept at 4 °C), and run at 100 V for 2 h in a cold room (4 °C).

6. After the transfer, stain the membrane with Ponceau S solution to visualize the molecular weight marker for the native gel.
7. Block the PVDF membrane with the blocking solution for 1 h at room temperature.
8. Add primary antibodies, and incubate overnight (for 12 h to 16 h) in the cold room at 4 °C.
9. Wash the blots with TBST three times (10 min each) at room temperature.
10. Incubate the membrane with the corresponding secondary antibody for 2 h at room temperature.
11. Wash with 1× TBST three times and proceed to signal visualization.

3.7 Chemiluminescent Detection

1. Develop the blots using the Thermo Scientific SuperSignal™ detection kit, which uses a luminol-based enhanced chemiluminescence (ECL) horseradish peroxidase (HRP) substrate, in a ChemiDoc MP system (Bio-Rad). Prepare the detection solution by mixing both components of the kit in a 1:1 ratio. Make approximately 0.1 mL detection solution for a 1 cm^2 membrane.
2. Take the membrane out of the TBST solution with tweezers, and remove excess buffer by holding the membrane in a vertical position, with the lower edge of the membrane touching a sheet of blotting paper.
3. Place the membrane on the developing tray.
4. Distribute the substrate evenly on the top of the membrane, incubate for 1 min, remove the solution, and insert the tray into the ChemiDoc MP machine for signal detection.
5. For re-blotting with a different antibody, the PVDF membranes can be stripped by boiling blots for 3 min in 5 mM EDTA in distilled water or by using commercially available stripping solutions.

4 Notes

1. Cells should be authenticated using short tandem repeat profiling and tested for mycoplasma before experiments.
2. All solutions employed for cell culture need to be prepared and maintained in sterile conditions by operating under a Class II Biological Safety Cabinet. Sterilize the Class II Biological Safety Cabinet with 70% ethanol before and after using it. All items placed in the Class II Biological safety cabinet must be sterilized with 70% ethanol.

3. Commercially available 10 × Tris-glycine native buffer buffers can also be used after 1× dilution.
4. Be cognizant that culture conditions (e.g., stress) and cell passage number may alter the cellular composition and the levels of epichaperomes.
5. From a 10 cm dish, we usually get 7 x 10^6 cells. For MDA-MB-468 cancer cells, this cell number yields approximately 7-8 mg/mL when lysed in 1 × native lysis buffer. The volume of the lysis buffer should be 1.5 × times the cell pellet volume.
6. Cell lysis and sample preparation for native-PAGE should be done at 4 °C to preserve the native epichaperome complexes. See Fig. 2 for the impact of sample handling and storage on epichaperomes. If lysates are not used immediately after preparation, then they should be stored at −80 °C for long-term storage. In our hands, storage at −20 °C for a week, without freeze-thaw cycles, is acceptable.
7. In addition to safety reasons, wearing of gloves is recommended at all times during the protocol to avoid contamination with keratin proteins and prevent degradation of proteins by proteases.
8. Instead of a gradient gel, a 7.5% gel can be used for the purpose of native-PAGE.
9. For a 4 to 20% gradient gel, the running time is very important and impacts the ability to resolve the epichaperome complexes.
10. APS should not be very old as it will negatively impact cross-linking time.
11. Instead of PVDF membrane, nitrocellulose membrane can also be used; however, in this case, the membrane cannot be boiled for stripping. A commercially available stripping solution should be used for re-probing with another antibody.
12. For the 4 to 12% Invitrogen precast gel, a roller is not recommended. In our hand, the gel can stretch out and generate a wavy pattern after the transfer to the PVDF membrane and subsequent protein detection (*see* Fig. 4).

Acknowledgments

This work was supported by the NIH (R01 CA172546, P01 CA186866, R56 AG061869, R01 AG067598, P01 AG014449, P01 AG017617, R01 AG074004, R56 AG072599, RF1 AG071805, and P30 CA08748). S.S. is supported by the Bright-Focus Foundation (Award ID: A2022020F).

References

1. Laemmli UK (1970) Cleavage of structural proteins during the assembly of the head of bacteriophage T4. Nature 227:680–685
2. Studier FW (1973) Analysis of bacteriophage T7 early RNAs and proteins on slab gels. J Mol Biol 79:237–248
3. Smith BJ (1984) SDS polyacrylamide gel electrophoresis of proteins. Methods Mol Biol 1:41–55
4. Wittig I, Braun H-P, Schägger H (2006) Blue native PAGE. Nat Protoc 1:418–428
5. Trepel J, Mollapour M, Giaccone G et al (2010) Targeting the dynamic HSP90 complex in cancer. Nat Rev Cancer 10:537–549
6. Prodromou C (2012) The 'active life' of Hsp90 complexes. Biochim Biophys Acta 1823:614–623
7. Saibil H (2013) Chaperone machines for protein folding, unfolding and disaggregation. Nat Rev Mol Cell Biol 14:630–642
8. Rodina A, Wang T, Yan P et al (2016) The epichaperome is an integrated chaperome network that facilitates tumour survival. Nature 538:397–401
9. Joshi S, Wang T, Araujo TLS et al (2018) Adapting to stress — chaperome networks in cancer. Nat Rev Cancer 18:562–575
10. Kishinevsky S, Wang T, Rodina A et al (2018) HSP90-incorporating chaperome networks as biosensor for disease-related pathways in patient-specific midbrain dopamine neurons. Nat Commun 9:4345
11. Pillarsetty N, Jhaveri K, Taldone T et al (2019) Paradigms for precision medicine in Epichaperome cancer therapy. Cancer Cell 36(559–573):e7
12. Inda MC, Joshi S, Wang T et al (2020) The epichaperome is a mediator of toxic hippocampal stress and leads to protein connectivity-based dysfunction. Nat Commun 11:319
13. Yan P, Patel HJ, Sharma S et al (2020) Molecular stressors engender protein connectivity dysfunction through aberrant N-glycosylation of a chaperone. Cell Rep 31:107840
14. Bolaender A, Zatorska D, He H et al (2021) Chemical tools for epichaperome-mediated interactome dysfunctions of the central nervous system. Nat Commun 12:4669
15. Ginsberg SD, Joshi S, Sharma S et al (2021) The penalty of stress - Epichaperomes negatively reshaping the brain in neurodegenerative disorders. J Neurochem 159:958–979
16. Joshi S, Gomes ED, Wang T et al (2021) Pharmacologically controlling protein-protein interactions through epichaperomes for therapeutic vulnerability in cancer. Commun Biol 4:1333
17. Ginsberg SD, Sharma S, Norton L et al (2023) Targeting stressor-induced dysfunctions in protein-protein interaction networks via epichaperomes. Trends Pharmacol Sci 44:20–33
18. Finka A, Goloubinoff P (2013) Proteomic data from human cell cultures refine mechanisms of chaperone-mediated protein homeostasis. Cell Stress Chaperones 18:591–605
19. Shemesh N, Jubran J, Dror S et al (2021) The landscape of molecular chaperones across human tissues reveals a layered architecture of core and variable chaperones. Nat Commun 12:2180
20. Sugita M, Wilkes DC, Bareja R et al (2021) Targeting the epichaperome as an effective precision medicine approach in a novel PML-SYK fusion acute myeloid leukemia. NPJ Precis Oncol 5:44

Chapter 15

Molecular Chaperone Receptors: An Update

Thiago J. Borges, Ayesha Murshid, Jimmy Theriault, and Stuart K. Calderwood

Abstract

Extracellular heat shock proteins (HSP) play important roles in cell signaling and immunity. Many of these effects are mediated by surface receptors expressed on a wide range of cell types, including immune cells. We have investigated the nature of such proteins by cloning candidate receptors into cells (CHO-K1) with the rare property of being null for HSP binding. Using this approach, we have discovered that mammalian and eukaryotic Hsp70 binds avidly to at least three classes of receptor including: (1) c-type lectin receptors (CLR), (2) scavenger receptors (SR) and (3) lectins. However, the structural nature of the receptor–ligand interactions is not currently clear. Hsp70 can bind to LOX-1 (a member of both the CLR and SR), with the c-type lectin binding domain (CTLD), to the SR family members SREC-I and FEEL-1/CLEVER-1/STABILIN-1, which by contrast have arrays of EGF-like repeats in their extracellular domains as well. In this chapter, we will discuss: (1) methods for the discovery of HSP receptors, (2) approaches to the study of individual receptors in cells that contain multiple such receptors and (3) methods for investigating HSP receptor function in vivo.

Key words Extracellular, Heat, Shock, Protein, Scavenger, Receptor, Immunity, Immune suppressive

1 Introduction

Heat shock proteins (HSP) play significant signaling roles in the extracellular microenvironment [1, 2]. HSP have been found in human serum, particularly after disease or stress [3, 4]. The 70 kilodalton heat shock protein (Hsp70) has been shown to be released from cells after acute stress as well as being secreted after exposure to a number of stimuli [5, 6]. Extracellular HSP may thus be able to play the role of danger signal (danger-activated molecular pattern or DAMP) [7]. In this context, they may interact with pattern recognition receptors (PRR) such as toll-like receptors (TLR) and activate pro-inflammatory signaling and transcription

This work was supported by NIH research grants RO-1CA047407, R01CA119045 and RO-1CA094397

Stuart K. Calderwood and Thomas L. Prince (eds.), *Chaperones: Methods and Protocols*, Methods in Molecular Biology, vol. 2693, https://doi.org/10.1007/978-1-0716-3342-7_15,

[8, 9]. Proteins including Hsp60, Hsp70, and Grp96 have been implicated as DAMPs [10, 11]. However, the interpretation of such experiments requires caution and careful control. Some HSP have the ability to bind to prokaryotic molecules that activate TLR signaling, such as lipopolysaccharides, and the inflammatory properties of HSP are influenced by tissue context [2, 12]. In addition, many members of the HSP family can participate in adaptive immunity by binding to antigenic peptides and transporting them into antigen-presenting cells (APCs) [13, 14]. HSP mediate the process of antigen cross-presentation [15] by facilitating the internalization of extracellular antigens and permitting their delivery to major histocompatibility complex (MHC) class I molecules [16]. HSP were also demonstrated to mediate antigen presentation in an MHC class II context [17]. MHC I-peptide complexes can then stimulate cognate T cell receptors on T lymphocytes and initiate the activation of clones of such powerful immune effectors [16, 17]. HSP may thus play a versatile role in anti-tumor immunity by activating innate and adaptive arms. In this context, HSP can additionally activate natural killer cells and lead to tumor cell killing [18–20] In contrast, some prokaryotic HSP like DnaK (Hsp70) and GroEL (Hsp60) have been shown to have anti-inflammatory effects by either modulating APCs [21, 22] and inducing regulatory T cells (Tregs) [23–26]. Thus, extracellular HSP can thus upregulate or downregulate immunity depending on the microenvironmental context.

Many studies have suggested that HSP activate immunity by binding to receptors on the cell surface [27–35]. HSP binding is saturable and competed for by unlabeled ligand properties of receptor-mediated signaling. However, this is where the consensus seems to end and some controversy exists as to the most significant HSP receptors. We have attempted to address this issue by screening the various contenders for binding to Hsp70, Hsp90, and DnaK.

2 Materials

Plasmids

pET23 *Hsp90a* plasmid.

pDEST™10.

AC-to-BAC Baculovirus transfection kit.

pCDNA3.1 eukaryotic expression vector.

Cell Lines

Chinese hamster ovary-K1 cells (CHO-K1).

A375 human melanoma.

MISA human breast carcinoma cells.

Sf9 insect cells.

DH10Bac competent cells.

MC38 cells stably expressing the MUC1 tumor antigen.

B16 melanoma cells.

B16 melanoma cells stably expressing the MUC1 tumor antigen.

Mouse Systems and Primary Murine Cells

Wild-type C57BL/6 and Tlr2-/-Tlr4-/- double knockout mice.

Primary mouse bone marrow dendritic cells were prepared from C57BL/6 as in text.

Splenocytes and/or lymph node cells (LNC) were isolated from mice immunized with Hsp70.PC fusion vaccine as described36.

Chromatography

Ni–NTA purification system (Qiagen).

10 ml Sephadex G-25 in PD10 column (Sigma-Aldrich).

5.0 Ml ADP–Agarose Column (Sigma-Aldrich).

20 ml DEAE-cellulose anion exchange column (Pierce Chemical).

Buffers and Reagents

Hypotonic buffer: 10 mM NaHCO3, 0.5 mM PMSF, pH 7.1.

Buffer D: 20 mM Tris-acetate, 20 mM NaCl, 15 mM b-mercaptoethanol, 3 mM MgCl2, 0.5 mM PMSF, pH 7.5.

ADP–agarose elution buffer: 3.0 mM ADP in buffer D.

FPLC buffer: 20 mM sodium mono- and diphosphate, 20 mM NaCl, pH 7.0.

DEAE-cellulose elution buffer: 150 mM NaCl in FPLC buffer.

Hsp70 binding buffer (PFNC): 0.5% FBS, 0.05% NaN3, and 1 mM CaCl2.

Hanks' buffered saline solution.

Cellfectin II Reagent (Thermo Fisher Scientific).

Sf900II serum-free medium (Thermo Fisher Scientific).

AcTEV protease (Thermo Fisher Scientific).

Recombinant murine GM-CSF (PeproTech).

Recombinant murine IL-4 (PeproTech).

FuGENE transfection reagent (Promega).

Antibodies

Anti-Hsp70 antibody (SPA-810, Assay Designs Inc.)

Anti-Myc antibody (clone 9E10, Stratagene)

Chromophores

Alexa Fluor 488 Microscale Protein Labeling Kit (Thermo Fisher Scientific)

shRNA to SREC-I

MISSION™ shRNA plasmids (shRNA) were purchased (Sigma-Aldrich, St. Louis, MI), and the *Lentivirus* generation and transduction were performed according to the manual of ViralPower™ Lentiviral Expression Systems (Invitrogen).

3 Methods

3.1 Screening for HSP Receptors

We have screened receptors for HSP binding in the context of cell surface expression, by expressing candidate receptors in cells null for Hsp70 binding. We screened several primary cells and established cell lines for lack of capacity to bind to Hsp70 (Table 1). Maintenance of established cell lines was previously described [37]. Human Umbilical Vein Endothelial Cells (HUVEC) were maintained in Endothelial Basal Medium-2 (EBM-2) supplemented with Clonetics™ SingleQuot® (Cambrex/Biowittaker). Isolation of peritoneal macrophages was carried out as previously described [38]. Briefly, peritoneal macrophages were isolated from 6-10-week-old C57BL/6 background mice. The mice were injected intraperitoneally with 3 ml of thioglycollate, and after 4 days peritoneal exudate cells were harvested by lavage with 10 ml of RPMI 1640 and cultured in RPMI 1640 medium supplemented with 10% heat-inactivated FBS and penicillin-streptomycin. Bone marrow-derived dendritic cells (BMDCs) were generated from the femur and tibiae of C57BL/6 mice. The bone marrow was flushed out and cultured in RPMI 1640 supplemented with 10% heat-inactivated fetal bovine serum (FBS), 40 ng/ml GM-CSF and 40 ng/ml of IL-4 for 6 days. On day 3, a third of the media was replaced by fresh growth media.

3.2 Alexa Fluor 488-Labeled Purified HSP70 Preparation

Human melanoma cells A375-MEL or mouse MISA cells were used as starting material for Hsp70 preparation because high endogenous HSC70 and/or HSP70 levels were detected in these cell types (J. Theriault and SK Calderwood, unpublished). In

Table 1
Binding to Hsp70

Cell type	Species	Hsp70 binding
THP1, monocyte	Human	+
RAW264.7, macrophage	Mouse	++
Primary macrophage	Mouse	++
Primary dendritic cell	Mouse	++
HEK293 embryonic kidney	Human	+
Vascular endothelial	Human	++
PC-3, prostate carcinoma	Human	+
HeLa, cervical carcinoma	Human	++
Hela S3 cervical carcinoma	Human	+
MCF7, mammary cancer	Human	+
IMR90, fibroblast	Human	–
K562, pluripotent leukemia	Human	–
A375, melanoma	Human	–
CHO K1, ovarian cells	Chinese hamster	–

addition, for some experiments, we used minced mouse liver as an abundant source of Hsp70. The Hsp70 purification protocol was based on previous studies [39]. Briefly, a 10 ml cell pellet of tumor cells or minced liver was homogenized in 40 ml hypotonic buffer by Dounce homogenization. The homogenate was then spun at 10,000xg for 30 min, and the supernatant was further treated for 60 min at 100,000xg. The sample buffer was changed to buffer D using a PD-10 desalting column (Amersham Biosciences). The material was then applied directly to a 5 ml ADP–agarose column pre-equilibrated with buffer D. Hsp70 was eluted from the ADP–agarose column with 3 mM ADP in buffer D. The sample buffer was then changed to FPLC buffer with PD-10 column. The supernatant was applied to a DEAE anion exchange column equilibrated with FPLC buffer (Amersham Biosciences). Hsp70 was eluted with the FPLC buffer containing 150 mM NaCl. Protein concentrations were determined by *Bradford* assay. Purified Hsp70 was then labeled with fluorophore Alexa Fluor 488 according to the manufacturer's instructions (Thermo Fisher Scientific). Intactness and purity of the labeled Hsp70 were checked by SDS-PAGE and Coomassie blue stain, and the presence of Hsp70 in the preparation was confirmed by Western blotting using a mouse monoclonal antibody specific for HSP70.

3.3 Alexa Fluor 488-Labeled Purified Hsp90 Preparation

Hsp90 alpha DNA was prepared by PCR amplification from the pET23 plasmid and cloned into pDEST™10 (*see* **Notes 1–3**). Overexpression of Hsp90 alpha in Sf9 insect cells was achieved according to the BAC-to-BAC transfection kit protocol of *Invitrogen* (Thermo Fisher Scientific). The transfer vector was transformed into DH10BAC competent cells containing bacmid DNA. Later, colonies containing recombinant bacmid were identified and prepared. The bacmid DNA was then transfected into Sf9 cells using Cellfectin II Reagent (Thermo Fisher Scientific) to make recombinant baculovirus according to the manufacturer's protocol. Sf9 cells were grown in Sf900II serum-free medium (Thermo Fisher Scientific) supplemented with 100 U/ml penicillin–streptomycin and 2 mM of L-glutamine in suspension cultures with continuous shaking at 150 rpm at 27 °C in a non-humidified environment. The insect cultures were infected in the log phase of growth with recombinant baculovirus. Cells were harvested 48 h postinfection and washed with Hank's buffered saline solution, and protein was purified using the Ni–NTA purification system according to the manufacturer's protocol (Invitrogen). AcTEV protease (Thermo Fisher Scientific) was used to cleave the 6× His tag from the fusion protein generated using pDEST™10 after purifying the recombinant protein on a nickel-chelating resin. Purified Hsp90 was then labeled with Alexa Fluor 488 as above. Intactness and purity of the labeled Hsp90 were checked by SDS-PAGE and Coomassie blue stain, and the presence of Hsp90 in the preparation was confirmed by immunoblot.

3.4 Alexa Fluor 488-Labeled Purified DnaK Preparation

Recombinant prokaryotic Hsp70 (DnaK) from *M. tuberculosis* was expressed and purified in a yeast system using *P. pastoris* (Gene ID: 885946). The gene was synthesized by DNA 2.0, followed by insertion into a pJ912 vector. Then, the construct was transformed into *Pichia* competent cells for protein expression. After screening with methanol induction, high-expression colonies were selected and scaled up to a 6 liter yeast expression system. Recombinant DnaK was purified with a Pharmacia FPLC system (GE Lifesciences). All chromatography runs were performed under clean 4 °C conditions. Protein preparations were quantitated with the Pierce BCA Protein assay kit (Thermo Fisher Scientific). Purified recombinant DnaK was then labeled with Alexa Fluor 488 as described for the other HSP proteins. Intactness and purity of the fluorescent-labeled DnaK were checked by SDS-PAGE, Coomassie blue stain, and Western blot.

3.5 HSP Binding Assay

Cells were first screened by binding to Alexa Fluor 488-labeled HSP in vitro and analyzed by flow cytometry. In addition, 2×10^5 non-trypsinized cells were washed twice in PFNC buffer and incubated with 150 nM Alexa Fluor 488-labeled BSA (negative control), Hsp70, Hsp90, or DnaK for 30 min on ice with gentle shaking. The cells were washed twice in PFNC buffer and Alexa 488-labeled protein binding was monitored by flow cytometry (BD LSR II).

Experiments utilizing flow cytometry were next confirmed by confocal fluorescence microscopy. Alexa Fluor 488-conjugated BSA, Hsp70, Hsp90, or DnaK were prepared as above. Cells were labeled with ligands for 20–30 min on ice. Cells were later washed with ice-cold stripping buffer to remove unbound HSP. Cells were then fixed with 4% paraformaldehyde and permeabilized with 0.1% TritonX-100. Fluorescence was then visualized using a Zeiss 510 confocal microscope (Carl Zeiss). Fluorophores were visualized using 488 nm excitation and a band-pass 505–530 emission filter for Alexa Fluor 488. Images were taken using a 63× numerical aperture (NA) 1.4 oil immersion objective lens.

To assay individual receptors for HSP binding, we selected CHO-K1 cells as null for HSP binding in the wild-type state. Cells were then transfected with expression plasmids for individual receptors following the protocol used for the study of LOX-1. 2.5×10^5 CHO-K1 cells that were transiently transfected with 5 ug of empty vector (pCDNA3) or pCDNA3 plasmids encoding Myc-tagged LOX-1, for 48 h using the SuPerfect (Qiagen) or FuGENE (Promega) transfection reagents according to the manufacturer's instructions. The expression of recombinant proteins was analyzed after transfection by SDS-PAGE and immunoblot with the mouse monoclonal anti-Myc antibody (clone 9E10, Stratagene). Cell lines were maintained by selection with neomycin and checked routinely for the expression of Myc-tagged product. In addition, we examined the cell surface location of the candidate receptors using antibodies to the extracellular domains of such proteins.

Using this approach, we have examined several candidate receptors. As previous studies had suggested a role for LOX-1 in immune responses to Hsp70, we began with the study of this protein and have confirmed that Hsp70 binds avidly to CHO--LOX-1 cells (Table 2) [30, 40]. Interestingly, DnaK (prokaryotic Hsp70) can also bind to LOX-1. This receptor has been assigned to at least two distinct protein families, the c-type lectins, and the scavenger receptors (SR) [40, 41]. C-type lectin receptors (CLR) are a large family of receptors characterized by the possession of a common binding domain – the Ca++ – dependent carbohydrate-binding motif (CTLD) [42, 43]. Binding to protein ligands can be inhibited using hapten sugars that differ between different CLR family members. In the case of LOX-1, fucoidan is a hapten sugar that interacts with its CTLD domain and inhibits Hsp70 binding to LOX-1 [30]. Scavenger receptors (SR) have been studied mostly in endothelial cells but are also expressed in dendritic cells and macrophages (J. Gong, A Murshid and SK Calderwood, in preparation). SRs are a group of proteins that are clustered according to their function in cells—their ability to interact with chemically modified

Table 2
Candidate HSP receptors

Receptor	Type	Expressed in	Hsp70	Hsp90	DnaK
TLR2	Signaling	APCs, etc.	–	ND	–
TLR4	Signaling	APCs, etc.	–	ND	–
CD14	Signaling	APCs, etc.	–	ND	ND
CD40	Signaling	APCs	–	ND	ND
CD91	Internalizing	Many	–	ND	–
LOX-1	Scavenger/CTL	APCs, endothelium	+++	++	+++
DC-SIGN	Scavenger/CTL	APCs	–	ND	–
Dectin-1	Scavenger	APCs	–	ND	+
CLEC-1	Scavenger	APCs	–	ND	ND
CLEC2	Scavenger	APCs	–	ND	ND
SREC-1	Scavenger	APCs	++	++	+
FEEL-1	Scavenger	APCs	++	++	–
SRA	Scavenger	APCs, endothelium, muscle	ND	ND	–
NKG2A	CTL	NK	++	ND	ND
NKG2C	CTL	NK	++	ND	ND
NKG2D	CTL	NK, T cell	++	ND	ND
mMGL2	Signaling	APCs	ND	ND	–
Siglec-E	Signaling	APCs, etc	++	ND	++

Binding to receptors was assayed in CHO transfectants with two exceptions, which are DC-SIGN, which was in K-562 (also HSP binding null), and the TLR, which was in HEK293 cells. APCs, antigen-presenting cells; ND, not done. The ability to compete with 25-fold XS cold HSP70 is indicated in the last column

proteins in the extracellular fluid, as exemplified by binding oxidized low-density lipoprotein and acetylated or maleylated bovine serum albumin (BSA) [44–47]. The binding of HSP to SR can initially be screened by competition assay using known SR binding proteins such as maleylated BSA, oxidized LDL, acetylated LDL, apolipoprotein B, or polyanions such as polyinosine [30]. In this approach, one ligand is labeled (fluorescently for flow cytometry), and the other one remains unlabeled. The competition assay is done with a constant concentration of labeled HSP and varying concentrations of unlabeled ligands such as mBSA and AcLDL. The basis for this interaction as well as HSP binding is not well understood as the extracellular domains of individual SR are highly divergent [47].

We screened members of these receptor families and confirmed that Hsp70 binds to LOX-1 and two other SR family members,

SREC-I and FEEL-1, when expressed in CHO-K1 (Table 2) [48–52]. DnaK was also found to bind to SREC-I, but not to FEEL-1. Others have subsequently shown that another SR family member—scavenger receptor A can interact with HSP family members which however fail to bind CD36, MARCO, and CLA-1 [30, 53]. Hsp70 also binds to other members of the CLR family such as NKG2D that are expressed in natural killer cells (Table 2) [48]. However, in our studies, Hsp70 failed to bind to a number of major CLR family members including Dectin-1, DC-SIGN, CLEC-1, and CLEC-2 (Table 2) [48]. Some studies also indicate a role for the LDL receptor-related protein (LRP) or CD91 in HSP binding [54]. LRP/CD91 contains four clusters of binding repeats that mediate association with at least 30 different ligands including apolipoprotein E, $\alpha_2$macroblobulin, pro-urokinase, and others [55, 56]. Most ligands bind specifically to two of these clusters of binding repeats within domains II and IV [55]. However, we could not detect Hsp70 binding to either domain II or IV when expressed in CHO-K1 cells, and in addition, LRP null cells appeared to bind Hsp70 as well as wild-type cells (Table 2) [29, 48, 57]. Endocytosis of the molecular chaperone calreticulin was also not decreased in CD91$^{-/-}$ cells casting some doubt on CD91 as a universal endocytic receptor for HSP [58]. By contrast scavenger receptor SRA has been shown to be required for a large proportion of gp96 and calreticulin uptake [28]. In the case of Hsp90, neither fucoidan (a LOX-1 agonist) nor α2-macroglobulin (LRP/CD91 agonist) was able to block the representation of a peptide bound to Hsp90 [59]. Nonetheless, others have shown that inactivation of CD91 can lead to loss of antigen representation ability in cells exposed to gp96/peptide complexes [60]. The large heat shock proteins Hsp110 and Grp170, which have potent immune properties, can bind to SRA and SREC-I and there is also some evidence for binding to LRP/CD91 [31]. There is additional evidence indicating a possible role for LRP/CD91 in Hsp70 binding to macrophages [30]. Inhibition by $\alpha_2$macroblobulin competition has often been used as a criterion for HSP binding to LRP/CD91 [31]. In addition, our unpublished experiments indicate that levels of SREC-I and LOX-1 are very low in unstimulated murine macrophages in which CD91 may play a significant role. The usage of HSP receptors could thus vary with the nature of the chaperone ligand, HSP posttranslational modifications (PTMs), immune cell type, and the activation state of the cell. Our studies in vivo indicate that TLR signaling is essential for SREC-I- expressing DC to traffic to afferent lymph nodes after vaccination with the Hsp70 vaccine [36]. Hsp70 itself can induce SREC-I expression in TLR pathway-proficient murine DC suggesting a feed-forward mechanism in which Hsp70 induces its own receptors (SR) and amplifies the immune effects of Hsp70-peptide complexes [36]. There are also indications that Hsp70 activates

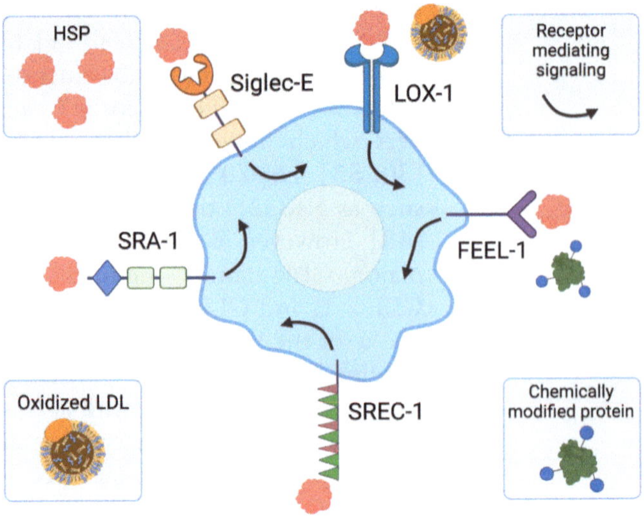

Fig. 1 Hsp70 receptors

signaling receptors such as TLR2, TLR4, and CD40 and may be involved in inducing inflammation and innate immunity [10, 11, 61]. However, using the CHO transfection system described above, we were unable to confirm the direct binding of Hsp70 to these molecules (Table 2). At least in the case of TLR2, indirect activation of this receptor downstream of both LOX-1 and SREC-I is observed after receptor binding to bacterial protein OMP1 [62]. Recently, Hsp70 was described to bind to members of the sialic acid-binding immunoglobulin-like lectins (Siglec) members [63]. Hsp70 binding to the human Siglec-5 dampened inflammatory responses while binding to Siglec-14 triggered an inflammatory pathway in monocytes [64]. Using our CHO-transfected system, we have found that both Hsp70 and DnaK can bind to Siglec-E—one of the main murine Siglec family members (Table 2).

Figure 1 shows a cartoon indicating that at least 5 HSP receptors exist and could potentially be co-expressed in a single cell. In addition, a range of other ligands can interact with such receptors. The existence of multiple receptors and their potential interaction, therefore, complicates the interpretation of experiments probing the function of extracellular HSP. This difficulty is exacerbated by the findings that, while SREC-I and LOX-1 have pro-immune functions, SRA-1 [30, 36, 53, 65] and Siglec-E [66, 67] appears to be inhibitory to the immune response by inhibiting the activity of TLR4. The receptors may thus have both additive and confounding effects. We recently described that Hsp90 induces the scavenger receptor Marco as well as Siglec-E in microglia [68]. Thus, HSP can both bind to and remodel the receptors on the cell surface, adding another layer of complexity to their role in shaping immune responses.

With this plethora of receptors, the nature of the HSP–receptor interaction is still not fully understood. However, the crystal structure of ligand-bound LOX-1 indicates that the ligand (oxidized LDL) binding surface is hydrophobic except for a basic spine composed of arginine residues essential for ligand binding [69, 70]. These positively charged arginine residues together with the hydrophobic residues appear to confer the specificity of LOX-1 for negatively charged lipids and lipoproteins [69]. LOX-1 binds to its ligands as a homodimer with an intramolecular disulfide bond [70]. It is not clear to what degree the ligand-binding properties of SREC-I and FEEL-1 resemble those of LOX-1. The extracellular domains of these two SRs do not contain CTLD but consist mostly of multiple EGF-like repeats [71]. These repeat domains have a length of approximately 40 amino acids and are characterized by a conserved arrangement of six cysteine residues found in EGF itself [72, 73]. It has been shown that at least four tandem repeats of EGF-like regions are required for the FEEL-1 homolog FEEL-2 (STABILIN-2) to bind the acidic lipid phosphatidylserine [74]. Therefore, much is left to be learned regarding the specificity and sequence requirements for HSP binding to these candidate receptors.

3.6 Studying HSP–SREC-I Interaction In Vivo

As we currently have not developed SREC-I knockout mice, we took the approach of knocking down SREC-I in dendritic cells by shRNA. Our studies had indicated that in mice with TLR2/TLR4 double knockdown, responses to the Hsp70 vaccine (Hsp70-FC) were strongly inhibited [36, 75]. We utilized this finding to study the significance of SREC-I in responses to Hsp70 in vivo. To address this issue, we first determined whether the decreased immunity in TLR knockout cells could be reversed by the adoptive transfer of wild-type DCs (WT DC) into *Tlr2*$^{-/-}$*Tlr4*$^{-/-}$ mice [36]. Indeed, the transfer of Hsp70-FC pulsed DC was able to induce immunity to tumors in naïve WT mice and support tumor-specific cytotoxicity [36]. We next showed that immunization of *Tlr2*$^{-/-}$*Tlr4*$^{-/-}$ mice with WT DC that had been pulsed with Hsp70-FC also increased the cytotoxic lymphocyte (CTL) activity and CTL frequency of such mice [36]. This implies that WT donor DCs were able to compensate for the endogenous, inactive DCs in *Tlr2*$^{-/-}$*Tlr4*$^{-/-}$ mice. Groups of such knockout mice were next immunized with DC generated from WT mice that had been infected with SREC-I shRNA or control shRNA constructs and then pulsed with Hsp70-FC. SREC-I knockdown inhibited the ability of WT DC to compensate for TLR knockout indicating that this is a viable approach for studying SREC-I function in vivo. These experiments, therefore, suggested that SREC-I is essential for antigen presentation by DC exposed to Hsp70.PC. These findings were confirmed by assessing the frequency of antigen-specific T cells induced using an MHC-class-I/peptide tetramer (iTAg). The 8-mer peptide (SAPDTRPA) is a dominant epitope from MUC1 that binds to C57BL/6 MHC class I, H-$2K^b$ [76]. MUC1 is one of the most prominent tumor antigens present in the Hsp70-FC vaccine [36]. The MUC1–8 iTAg was used to identify and assess the tetramer-positive T cells. The numbers of MUC1–8 iTAg-positive T cells from *Tlr2*$^{-/-}$*Tlr4*$^{-/-}$ immunized with SREC-I knockdown DCs were significantly decreased compared with those from mice transferred with DCs infected with the control virus. Immunization of *Tlr2*$^{-/-}$*Tlr4*$^{-/-}$ mice either with Hsp70-FC pulsed DCs after SREC-I knockdown or after infection with the control virus resulted in 1.45% and 3.51% CD8 T cells positive for MUC1–8 indicating that SREC-I plays an important role in the induction of antigen-specific T cells by Hsp70-based vaccines.

3.7 shRNA Directed Against SREC-I

MISSION™ shRNA plasmids (shRNA) were purchased (Sigma-Aldrich, St. Louis, MI) and the *Lentivirus* generation and transduction were performed according to the manual of ViralPower™ Lentiviral Expression Systems (Invitrogen). The effectiveness of shRNA for knockdown of murine SREC-I (five clones: TRCN0000067873; TRCN0000067874; TRCN0000067875; TRCN0000067876 and TRCN0000067877) was examined through real-time PCR and immunoblot. The most effective construct for mSREC-I mRNA knockdown (TRCN0000067875) was used. The *Lentivirus* (insert sequence was 5'–CCGCAGGTATG-CACGCGT-3′ which does not target any mouse genes but will activate RISC and the RNAi pathway in the cells) was used as a negative control.

Six-day-old DCs were collected, purified, and placed in 96-well round bottom plates with 1×10^6 cells per well for O/N culture in a medium containing GM-CSF and IL-4. On the second day, half of the medium (100 μl) was removed, and 100 μl *Lentivirus* supernatant (1 virus: 1 DC) was added. After 20 h, 150 μl of medium from each well was replaced by fresh GM-CSF + IL-4 medium for additional culture. DC infected with SREC-I shRNA, control constructs, or uninfected DC were collected for SREC1 expression and T cell stimulation assay. Significant SREC-I knockdown at the protein level was achieved in 90–100 h.

4 Notes

1. *Purified HSP.* Most heat shock proteins are encoded by multigene families. This is particularly true for Hsp70 in which there are, depending on species, at least 12 closely related gene family members each of which may bind to cells with differing affinities [42]. This complication could be remedied using recombinant proteins. However, if recombinant proteins are

produced in *E. coli*, great care must be taken to remove contaminating PAMPs such as lipopolysaccharides or peptidoglycans [12]. We have prepared recombinant HSP using the baculovirus/sf9 system to avoid these problems. However, LPS in particular can be introduced into protein preparations from contaminated glassware, and we have been very careful to assay all batches of HSP by the *Limulus* assay.

2. HSP bind with only moderate affinity to the receptors uncovered so far, and these are promiscuous in terms of binding partners. In most cases, the structural basis for HSP–receptor binding is not known. Thus, much needs to be learned regarding HSP–receptor interaction. Future in vitro studies should address questions about the three-dimensional structure of HSP–receptor complexes and the exact affinities of the interactions.
3. *Many* cells express multiple HSP receptors and great care must be taken to isolate the properties of individual receptors. It seems likely that other yet unknown receptors exist. Also, how these receptors interact affecting the binding and signaling triggered by HSP still needs to be determined.

References

1. Calderwood SK, Mambula SS, Gray PJ, Theriault JR (2007) Extracellular heat shock proteins in cell signaling. FEBS Lett 581:3689–3694
2. Calderwood SK, Gong J, Murshid A (2016) Extracellular HSPs: the complicated roles of extracellular in immunity. Front Immunol 7: 159
3. Pockley AG (2002) Heat shock proteins, inflammation, and cardiovascular disease. Circulation 105:1012–1017
4. Pockley AG, Shepherd J, Corton JM (1998) Detection of heat shock protein 70 (Hsp70) and anti-Hsp70 antibodies in the serum of normal individuals. Immunol Investig 27: 367–377
5. Mambula SS, Calderwood SK (2006) Heat induced release of Hsp70 from prostate carcinoma cells involves both active secretion and passive release from necrotic cells. Int J Hyperth Off J Eur Soc Hyperthermic Oncol North Am Hyperth Gr 22:575–585
6. Mambula SS, Calderwood SK (2006) Heat shock protein 70 is secreted from tumor cells by a nonclassical pathway involving lysosomal endosomes. J Immunol 177:7849–7857
7. Matzinger P (2002) The danger model: a renewed sense of self. Science 296:301–305
8. Vabulas RM et al (2001) Endocytosed HSP60s use toll-like receptor 2 (TLR2) and TLR4 to activate the toll/interleukin-1 receptor signaling pathway in innate immune cells. J Biol Chem 276:31332–31339
9. Vabulas RM et al (2002) HSP70 as endogenous stimulus of the toll/interleukin-1 receptor signal pathway. J Biol Chem 277:15107–15112
10. Asea A et al (2002) Novel signal transduction pathway utilized by extracellular HSP70: role of toll-like receptor (TLR) 2 and TLR4. J Biol Chem 277:15028–15034
11. Asea A et al (2000) HSP70 stimulates cytokine production through a CD14-dependant pathway, demonstrating its dual role as a chaperone and cytokine. Nat Med 6:435–442
12. Henderson B et al (2010) Caught with their PAMPs down? The extracellular signalling actions of molecular chaperones are not due to microbial contaminants. Cell Stress Chaperones 15:123–141
13. Singh-Jasuja H et al (2000) Cross-presentation of glycoprotein 96-associated antigens on major histocompatibility complex class I molecules requires receptor-mediated endocytosis. J Exp Med 191:1965–1974
14. Srivastava P (2002) Interaction of heat shock proteins with peptides and antigen presenting

cells: chaperoning of the innate and adaptive immune responses. Annu Rev Immunol 20: 395–425
15. Rock KL (2003) The ins and outs of cross-presentation. Nat Immunol 4:941–943
16. Murshid A, Gong J, Calderwood SK (2010) Heat shock protein 90 mediates efficient antigen cross presentation through the scavenger receptor expressed by endothelial cells-I. J Immunol 185:2903–2917
17. Murshid A, Gong J, Calderwood SK (2014) Hsp90-peptide complexes stimulate antigen presentation through the class II pathway after binding scavenger receptor SREC-I. Immunobiology 219:924–931
18. Multhoff G (2002) Activation of natural killer cells by heat shock protein 70. Int J Hyperth Off J Eur Soc Hyperthermic Oncol North Am Hyperth Gr 18:576–585
19. Multhoff G, Hightower LE (1996) Cell surface expression of heat shock proteins and the immune response. Cell Stress Chaperones 1: 167–176
20. Shevtsov M et al (2019) Ex vivo Hsp70-activated NK cells in combination with PD-1 inhibition significantly increase overall survival in preclinical models of glioblastoma and lung cancer. Front Immunol 10:454
21. Borges TJ et al (2013) Extracellular Hsp70 inhibits pro-inflammatory cytokine production by IL-10 driven down-regulation of C/EBPβ and C/EBPδ. Int J Hyperth 29:455–463
22. Borges TJ et al (2018) March1-dependent modulation of donor MHC II on CD103+ dendritic cells mitigates alloimmunity. Nat Commun 9:3482
23. Borges TJ et al (2012) The anti-inflammatory mechanisms of Hsp70. Front Immunol 3:1–12
24. van Herwijnen MJC et al (2012) Regulatory T cells that recognize a ubiquitous stress-inducible self-antigen are long-lived suppressors of autoimmune arthritis. Proc Natl Acad Sci 109:14134–14139
25. Wendling U et al (2000) A conserved mycobacterial heat shock protein (hsp) 70 sequence prevents adjuvant arthritis upon nasal administration and induces IL-10-producing T cells that cross-react with the mammalian self-hsp70 homologue. J Immunol 164:2711–2717
26. Van Eden W, Van Der Zee R, Prakken B (2005) Heat-shock proteins induce T-cell regulation of chronic inflammation. Nat Rev Immunol 5: 318–330
27. Berwin B, Delneste Y, Lovingood RV, Post SR, Pizzo SV (2004) SREC-I, a type F scavenger receptor, is an endocytic receptor for calreticulin. J Biol Chem 279:51250–51257
28. Berwin B et al (2003) Scavenger receptor-a mediates gp96/GRP94 and calreticulin internalization by antigen-presenting cells. EMBO J 22:6127–6136
29. Binder RJ, Han DK, Srivastava PK (2000) CD91: a receptor for heat shock protein gp96. Nat Immunol 1:151–155
30. Delneste Y et al (2002) Involvement of LOX-1 in dendritic cell-mediated antigen cross-presentation. Immunity 17:353–362
31. Facciponte JG, Wang X-Y, Subjeck JR (2007) Hsp110 and Grp170, members of the Hsp70 superfamily, bind to scavenger receptor-a and scavenger receptor expressed by endothelial cells-I. Eur J Immunol 37:2268–2279
32. Gross C, Hansch D, Gastpar R, Multhoff G (2003) Interaction of heat shock protein 70 peptide with NK cells involves the NK receptor CD94. Biol Chem 384:267–279
33. Kettner S et al (2007) EWI-2/CD316 is an inducible receptor of HSPA8 on human dendritic cells. Mol Cell Biol 27:7718–7726
34. Sondermann H, Becker T, Mayhew M, Wieland F, Hartl FU (2000) Characterization of a receptor for heat shock protein 70 on macrophages and monocytes. Biol Chem 381: 1165–1174
35. Whittall T et al (2006) Interaction between the CCR5 chemokine receptors and microbial HSP70. Eur J Immunol 36:2304–2314
36. Gong J et al (2009) T cell activation by heat shock protein 70 vaccine requires TLR signaling and scavenger receptor expressed by endothelial Cells-1. J Immunol 183:3092–3098
37. Thériault JR, Mambula SS, Sawamura T, Stevenson MA, Calderwood SK (2005) Extracellular HSP70 binding to surface receptors present on antigen presenting cells and endothelial/epithelial cells. FEBS Lett 579:1951–1960
38. Mambula SS, Sau K, Henneke P, Golenbock DT, Levitz SM (2002) Toll-like receptor (TLR) signaling in response to aspergillus fumigatus. J Biol Chem 277:39320–39326
39. Peng P, Ménoret A, Srivastava PK (1997) Purification of immunogenic heat shock protein 70-peptide complexes by ADP-affinity chromatography. J Immunol Methods 204:13–21
40. Mehta JL, Chen J, Hermonat PL, Romeo F, Novelli G (2006) Lectin-like, oxidized low-density lipoprotein receptor-1 (LOX-1): a critical player in the development of atherosclerosis and related disorders. Cardiovasc Res 69: 36–45

41. Chen M, Masaki T, Sawamura T (2002) LOX-1, the receptor for oxidized low-density lipoprotein identified from endothelial cells: implications in endothelial dysfunction and atherosclerosis. Pharmacol Ther 95:89–100
42. Zelensky AN, Gready JE (2005) The C-type lectin-like domain superfamily. FEBS J 272: 6179–6217
43. Drickamer K (1999) C-type lectin-like domains. Curr Opin Struct Biol 9:585–590
44. Adachi H, Tsujimoto M (2006) Endothelial scavenger receptors. Prog Lipid Res 45:379–404
45. Rigotti A (2000) Scavenger receptors and atherosclerosis. Biol Res 33:97–103
46. van Berkel TJC et al (2005) Scavenger receptors: friend or foe in atherosclerosis? Curr Opin Lipidol 16:525–535
47. Krieger M (1997) The other side of scavenger receptors: pattern recognition for host defense. Curr Opin Lipidol 8:275–280
48. Thériault JR, Adachi H, Calderwood SK (2006) Role of scavenger receptors in the binding and internalization of heat shock protein 70. J Immunol 177:8604–8611
49. Adachi H, Tsujimoto M (2002) Characterization of the human gene encoding the scavenger receptor expressed by endothelial cell and its regulation by a novel transcription factor, endothelial zinc finger protein-2. J Biol Chem 277:24014–24021
50. Politz O et al (2002) Stabilin-1 and -2 constitute a novel family of fasciclin-like hyaluronan receptor homologues. Biochem J 362:155–164
51. Murshid A, Borges TJ, Calderwood SK (2015) Emerging roles for scavenger receptor SREC-I in immunity. Cytokine 75:256–260
52. Murshid A, Borges TJ, Lang BJ, Calderwood SK (2016) The scavenger receptor SREC-I cooperates with toll-like receptors to trigger inflammatory innate immune responses. Front Immunol 7:226
53. Wang X-Y, Facciponte J, Chen X, Subjeck JR, Repasky EA (2007) Scavenger receptor-a negatively regulates antitumor immunity. Cancer Res 67:4996–5002
54. Basu S, Binder RJ, Ramalingam T, Srivastava PK (2001) CD91 is a common receptor for heat shock proteins gp96, hsp90, hsp70, and calreticulin. Immunity 14:303. https://doi.org/10.1016/S1074-7613(01)00111-X
55. Herz J, Strickland DK (2001) LRP: a multifunctional scavenger and signaling receptor. J Clin Invest 108:779–784
56. Newton CS et al (2005) Platelet-derived growth factor receptor-beta (PDGFR-beta) activation promotes its association with the low density lipoprotein receptor-related protein (LRP): evidence for co-receptor function. J Biol Chem 280:27872–27878
57. Obermoeller-McCormick LM et al (2001) Dissection of receptor folding and ligand-binding property with functional minireceptors of LDL receptor-related protein. J Cell Sci 114:899–908
58. Walters JJ, Berwin B (2005) Differential CD91 dependence for calreticulin and pseudomonas exotoxin-a endocytosis. Traffic 6:1173–1182
59. Kurotaki T et al (2007) Efficient cross-presentation by heat shock protein 90-peptide complex-loaded dendritic cells via an endosomal pathway. J Immunol 179:1803–1813
60. Binder RJ, Srivastava PK (2005) Peptides chaperoned by heat-shock proteins are a necessary and sufficient source of antigen in the cross-priming of CD8+ T cells. Nat Immunol 6:593–599
61. Becker T, Hartl F-U, Wieland F (2002) CD40, an extracellular receptor for binding and uptake of Hsp70-peptide complexes. J Cell Biol 158:1277–1285
62. Jeannin P et al (2005) Complexity and complementarity of outer membrane protein a recognition by cellular and humoral innate immunity receptors. Immunity 22:551–560
63. Calderwood SK, Murshid A (2015) Siglecs take a TOLL on inflammation: deciphering the Hsp70 riddle. EMBO J 34:2733–2734. https://doi.org/10.15252/embj.201593172
64. Fong JJ et al (2015) Immunomodulatory activity of extracellular Hsp70 mediated via paired receptors Siglec-5 and Siglec-14. EMBO J 34: 2775–2788. https://doi.org/10.15252/embj
65. Kottke T et al (2009) Antitumor immunity can be uncoupled from autoimmunity following heat shock protein 70-mediated inflammatory killing of normal pancreas. Cancer Res 69: 7767–7774
66. Spence S et al (2015) Targeting Siglecs with a sialic acid-decorated nanoparticle abrogates inflammation. Sci Transl Med 7:303ra140–303ra140
67. Chen GY et al (2014) Broad and direct interaction between TLR and Siglec families of pattern recognition receptors and its regulation by Neu1. elife 3:e04066
68. Okusha Y et al (2022) Extracellular Hsp90α stimulates a unique innate gene profile in microglial cells with simultaneous activation of Nrf2 and protection from oxidative stress.

Cell Stress Chaperones 27:461. https://doi.org/10.1007/s12192-022-01279-9

69. Sawamura T (2005) LOX-1 unlocked. Structure (London, England: 1993) 13:834–835
70. Ohki I et al (2005) Crystal structure of human lectin-like, oxidized low-density lipoprotein receptor 1 ligand binding domain and its ligand recognition mode to OxLDL. Structure 13:905. https://doi.org/10.1016/j.str.2005.03.016
71. Plüddemann A, Neyen C, Gordon S (2007) Macrophage scavenger receptors and host-derived ligands. Methods 43:207–217
72. Appella E, Weber IT, Blasi F (1988) Structure and function of epidermal growth factor-like regions in proteins. FEBS Lett 231:1–4
73. Shibata M et al (2004) Type F scavenger receptor SREC-I interacts with advillin, a member of the gelsolin/villin family, and induces neurite-like outgrowth. J Biol Chem 279:40084–40090
74. Park S-Y, Kim S-Y, Jung M-Y, Bae D-J, Kim I-S (2008) Epidermal growth factor-like domain repeat of stabilin-2 recognizes phosphatidylserine during cell corpse clearance. Mol Cell Biol 28:5288–5298
75. Enomoto Y et al (2006) Enhanced immunogenicity of heat shock protein 70 peptide complexes from dendritic cell-tumor fusion cells. J Immunol 177:5946–5955
76. Apostolopoulos V et al (2002) Crystal structure of a non-canonical low-affinity peptide complexed with MHC class I: a new approach for vaccine design. J Mol Biol 318:1293–1305

Chapter 16

A Novel Heat Shock Protein 70-Based Vaccine Prepared from DC Tumor Fusion Cells: An Update

Desheng Weng, Stuart K. Calderwood, and Jianlin Gong

Abstract

We have developed an enhanced molecular chaperone-based vaccine through rapid isolation of Hsp70 peptide complexes after the fusion of tumor and dendritic cells (Hsp70.PC-F). In this approach, the tumor antigens are introduced into the antigen-processing machinery of dendritic cells through the cell fusion process, and thus we can obtain antigenic tumor peptides or their intermediates that have been processed by dendritic cells. Our results show that Hsp70.PC-F has increased immunogenicity compared to preparations from tumor cells alone and therefore constitutes an improved formulation of the chaperone protein-based tumor vaccine.

Key words Heat shock proteins70 (Hsp70), Dendritic cells (DC), Cell fusion, Extraction of Hsp70 peptide complexes (Hsp70.PC), Tumor vaccine

1 Introduction

The heat shock proteins 70 (Hsp70) family is intrinsic to cellular life, permitting client proteins to perform essential enzymatic, signaling, and structural functions within the tightly crowded milieu of the cell and working to avert the catastrophe of protein aggregation during stress [1–5]. There are at least twelve members of the human Hsp70 family, including proteins expressed in the cytoplasm, endoplasmic reticulum, and mitochondria [1, 5–7]. For molecular chaperone function, Hsp70 family members are equipped with two major functional domains, including a carboxy-terminal region that binds peptides and denatured proteins and an N-terminal ATPase domain that controls the opening and closing of the peptide binding domain [6, 8]. These two domains play important roles in the functions of Hsp70 in tumor immunity, mediating the acquisition of cellular antigens and their delivery to immune effector cells [9, 10]. Hsp70 expression becomes dysregulated in many types of cancer leading to elevated

Stuart K. Calderwood and Thomas L. Prince (eds.), *Chaperones: Methods and Protocols*, Methods in Molecular Biology, vol. 2693, https://doi.org/10.1007/978-1-0716-3342-7_16,

Hsp70 levels under nonstress conditions that protect emerging cancer cells from the apoptosis that accompanies many steps in transformation but also create an opportunity for vaccine design [7, 11–14].

The molecular chaperone-based tumor vaccine was pioneered by Srivastava et al. who prepared autologous vaccines in mice and human patients with the direct aim of targeting the unique antigens that characterize each neoplasm [15–19]. In this approach, Hsp70 peptide complexes (Hsp70.PC) are isolated from the patients' tumors by affinity chromatography using ATP–agarose and formulations of Hsp70 applied in a multidose regimen. The aim is for Hsp70.PC to facilitate antigen cross-presentation to the patient's T cells through host APC and for the unique mixture of peptides from the individual tumor to induce antitumor immunity. Despite immunologic and clinical responses obtained in a subset of patients with malignant tumors in the early phase I and/or II trials with molecular chaperone GP96.PC (vitespen) purified from patient-derived tumors [19–22], the randomized phase III trials, however, showed mixed results [23, 24].

We have designed an enhanced molecular chaperone-based vaccine, employing rapid isolation of Hsp70 peptide complexes from fusions of tumor and dendritic cells (Hsp70.PC-F). In our animal studies, Hsp70.PC-F vaccines show superior immunological properties such as enhanced induction of CTL against tumor cells and stimulation of DC maturation over counterparts from tumor cells [25, 26]. More importantly, immunization of mice with Hsp70.PC-F resulted in a T cell-mediated immune response including a significant increase of CD8 T cells and induction of effector and memory T cells able to break T cell unresponsiveness to a non-mutated tumor antigen and provide protection of mice against challenge with tumor cells. By contrast, the immune response to vaccination with Hsp70.PC derived from tumor cells alone is muted against such non-mutated tumor antigen. Hsp70.PC-F complexes differed from those derived from tumor cells in a number of key manners, most notably, enhanced association with immunologic peptides. In addition, the molecular chaperone Hsp90 was found to be associated with Hsp70.PC-F as indicated by co-immunoprecipitation, suggesting its ability to carry an increased repertoire of antigenic peptides by the two chaperones. These experiments indicate that Hsp70.PC derived from DC tumor fusion cells have increased their immunogenicity and therefore constitute an improved formulation of chaperone protein-based tumor vaccine.

The rationale for the extraction of Hsp70.PC from DC tumor fusion cells is based on the observation that DC are the most potent antigen-presenting cells [27, 28]. The fusion of DC and tumor cells through chemical [29–46], physical [32, 47–56], or biological [57, 58] means creating a heterokaryon which combines

DC-derived costimulatory molecules, efficient antigen-processing and presentation machinery, and an abundance of tumor-derived antigens. The DC and tumor cells become one hybrid cell sharing a unified cytoplasm. The integration of cytoplasm from DC and tumor cells renders the tumor antigens endogenous to the DC heterokaryon and, therefore, facilitates the entry of tumor antigens into the DC endogenous pathway of antigen-processing and presentation machinery [35, 59, 60]. It is likely that the antigen-processing machinery from DC can sort or select the immunogenic peptides to be processed and presented and work much more efficiently than that from tumor cells, thus increasing the quality and quantity of the HSP-associated complexes.

2 Materials

2.1 Isolation of Tumor Cells from Patient-Derived Solid Sample or Malignant Fluid

DNase (0.1 mg/ml, Sigma-Aldrich, Saint Louis, MO).

Collagenase (1 mg/ml, Worthington Biochemical Corporation, Lakewood, NJ).

Ca^{2+}/Mg^{2+}-free Hanks balanced salt solution (HBSS medium, Mediatech Inc., Manassas, VA).

A sterile 50 μm nylon mesher (Sigma-Aldrich, Saint Louis, MO).

Heat-inactivated human AB serum (Sigma-Aldrich, Saint Louis, MO).

RPMI 1640 medium (Mediatech, Manassas, VA).

L-glutamine (2 mM, Mediatech, Manassas, VA).

Penicillin and streptomycin (100 units/ml and 100 μg/ml) (Mediatech Inc., Manassas, VA).

2.2 Generation of DC from Human Peripheral Blood Monocytes

Ficoll density gradient centrifugation (Ficoll-Paque™ plus, GE Healthcare Bio-Sciences AB, Sweden).

Granulocyte–macrophage colony-stimulating factor (hGM-CSF, 1000 units/ml) (Genzyme, Framingham, MA).

Interleukin-4 (hIL-4, 500 units/ml) (R&D Systems, Minneapolis, MN).

2.3 Preparation of DC Tumor Fusions

Polyethylene glycol (PEG, 50% MW1450) (Sigma-Aldrich, Saint Louis, MO).

2.4 Preparation of Hsp70.PC Extraction from DC Tumor Fusions

Tris-HCl (pH 7.4, 50 mM) (Boston Bioproduct, Ashland, MA).

NaCl (50 mM) (Sigma-Aldrich, Saint Louis, MO).

Nonidet P-40 (NP40, 1%, Sigma-Aldrich, Saint Louis, MO).

Protease inhibitor cocktail tablets, cOmplete Mini (Roche, Mannheim, Germany).

Sodium orthovanadate ($NaVO_4$, 1 mM) (Boston Bioproduct, Ashland, MA).

Antibody against human Hsp70 (5C1A12, developed by ProMab Biotechnologies, Inc., Albany, CA).

Dye Reagent Concentrate for protein assay (Bio-Rad, Hercules, CA).

Protein A Sepharose (GE Healthcare, Waukesha, WI).

Protein G Sepharose (GE Healthcare, Waukesha, WI).

2.5 Measurement of Levels of Endotoxin

Limulus amebocyte lysate (LAL kit, Cambrex Bio Science Inc., Walkersville, MD).

3 Methods

Cell fusion between DC and tumor cells can be achieved through chemical, physical, or biological means. In our laboratory, we use PEG to fuse DC and tumor cells. We have used the following protocol to prepare Hsp70.PC extracts from DC tumor fusion cells (Fig. 1).

3.1 Generation of DC from Human Peripheral Blood Monocytes

DC can be generated from human peripheral blood monocytes (PBMC) derived from patients or healthy donors. We usually use Ficoll to separate PBMC and culture these cells in a medium containing hGM-CSF. The protocol is based on a previously described method [61–63] with modifications:

1. Peripheral blood mononuclear cells (PBMC) obtained from patients or leucopaks are transferred into a 50 ml centrifuge tube and sedimented at low speed.
2. The serum on the top of the tube is collected into a clear tube as a serum for cell culture. The blood cells at the bottom of the tube are resuspended with RPMI 1640 medium without serum (1:2 dilution).
3. The blood cells are gently laid on top of the tube containing a Ficoll density gradient.
4. Tubes are centrifuged at 1800 rpm for 20 min at room temperature.
5. After Ficoll density gradient centrifugation, the cells in the interface layer are collected into another tube with RPMI 1640 and 2% serum.
6. Cells are washed twice with serum-free medium, and the cells are counted.

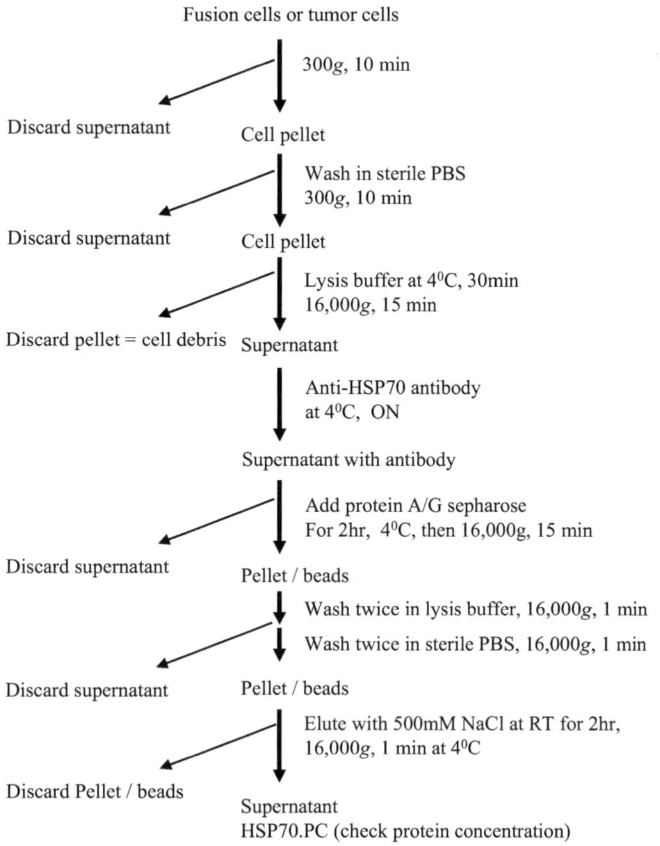

Fig. 1 Procedures for HSP70.PC purification

7. Culture 1 × 10^6 cells /ml in RPMI 1640 containing 5% human serum, 2 mM L-glutamine, 100 U/ml penicillin, and 100 μg/ ml streptomycin for 1 h in a humidified CO_2 incubator.
8. After 1 h culture, gently wash and remove the non-adherent cells. The adherent fraction is cultured in RPMI/AIM-V (1:1) medium with 1% of human serum and 1000 U/ml hGM-CSF and 500 U/ml hIL-4 for 5 days.
9. On day 3 of culture, cell clusters appear. Fresh medium with 1000 U/ml of hGM-CSF is added if the color of the medium becomes yellow.
10. On day 5 of culture, the loosely adherent cell or cell clusters are collected by gently dislodging the cells by pipetting and then the cells are counted (most cells are immature DC).

3.2 Preparation of Tumor Cells

Tumor cells can be either freshly isolated from tumor samples or obtained from vials of frozen cell lines. The method described here is used to isolate and culture tumor cells from patient-derived breast or ovarian cancer sample under a sterile condition.

1. The resected human tumor sample is weighed, minced into small pieces (1–3 mm), and digested in HBSS solution containing 1 mg/ml collagenase, 0.1 mg/ml DNase, 100 U/ml penicillin, and 100 μg/ml streptomycin.
2. The digested tumor tissue is then mashed through a sterile 50-μm nylon mesher under sterile conditions in a tissue culture hood.
3. Cells are washed twice with cold HBSS solution.
4. Single-tumor cell suspensions are obtained by passing through a cell strainer, and the tumor cells are counted.
5. Culture the tumor cells in high glucose DMEM medium containing 10% human serum and antibiotics. Remove the non-adherent dead cells.
6. Incubate the cells at 37 °C for 2 to 3 days. Cells are ready for fusion when they are in the logarithmic phase of growth.

3.3 Cell Fusion

1. DC generated from PBMC are cultured in 1000 U/ml hGM-CSF medium for 5 days.
2. Tumor cells are maintained in DMEM supplemented with 10% heat-inactivated FCS, 2 mM L-glutamine, 100 U/ml penicillin, and 100 μg/ml streptomycin.
3. The DC are mixed with tumor cells at a 10:1 ratio, and the mixture is washed once with serum-free medium followed by low-speed spin (500 rpm) to obtain cell pellets.
4. The mixed cell pellets are gently resuspended in pre-warmed 50% PEG solution (1 ml per 1–5 × 108 cells) for 5 min at room temperature.
5. The PEG solution is diluted by slow addition and mixing of 1, 2, 4, 8, and 16 ml warm serum-free medium within 10 min.
6. The cell pellets obtained after centrifugation at 1350 rpm are resuspended in RPMI 1640 medium supplemented with 10% heat-inactivated FCS, 2 mM L-glutamine, 10 mM nonessential amino acids, 1 mM sodium pyruvate, 10% NCTC 109, 100 U/ml penicillin, 100 μg/ml streptomycin, and 500 U/ml hGM-CSF and further cultured for 5 days.
7. After 5 days, DC tumor fusion cells are loosely adherent to the culture dish, whereas tumor–tumor fusions and unfused tumor cells are attached firmly to the dish. The loosely adherent fusion cells are obtained first by gentle pipetting.
8. The fusion efficiency is determined by the dual expression of tumor antigens such as MUC1 and DC markers (MHC class II molecules or costimulatory molecules).

3.4 Extraction of Hsp70 Peptide Complexes (Hsp70.PC) from DC Tumor Fusion Cell Products

1. DC tumor fusion cells are collected and counted.
2. Resuspend the cell pellets with lysis buffer (50 mM Tris-HCl, pH 8.0, containing 50 mM NaCl, 1% Nonidet P-40, 1 mM PMSF) (1 ml lysis buffer for 2×10^7 cells) on ice for 30 min.
3. Centrifuge the cell lysate at 13,000 rpm for 15 min at 4 °C.
4. After centrifuge, collect the supernatant into a clear tube.
5. Check protein concentration with standard procedure (Bio-Rad Bradford Protein Assay Kit):
   (i) Prepare dye reagent by diluting 1 part Dye Reagent Concentrate with 4 parts distilled, deionized water.
   (ii) Prepare three to five dilutions of a protein standard, which is representative of the protein solution to be tested.
   (iii) Pipette 100 μl of each standard and sample solutions into 5.0 ml of diluted dye reagent.
   (iv) Incubate the test samples at room temperature for at least 5 min. Absorbance will increase over time. Samples should be incubated at room temperature for no more than 15 min.
   (v) Measure the absorbance at 595 nm. Calculate the protein concentration based on the standard curve.
6. The lysates are clarified by centrifugation, and the aqueous phase is collected and incubated with mAb against human Hsp70 at a concentration of 1:100, rotating through overnight at 4 °C.
7. For protein analysis, the immunoprecipitates are dissolved in Laemmli SDS sample buffer (0.1 Tris-HCl, 4% SDS, 20% glycerol, 0.05% bromphenol blue, 5% 2-ME) and analyzed by immunoblotting.

   For binding of the immune complex

   (i) Mix Protein A Sepharose and Protein G Sepharose at a 1:1 ratio followed by washing with lysis buffer once.
   (ii) Spin down at 13,000 rpm for 1 min at 4 °C, discard the supernatant, and resuspend the beads with 250 μl lysis buffer.
   (iii) Pipette 100 μl A/G mixture beads into sample tubes and incubate for 2 h at 4 °C.
   (iv) After incubation, spin down at 13,000 rpm for 1 min at 4 °C. Remove supernatant.
   (v) Wash beads with 0.5 ml of lysis buffer for 5 min (rotate at 4 °C) followed by centrifugation at 13,000 rpm for 1 min at 4 °C.
   (vi) Wash beads with 0.5 ml of sterile PBS for 5 min (rotate at 4 °C) followed by centrifugation at 13,000 rpm for 1 min at 4 °C.

8. After extensive wash with lysis buffer, the immunoprecipitates are eluted with sterile high salt elution buffer.
   (i) Elute the proteins with (500 mM NaCl, 100 μl) at RT for 2 h.
   (ii) Centrifuge at 13,000 rpm for 1 min at 4 °C.
   (iii) Collect the supernatant and measure the protein concentration by Bradford protein assay.
9. The Hsp70.PC preparations are checked by limulus amebocyte lysate assay to ensure minimal contamination with endotoxins, aliquoted into 1.5 ml eppendorf tubes and stored at −80 °C.

4 Notes

4.1 Cell Fusion

The PEG solution is diluted by gradual addition and progressive mixing of 1, 2, 4, 8, and 16 ml warm serum-free medium. The cell pellets obtained after centrifugation at 1350 rpm are resuspended in a medium containing 10% heat-inactivated FCS and GM-CSF. The variable factor for cell fusion is the length of time the cells are exposed to PEG. We have found that there is some difference in the sensitivity of cells to PEG. It is desirable to perform a dose–response test to evaluate the conditions of PEG fusion for each type of tumor cell and to determine the optimal exposure time. Unlike electrofusion, DC tumor fusion by PEG is an active and evolving process, and it is thus likely that the larger the initial contact surface between cells, the faster the integration of these cells. Fusion efficiency is lowest immediately after the fusion process is initiated, and 1 week culture results in more than a ten-fold increase in efficiency (Gong J., unpublished data). In addition, short-term culture will give the fusion cells sufficient time to integrate and display the antigen in the context of MHC molecules.

4.2 Extraction of Hsp70 Peptide Complexes

For protein concentration measurement, Bradford dye reagent absorbance will increase over time. Samples should incubate at room temperature for at least 5 mi, but no more than 15 min. After protein A/G sepharose binding with Hsp70 immunoglobulin, the beads should be gently mixed with lysis buffer to wash off nonspecific interactions. However, the use of a vortex should be avoided since it may break the binding of sepharose beads and immunoglobulins. Background caused by actin contamination can be avoided by adding 10 mM ATP to the lysis buffer. All steps should be performed at 4 °C to reduce proteolysis and denaturation of antigens. This is especially important for the binding step which is typically incubated overnight (or at least 2 h) at 4 °C.

References

1. Lindquist S, Craig EA (1988) The heat shock proteins. Ann Rev Genet 22:631–637
2. Georgopolis C, Welch WJ (1993) Role of the major heat shock proteins as molecular chaperones. Ann Rev Cell Biol 9:601–634
3. Richter K, Haslbeck M, Buchner J (2010) The heat shock response: life on the verge of death. Mol Cell 40(2):253–266
4. Lang BJ et al (1869) Heat shock proteins in cell signaling and cancer. Biochim Biophys Acta Mol Cell Res 2022(3):119187
5. Lang BJ et al (2021) The functions and regulation of heat shock proteins; key orchestrators of proteostasis and the heat shock response. Arch Toxicol 95(6):1943–1970
6. Bukau B, Horwich AL (1998) The Hsp70 and Hsp60 chaperone machines. Cell 92(3): 351–366
7. Tang D et al (2005) Expression of heat shock proteins and HSP messenger ribonucleic acid in human prostate carcinoma in vitro and in tumors in vivo. Cell Stress Chaperones 10:in press
8. Kityk R et al (2015) Pathways of allosteric regulation in Hsp70 chaperones. Nat Commun 6:8308
9. Noessner E et al (2002) Tumor-derived heat shock protein 70 peptide complexes are cross-presented by human dendritic cells. J Immunol 169(10):5424–5432
10. Srivastava PK, Amato RJ (2001) Heat shock proteins: the 'Swiss Army Knife' vaccines against cancers and infectious agents. Vaccine 19(17–19):2590–2597
11. Nylandsted J, Brand K, Jaattela M (2000) Heat shock protein 70 is required for the survival of cancer cells. Ann N Y Acad Sci 926:122–125
12. Cornford PA et al (2000) Heat shock protein expression independently predicts clinical outcome in prostate cancer. Cancer Res 60(24): 7099–7105
13. Clark PR, Menoret A (2001) The inducible Hsp70 as a marker of tumor immunogenicity. Cell Stress Chaperones 6(2):121–125
14. Calderwood SK, Gong J (2016) Gong, *heat shock proteins promote cancer: it's a protection racket*. Trends Biochem Sci 41:311
15. Srivastava P (2002) Interaction of heat shock proteins with peptides and antigen presenting cells: chaperoning of the innate and adaptive immune responses. Annu Rev Immunol 20: 395–425
16. Srivastava P (2003) Hypothesis: controlled necrosis as a tool for immunotherapy of human cancer. Cancer Immun 3:4
17. Srivastava PK (2000) Immunotherapy of human cancer: lessons from mice. Nat Immunol 1(5):363–366
18. Belli F et al (2002) Vaccination of metastatic melanoma patients with autologous tumor-derived heat shock protein gp96-peptide complexes: clinical and immunologic findings. J Clin Oncol 20(20):4169–4180
19. Mazzaferro V et al (2003) Vaccination with autologous tumor-derived heat-shock protein gp96 after liver resection for metastatic colorectal cancer. Clin Cancer Res 9(9):3235–3245
20. Parmiani G et al (2006) Heat shock proteins gp96 as immunogens in cancer patients. Int J Hyperth 22(3):223–227
21. Belli F et al (2002) Vaccination of metastatic melanoma patients with autologous tumor-derived heat shock protein gp96-peptide complexes: clinical and immunologic findings. J Clin Oncol 20(20):4169–4180
22. Pilla L et al (2006) A phase II trial of vaccination with autologous, tumor-derived heat-shock protein peptide complexes Gp96, in combination with GM-CSF and interferon-alpha in metastatic melanoma patients. Cancer Immunol Immunother 55(8):958–968
23. Testori A et al (2008) Phase III comparison of vitespen, an autologous tumor-derived heat shock protein gp96 peptide complex vaccine, with physician's choice of treatment for stage IV melanoma: the C-100-21 study group. J Clin Oncol 26(6):955–962
24. Wood C et al (2008) An adjuvant autologous therapeutic vaccine (HSPPC-96; vitespen) versus observation alone for patients at high risk of recurrence after nephrectomy for renal cell carcinoma: a multicentre, open-label, randomised phase III trial. Lancet 372(9633):145–154
25. Enomoto Y et al (2006) Enhanced immunogenicity of heat shock protein 70 peptide complexes from dendritic cell-tumor fusion cells. J Immunol 177(9):5946–5955
26. Weng D et al (2013) Immunotherapy of radioresistant mammary tumors with early metastasis using molecular chaperone vaccines combined with ionizing radiation. J Immunol 191(2):755–763
27. Steinman RM (1991) The dendritic cell system and its role in immunogenicity. Annu Rev Immunol 9:271–296
28. Steinman RM (2001) Dendritic cells and the control of immunity: enhancing the efficiency of antigen presentation. Mt Sinai J Med 68(3): 106–166
29. Gong J et al (1997) Induction of antitumor activity by immunization with fusions of

dendritic and carcinoma cells. Nat Med 3(5): 558–561

30. Gong J et al (2002) Immunization against murine multiple myeloma with fusions of dendritic and plasmacytoma cells is potentiated by interleukin 12. Blood 99(7):2512–2517
31. Liu Y et al (2002) Engineered fusion hybrid vaccine of IL-4 gene-modified myeloma and relative mature dendritic cells enhances antitumor immunity. Leuk Res 26(8):757–763
32. Lindner M, Schirrmacher V (2002) Tumour cell-dendritic cell fusion for cancer immunotherapy: comparison of therapeutic efficiency of polyethylen-glycol versus electro-fusion protocols. Eur J Clin Investig 32(3):207–217
33. Homma S et al (2001) Preventive antitumor activity against hepatocellular carcinoma (HCC) induced by immunization with fusions of dendritic cells and HCC cells in mice. J Gastroenterol 36(11):764–771
34. Cao X et al (1999) Therapy of established tumour with a hybrid cellular vaccine generated by using granulocyte-macrophage colony-stimulating factor genetically modified dendritic cells. Immunology 97(4):616–625
35. Wang J et al (1998) Eliciting T cell immunity against poorly immunogenic tumors by immunization with dendritic cell-tumor fusion vaccines. J Immunol 161(10):5516–5524
36. Hayashi T et al (2002) Immunogenicity and therapeutic efficacy of dendritic-tumor hybrid cells generated by electrofusion. Clin Immunol 104(1):14–20
37. Xia J et al (2003) Prevention of spontaneous breast carcinoma by prophylactic vaccination with dendritic/tumor fusion cells. J Immunol 170(4):1980–1986
38. Kao JY et al (2003) Tumor-derived TGF-beta reduces the efficacy of dendritic cell/tumor fusion vaccine. J Immunol 170(7):3806–3811
39. Takeda A et al (2003) Immature dendritic cell/tumor cell fusions induce potent antitumour immunity. Eur J Clin Investig 33(10):897–904
40. Zhang JK et al (2003) Antitumor immunopreventive and immunotherapeutic effect in mice induced by hybrid vaccine of dendritic cells and hepatocarcinoma in vivo. World J Gastroenterol 9(3):479–484
41. Li J et al (2001) Purified hybrid cells from dendritic cell and tumor cell fusions are superior activators of antitumor immunity. Cancer Immunol Immunother 50(9):456–462
42. Xia D, Chan T, Xiang J (2005) Dendritic cell/myeloma hybrid vaccine. Methods Mol Med 113:225–233
43. Homma S et al (2005) Cancer immunotherapy by fusions of dendritic and tumour cells and rh-IL-12. Eur J Clin Investig 35(4):279–286
44. Kao JY et al (2005) Superior efficacy of dendritic cell-tumor fusion vaccine compared with tumor lysate-pulsed dendritic cell vaccine in colon cancer. Immunol Lett 101(2):154–159
45. Ogawa F, Iinuma H, Okinaga K (2004) Dendritic cell vaccine therapy by immunization with fusion cells of interleukin-2 gene-transduced, spleen-derived dendritic cells and tumour cells. Scand J Immunol 59(5): 432–439
46. Akasaki Y et al (2001) Antitumor effect of immunizations with fusions of dendritic and glioma cells in a mouse brain tumor model. J Immunother 24(2):106–113
47. Scott-Taylor TH et al (2000) Human tumour and dendritic cell hybrids generated by electrofusion: potential for cancer vaccines. Biochim Biophys Acta 1500(3):265–279
48. Tanaka H et al (2002) Therapeutic immune response induced by electrofusion of dendritic and tumor cells. Cell Immunol 220(1):1–12
49. Siders WM et al (2003) Induction of specific antitumor immunity in the mouse with the electrofusion product of tumor cells and dendritic cells. Mol Ther 7(4):498–505
50. Jantscheff P et al (2002) Cell fusion: an approach to generating constitutively proliferating human tumor antigen-presenting cells. Cancer Immunol Immunother 51(7):367–375
51. Goddard RV et al (2003) In vitro dendritic cell-induced T cell responses to B cell chronic lymphocytic leukaemia enhanced by IL-15 and dendritic cell-B-CLL electrofusion hybrids. Clin Exp Immunol 131(1):82–89
52. Marten A et al (2003) Allogeneic dendritic cells fused with tumor cells: preclinical results and outcome of a clinical phase I/II trial in patients with metastatic renal cell carcinoma. Hum Gene Ther 14(5):483–494
53. Trevor KT et al (2004) Generation of dendritic cell-tumor cell hybrids by electrofusion for clinical vaccine application. Cancer Immunol Immunother 53(8):705–714
54. Suzuki T et al (2005) Vaccination of dendritic cells loaded with interleukin-12-secreting cancer cells augments in vivo antitumor immunity: characteristics of syngeneic and allogeneic antigen-presenting cell cancer hybrid cells. Clin Cancer Res 11(1):58–66
55. Trefzer U et al (2005) Tumour-dendritic hybrid cell vaccination for the treatment of patients with malignant melanoma: immunological effects and clinical results. Vaccine 23(17–18):2367–2373
56. Shimizu K et al (2004) Comparative analysis of antigen loading strategies of dendritic cells for tumor immunotherapy. J Immunother 27(4): 265–272

57. Phan V et al (2003) A new genetic method to generate and isolate small, short-lived but highly potent dendritic cell-tumor cell hybrid vaccines. Nat Med 9(9):1215–1219
58. Hiraoka K et al (2004) Enhanced tumor-specific long-term immunity of hemagglutinating [correction of hemaggluttinating] virus of Japan-mediated dendritic cell-tumor fused cell vaccination by coadministration with CpG oligodeoxynucleotides. J Immunol 173(7): 4297–4307
59. Koido S et al (2004) Dendritic cells fused with human cancer cells: morphology, antigen expression, and T cell stimulation. Clin Immunol 113(3):261–269
60. Galea-Lauri J et al (2002) Eliciting cytotoxic T lymphocytes against acute myeloid leukemia-derived antigens: evaluation of dendritic cell-leukemia cell hybrids and other antigen-loading strategies for dendritic cell-based vaccination. Cancer Immunol Immunother 51(6): 299–310
61. Gong J et al (2000) Fusions of human ovarian carcinoma cells with autologous or allogeneic dendritic cells induce antitumor immunity. J Immunol 165(3):1705–1711
62. Gong J et al (2000) Activation of antitumor cytotoxic T lymphocytes by fusions of human dendritic cells and breast carcinoma cells. Proc Natl Acad Sci USA 97(6):2715–2718
63. Koido S et al (2005) Assessment of fusion cells from patient-derived ovarian carcinoma cells and dendritic cells as a vaccine for clinical use. Gynecol Oncol 99(2):462–471

Chapter 17

Methods to Assess the Impact of Hsp90 Chaperone Function on Extracellular Client MMP2 Activity

SarahBeth D. Votra, Deema Alsalih, and Dimitra Bourboulia

Abstract

Secreted, or extracellular, heat shock protein 90 (eHsp90) is considered a recent discovery in eukaryotes. Over the last two decades, studies have provided significant supporting evidence that implicates eHsp90 both in normal cellular processes such as wound healing and in the development of human pathologies and diseases including fibrosis and cancer. In the early 2000s, Eustace et al. demonstrated that eHsp90 promotes the invasion of breast cancer cells by binding to and regulating the activity of an extracellular matrix (ECM) remodeling enzyme, the matrix metalloproteinase 2 or MMP2. Interestingly, inside mammalian cells, Hsp90 is an essential chaperone that interacts with hundreds of newly synthesized proteins, known as "clients," that require Hsp90's assistance to perform their function. Several methods are routinely used to characterize the role and impact of Hsp90 on a client protein's functionality in vitro and in vivo. However, the mechanistic role of eHsp90 is less well-defined since, so far, only a handful of extracellular client proteins have been identified. Here, we describe methods to characterize the impact of the secreted chaperone on MMP2 activity, the most characterized extracellular client of eHsp90. The procedures described here can be applied and adapted to characterize other extracellular clients, particularly members of the MMP family.

Key words Extracellular heat shock protein 90 (eHsp90) , MMP2, Interaction, Enzyme kinetics, ECM

1 Introduction

Heat shock protein 90 (Hsp90) is an essential molecular chaperone in eukaryotes and the central building block of a multimolecular assembly known as "the Hsp90 chaperone machinery" [1–3]. Hsp90 utilizes its key ATP binding and hydrolysis mechanism to perform its chaperone function [4, 5]. Hsp90 interacts with over 300 different proteins known as "clients" ensuring their proper folding, activation, maturation, and stability. A number of regulatory elements facilitate, adapt, and fine-tune Hsp90 function including association with a group of auxiliary proteins named as "co-chaperones" as well as by post-translational modifications [6–

Stuart K. Calderwood and Thomas L. Prince (eds.), *Chaperones: Methods and Protocols*, Methods in Molecular Biology, vol. 2693, https://doi.org/10.1007/978-1-0716-3342-7_17,

8]. While processes that regulate the intracellular Hsp90 have been extensively studied, mechanisms that control the secreted form of Hsp90 (extracellular or eHsp90) are less characterized.

eHsp90 has been studied particularly in the context of pathophysiological processes such as wound healing, tumor angiogenesis, and metastasis (reviewed in [9–12]). In our laboratory, we have been investigating eHsp90 at the cellular level, in the context of tumor cell migration and invasion, key early steps of tumor metastasis [13]. MMPs are a large family of zinc-dependent endopeptidases that cleave all components of the extracellular matrix (ECM) [14]. Proteolysis of the ECM is essential for tumor cells to spread locally and through lymphatic and blood vessels to disseminate to distant sites, where they could establish metastatic islands. Gelatinases MMP2 (type A, 72 kDa) and MMP9 (type B, 92 kDa, also an eHsp90 client) degrade collagen, elastin, fibronectin, gelatin, and laminin [14]. Our studies, as well as work from others, have demonstrated that eHsp90's proinvasive properties are due to its association with MMP2 [15–18]. Formation of the MMP2:Hsp90 complex protects the protease from autolysis and, consequently, increases the extracellular levels of a proteolytically active MMP2 pool. MMP catalytic activity is inhibited by their specific endogenous inhibitors, the tissue inhibitors of metalloproteinases (TIMPs) (reviewed in [14]). Interestingly, our recent studies have demonstrated that TIMP2 functions as an extracellular co-chaperone of eHsp90, thus, contributing broadly to the regulation of the tumor microenvironment, controlling both MMP proteolytic activity and Hsp90 chaperone function [15, 19, 20].

MMPs are released by cells in an inactive form (zymogens) and require further proteolytic processing to become active enzymes extracellularly [14, 21]. proMMP2 zymogen has a molecular weight of 72 kDa (proMMP2) [22–25]. Following secretion, proMMP2 is sequentially processed to an intermediate 64 kDa and then to a fully active 62 kDa protein fragment at the cell surface. Gelatin zymography is a biochemical method that is routinely used to determine processing and the gelatinolytic activity of MMP2 and MMP9 as originally described by Kleiner and Stetler-Stevenson, 1994 [26]. Here, we test the gelatinase activity or recombinant active MMP2 (62 kDa) following incubation of the protease with or without Hsp90α at 37 °C for a set period of time. Active MMP2 will target and proteolytically cleave other active MMP2 molecules (autolysis) resulting in degradation, loss of activity, and processing to smaller fragments. Gelatin zymography will verify only the presence of active MMP2 fragments, those that have the catalytic domain intact and able to degrade gelatin incorporated in the polyacrylamide gel. Processed fragments that have no activity are not observed in this assay. These may only be detected by western blotting.

In this chapter, we describe detailed procedures for characterizing the interaction between Hsp90α chaperone and active enzyme MMP2 as well as the impact of this interaction on MMP2 activity using a number of techniques. These methods have been optimized or modified in our laboratory and can be used to test other members of the MMP family particularly gelatinase MMP9.

2 Materials

2.1 Hsp90α:MMP2 Complex Formation

1. Ice/ice box/racks.
2. 1 ml and 0.5 ml microtubes.
3. Microtube rotator placed in a 4 °C chiller cabinet or cold room.
4. Bacterial incubator set at 37 °C.
5. Recombinant human 62 kDa MMP2 (active) (Sigma-Aldrich, SRP3118).
6. Recombinant human Hsp90α-his_6 (produced in-house), untagged or commercially available.
7. TIMP2 reconstitution buffer (50 mM Tris–HCl 7.4, 150 mM NaCl, 5 mM CaCl2, 0.05% Brij-35).

2.2 Immunoblotting (Bio-Rad Criterion System)

1. Protein loading buffer (5×, 15 ml): 4.7 mL 1 M Tris–HCl pH 6.8, 4.5 ml glycerol, 1.5 g SDS, 0.75 ml 1% bromophenol blue, 0.75 ml beta-mercaptoethanol, ~4.3 ml dH_20.
2. 10xTBS (1 L): 24.2 g Trizma (Sigma), 80 g NaCl, 13 mL HCl, 1 L dH_20.
3. 1xTBST (1×, 1 L): 900 mL dH_20, 100 mL 10 × TBS, 1 mL Tween 20 (working solution).
4. 10× running buffer: 121.2 g Trizma (Sigma), 576.8 g glycine, 20 g SDS, 3.2 L dH_20.
5. 10× transfer buffer: 800 mL 5× transfer buffer (Bio-Rad), 800 mL ethanol, 2.4 L dH_20. 4–20% Criterion™ Tris–HCl polyacrylamide gel (Bio-Rad).
6. Ponceau S solution (Sigma).
7. Dried skimmed milk.
8. Protran BA85, 0.45 μm nitrocellulose membrane (Whatman).
9. ThermoFisher™ Pierce™ ECL 2 Western Blotting Substrate (Thermo Scientific). ECL2 or SuperSignal™ West Femto Maximum Sensitivity Substrate (Thermo Scientific).
10. Mouse anti-his_6 antibody (HIS.H8), Cat# MA1–21315, (Thermo Scientific).
11. Rat anti-Hsp90 mAb (16F1), Cat# ADI-SPA-835F (Enzo Life Sciences).

12. Rabbit anti-MMP2 (D8N9Y) mAb, Cat# 13132 (Cell Signaling Technology).
13. HRP-conjugated secondary antibodies (Santa Cruz, anti-rabbit Cat #sc-2004, anti-mouse Cat# sc-2005, anti-rat Cat# sc-2006).
14. X-ray film, X-ray cassette, and X-ray film developing machines.

2.3 Gelatin Zymography *(ThermoFisher Scientific Protein Gel Electrophoresis Chambers, Empty Gel Cassettes, and Buffers)*

1. Electrophoresis chamber system (XCell SureLock Mini-Cell, ThermoFisher).
2. Empty gel cassettes, mini (#NC2010, Novex) 1.0 mm thickness and 10-well (#NC3010, Novex) or 12-well combs (#NC3012, Novex).
3. Horizontal rocker shaker.
4. Bacterial incubator set at 37 °C.
5. Gelatin from porcine type A (Sigma G2500): Prepare a stock at 10 mg/ml (see methods).
6. 10× running buffer: 121.2 g Trizma (Sigma), 576.8 g glycine, 20 g SDS, 3.2 L dH_20.
7. 1.5 M Tris–HCl pH 8.8 (Teknova, #T1588).
8. 1 M Tris–HCl pH 6.8 (Fisher Scientific, #50-103-1384).
9. 10% SDS.
10. 30% acrylamide/Bis solution, 29:1 (Bio-Rad, #1610156).
11. Ammonium persulfate (APS) 10% in H_2O (Sigma-Aldrich, #A3678).
12. TEMED (Sigma-Aldrich, #110–18-9).
13. 70% EtOH.
14. dH_20.
15. 72 kDa recombinant proMMP2 (Sigma-Aldrich, PF037-10 μg).
16. 5× protein loading buffer (see recipe above; DO NOT ADD reducing agent).
17. Novex™ Zymogram Renaturing Buffer (10×) Invitrogen™.
18. Novex™ Zymogram Developing Buffer (10×) Invitrogen™.
19. Coomassie Brilliant Blue R-250 (Bio-Rad # 1610436).
20. Destaining solution (10% acetic acid, 50% MeOH, 40% water).

2.4 MMP2 Fluorometric Enzyme Activity Assay

1. Thermo Scientific™ 96 Well Black/Clear Bottom Plate, TC Surface (Thermo Scientific™ 165,305).
2. 62 kDa active MMP2 (Sigma-Aldrich, SRP3118).
3. Fluorogenic MMP2 substrate peptide (Dabcyl-GPLGMRGK (5FAM)-NH2) (BioZyme Inc. #PEPDAB011).

4. Fluorescent peptide buffer: 50 mM Tris–HCl, pH 7.5, 150 mM NaCl, 2 mM $CaCl_2$, 5 μM $ZnSO_4$, and 0.01% Brij-35.
5. Fluorometer, fluorescent plate reader.

2.5 Statistic Software

1. Prism 7 (GraphPad Software, Inc.).

3 Methods

Fully active MMP2 is highly proteolytic and also able to self-degrade over time by autolysis at 37 °C. In our hands, the enzyme activity of active MMP2 is almost completely lost following 1 h incubation at 37 °C. Therefore, the length of protein incubation time at 37 °C has to be predetermined according to the specifics of the experiment and the questions to be answered.

Here, we tested the role of Hsp90α on active MMP2 stability, i.e., resistance to autodegradation and loss of enzymatic activity. In summary, proteins are mixed and incubated at 37 °C. Small volumes are picked from the protein master mix at specific time points (e.g., 0, 25 min, 45 min, 60 min). Protein samples are chilled on ice and briefly spun down. If a number of analyses are to be performed, protein samples can be split, for example, (A) for immunoblotting, (B) for gelatin zymography, and (C) for MMP2 fluorogenic substrate degradation. NOTE: It is essential to pre-calculate the amount and volume of protein needed for each of these analyses. For example, 10-50 ng of protein is a sufficient amount of protein for immunoblotting, and 1-4 ng of MMP2 is enough to determine the presence and activity of MMP2 by gelatin zymography or/and using a fluorogenic peptide substrate to assess proteolytic activity in a quantitative way using enzyme kinetic. These assays are also performed in [15, 20, 27].

Here, we describe a method where recombinant MMP2 and Hsp90α proteins are mixed at 1:1 or 1:2 molar ratios and different aliquots are picked from a master mix during protein incubation at 37 °C.

3.1 *In Vitro* Hsp90α: Active MMP2 Complex Formation

This protocol describes steps taken to prepare complexes of bacterially purified human Hsp90α (produced in-house, with or without a tag) with active MMP2 enzyme (Sigma-Aldrich, #SRP3118).

1. Prepare protein stocks at 100 μg/ml (containing 0.1% BSA) in TIMP2 reconstitution buffer.
2. On ice, prepare two sets of protein samples: One contains 1 μg of MMP2 while the second contains 1 μg of MMP2 mixed with 1 μg of Hsp90α. The total volume was made up of 4 μl of

TIMP2 reconstitution buffer (protein concentration at 250 μg/ml).

3. Mix well and let them rotate at 4 °C for 1 h.
4. Briefly spin down and incubate at 37 °C.
5. Collect protein aliquots at different time points or the end of the experiment (e.g., 0-45mins), and immediately chill them on ice.
6. When all samples are collected, dilute samples to 25 μg/ml in TIMP2 reconstitution buffer. Use these stocks for the following analyses.

3.2 Protein Detection: Immunoblotting

1. Protein samples are prepared using 5× protein loading buffer and heated for 5 mins at 98 °C. Consider loading ~10-20 ng of protein-containing sample for each time point.
2. Samples are loaded into criterion Tris–HCl precast gels (Bio-Rad) with PageRuler Prestained Protein Ladder (ThermoScientific).
3. Samples are separated by SDS-PAGE at 200 V until ladder fragments above 50 kDa are well resolved.
4. Proteins are transferred onto nitrocellulose membrane via Trans-Blot Turbo (Bio-Rad) semidry transfer system.
5. The transferred membrane is then blocked in a 5% milk solution containing Tris-buffered saline with 0.1% Tween 20 (TBST) for 1 h at room temperature on a rocker.
6. Samples are incubated with antibodies diluted in 5% milk solution rocking for 1 h at room temperature or overnight at 4 °C.
7. Samples were subsequently washed 3 times for a minimum of 5 min in 1xTBST.
8. Samples are incubated with corresponding secondary antibodies for 1 h at room temperature.
9. Nitrocellulose membranes are washed with TBST 3 times for a minimum of 5 min.
10. The membranes are treated with western blotting substrate ECL2 (Thermo) as described by the manufacturer.
11. Blots are then visualized via exposure to x-ray film (Fig. 1).

Note: Data indicates that, in vitro, *Hsp90α protects active MMP2 from autodegradation.*

3.3 MMP2 Gelatinolytic Activity: Gelatin Zymography

1. Dissolve gelatin powder at 10 mg/ml in 10mls of dH_20 containing 2% SDS:
   1.1. Warm up the 2% SDS solution for at least 30 min at 50 °C.
   1.2. Add gelatin powder, and mix immediately (do not let the gelatin cool before it has been thoroughly mixed within the solution).

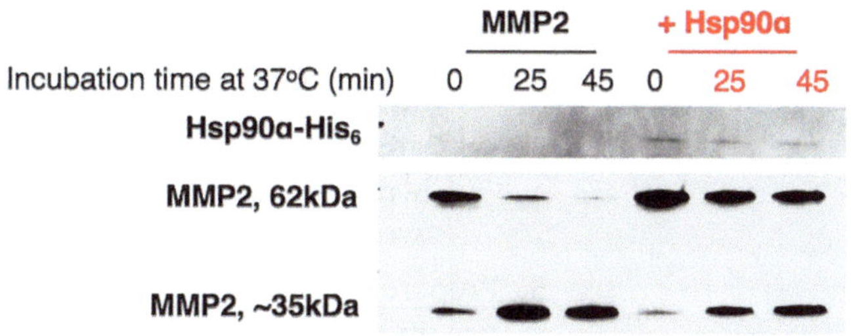

Fig. 1 Detection of recombinant MMP2 with and without Hsp90α by immunoblotting. Proteins were incubated for the indicated periods of time at 37 °C. Proteins were detected using MMP2-, His6-, or Hsp90-specific antibodies. Two bands appear for MMP2. The 62 kDa protein indicates that the full-length active enzyme is present. The 35 kDa fragment indicates that the protein is proteolytically processed over time

1.3. Beware of the gelatin solidifying before it fully dissolves in the solution.

1.4. Prepare 1 ml aliquots and freeze at −80 °C until use.

2. Prepare 1× protein running buffer (~1000 ml).
3. Prepare one empty gel cassette (mini-gel). Place vertically and hold stably in place.
4. Prepare separation gel: The total volume of 9.018 mL is enough for 1 mini-gel. To prepare the separation gel, the following should be added in a 15 ml tube in this particular order:
   - 3.30 mL dH_2O.
   - 2.25 ml 1.5 M Tris–HCl pH 8.8.
   - 900 μL of 10 mg/mL gelatin in 2% SDS (if gelatin is frozen, liquify by warming up at 40 °C for 5-10 min). Mix well, gelatin needs to be fully dissolved.
   - 2.4 ml acrylamide-Bis (30% 29:1). Mix well.
   - 150 μl 10% APS.
   - 18 μl TEMED (add only when you are ready to load the gel solution in the empty gel cassette).

4.1. Mix the gel immediately upon adding the TEMED, and pour it into the cassette without delay. Can be pipetted on top of the empty cassette using a 5 ml pipette.

4.2. Pour 6–8 mL to fill ~4/5 of the cassette.

4.3. Gently add 1 ml (enough to cover up the surface of the gel) of 70% ethanol on top of the gel. It prevents contact with air, which slows down polymerization at the upper part and results in a heterogeneous gel. Alcohol also provides a sharp border for the gel.

4.4. After the gel has polymerized (no more than 30 min), pour the alcohol out, and gently wash the top of the gel with dH_2O. If you are not going to use the gel immediately, fill the remaining space on top of the gel with water, wrap the top of the gel with a wet paper towel and Saran wrap, and store at 4 °C.

5. Prepare stacking gel (the total volume of 10 mL is enough for two mini-gel). For one mini-gel, use one 10-well or 12-well comb. In a 12 ml tube, add the following in this order:
   - 3.675 ml dH2O.
   - 100 μl 10% SDS.
   - 625 μl 1 M Tris–HCl pH 6.8.
   - 500 μl acrylamide-Bis (30% 29:1).
   - 100 μl 10% APS.
   - 12.5 μl TEMED (again, this is added only when ready to load on the cassette).

   5.1. Load ~3-4 ml of stacking gel on the cassette.

   5.2. Add the comb inside the cassette. Make sure there are no trapped bubbles.

   5.3. After the gel polymerizes (~15–20 min), remove the comb and rinse gently with water. Make sure no trapped pieces of acrylamide are left. Dry to remove the excess water.

6. Prepare protein samples for loading (**NO BOIL, NO REDUCING AGENT**).
7. Protein samples are diluted at 200 ng/ml in TIMP2 reconstitution buffer so that you will be able to load 2 ng of MMP2-containing sample.
8. Samples are mixed with protein loading buffer (max load 25 μl per well).
9. Prepare control proMMP2 to load 2 ng.
10. Mix well and spin down. Keep samples at room temperature before loading.
11. Prepare the electrophoresis chamber system (XCell SureLock Mini-Cell, ThermoFisher). Lock in the cassette and add protein running buffer.
12. Load 7 μl of the ladder followed by the samples. The last well should contain 2 ng of proMMP2 (zymogen) control.
13. Gels are run at 140 V for the first 10 min, and the voltage is increased by 10 V every 10 min up to 200 V.
14. The gel is run until the ladder bands over 50 kDa are well separated. Expect proMMP2 at 72 kDa, and active MMP2 at 62 kDa.

15. Following completion, remove, trim, and immerse gel in 100 ml of 1× renaturing buffer.
16. Gels are rocked on a horizontal shaker for 30 min at room temperature.
17. Buffer is then removed and replaced with 100 ml of 1× developing buffer.
18. Rock for another 30 min.
19. Replace buffer with 100 ml of fresh 1× developing buffer.
20. Rock horizontally overnight at 37 °C.
21. Next day remove the developing solution, and add 100 ml of Coomassie Brilliant Blue R-250 (Bio-Rad).
22. Rock for 30-60mins at room temperature to stain the gel.
23. Wash gel with a destaining solution for ~15 mins.
24. Observe clearly identifiable MMP2 bands (white bands) against a blue background.
25. Replace with dH_2O and scan images (Fig. 2).

Note: Data demonstrate that the activity of MMP2 is lost over time in the absence of Hsp90α significantly faster than in the presence of Hsp90α. ImageJ is used to quantify band intensities. CTR; control.

3.4 MMP2 Fluorometric Enzyme Activity Assay

During our optimization studies, MMP2 and Hsp90α were incubated at 37 °C at different molar ratios (MMP2:Hsp90α), e.g., 1:1, 1:2, and 1:4 (Fig. 3). The following protocol describes an example of protein ratio ~ 1:2.

1. After incubation at 37 °C is complete and proteins are chilled, dilute protein samples further down to 25 ng/ml in fluorescent peptide buffer.
2. Fluorogenic peptide (Dabcyl-GPLGMRGK(5FAM)-NH2) is diluted in DMSO to a concentration of 10 mM.
3. Further dilutions were made to 10 μM in fluorescent peptide buffer.

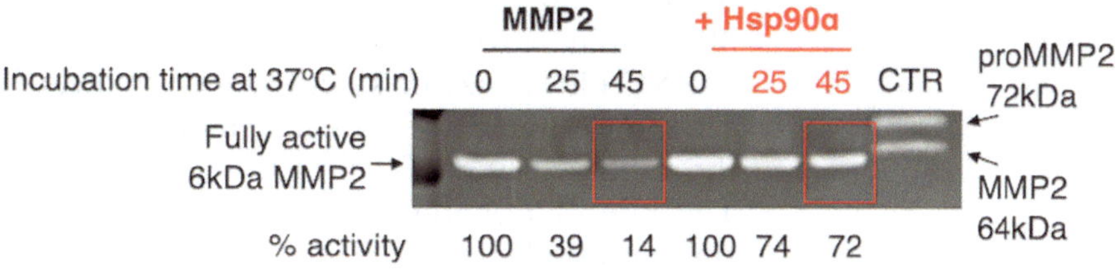

Fig. 2 Hsp90α stabilizes active MMP2. Gelatin zymography showing 62 kDa active MMP2 and the ability to degrade gelatin with and without Hsp90α. MMP2 proteolytic activity is prolonged in the presence of Hsp90α, as demonstrated by the increased band intensity of MMP2 + Hsp90α (72%) compared to MMP2 alone (14%) after 45 min incubation at 37 °C

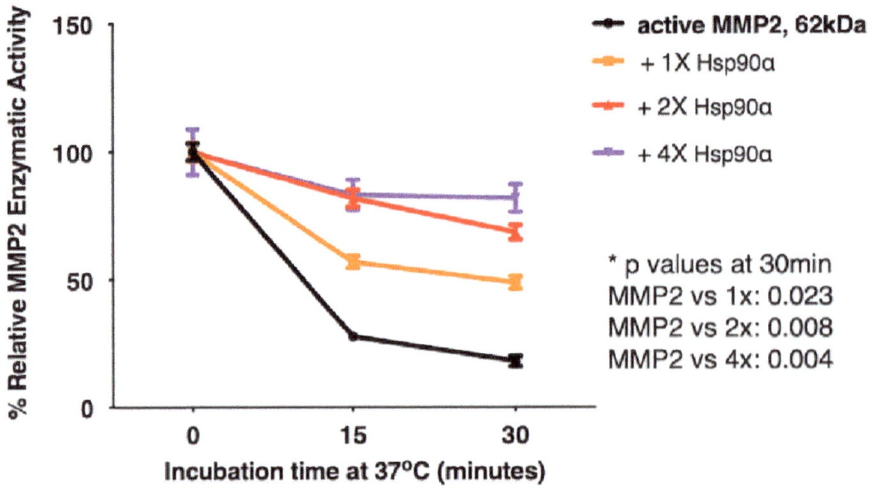

Fig. 3 Active MMP2 enzymatic activity measured by a fluorometric activity assay measuring 62 kDa MMP2-mediated degradation of fluorogenic substrate peptide. Hsp90α was added at different molar ratios as shown on the graph. The complexes were incubated at 37 °C. p values are indicated. The stability of MMP2 is increased with increasing amounts of Hsp90α present

4. Add diluted protein samples and peptide substrate 1:1 (the total volume is 100 μl/well) in a 96-well optical bottom plate. Specifically:
   - 4.1. All samples are brought to room temperature, and all work is continued at room temperature.
   - 4.2. As blanks, prepare 100 μl of fluorescent peptide buffer and 100 μl of 1:1000 DMSO in fluorescent peptide buffer.
   - 4.3. Load 50 μl of each of the samples into two wells of the 96-well plate.
   - 4.4. Load 50 μl of fluorescent peptide (10 μM) on the samples.
   - 4.5. Cover the plate with aluminum foil.
   - 4.6. Load the plate into a fluorometer, e.g., SpectraMax i3, plate reader (Molecular Devices).
   - 4.7. Analyze samples at an excitation and emission wavelength of 485 and 530 nm respectively, every 5 min for 30–60 min, at room temperature.

     NOTE: If enzyme kinetics are desirable, take fluorescence measurements for 1 h, at 5 min intervals. Determine the initial velocities of the reaction for each measurement. The rate of reaction is then plotted in relation to the amount of substrate.
   - 4.8. Following linear regression, experiments are either modeled to the Michaelis–Menten approach or changes seen in the percentage activity of an MMP2 bound complex with respect to a sole MMP2 control (Fig. 4).

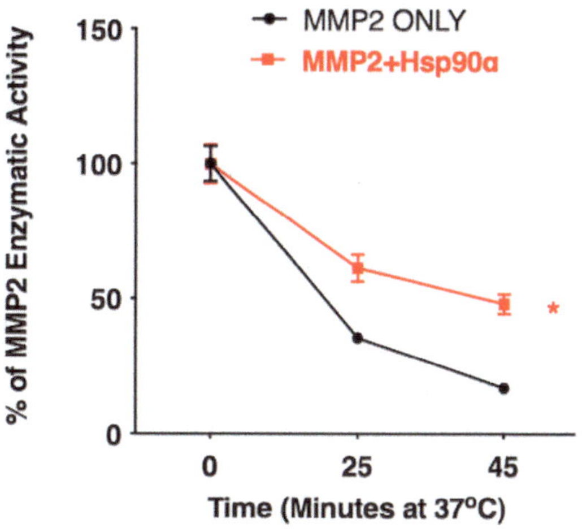

Fig. 4 The same assay as in Fig. 3. Here, Hsp90α-his_6 was added at a molar ratio 2:1 (Hsp90α:MMP2), and the complexes were incubated at 37 °C. Representative of three independent experiments *$p < 0.05$ at 45 min (MMP2 vs MMP2+ Hsp90α)

4.9. Import data in Prism 7 (GraphPad Software, Inc.) and perform statistics.

Note: Data demonstrate that the activity of MMP2 is lost over time in the absence of Hsp90α as measured by fluorescence released following the degradation of the peptide substrate.

Acknowledgments

We would like to thank Dr. Alexander J. Baker-Williams for critical discussions and technical support. Dimitra Bourboulia is supported by the National Institute of General Medical Sciences of the National Institutes of Health under Award Number R01GM139932. The content is solely the responsibility of the authors and does not necessarily represent the official views of the National Institutes of Health. Additional funding sources include the SUNY Upstate Medical University, the Upstate Foundation, the SUNY Research Foundation, and the Carol M. Baldwin Breast Cancer Research fund grant (D.B.).

References

1. Biebl MM, Buchner J (2019) Structure, function, and regulation of the Hsp90 machinery. Cold Spring Harb Perspect Biol 11(9):106
2. Schopf FH, Biebl MM, Buchner J (2017) The HSP90 chaperone machinery. Nat Rev Mol Cell Biol 18(6):345–360
3. Taipale M, Jarosz DF, Lindquist S (2010) HSP90 at the hub of protein homeostasis: emerging mechanistic insights. Nat Rev Mol Cell Biol 11(7):515–528
4. Hessling M, Richter K, Buchner J (2009) Dissection of the ATP-induced conformational

cycle of the molecular chaperone Hsp90. Nat Struct Mol Biol 16(3):287–293
5. Prodromou C (2012) The 'active life' of Hsp90 complexes. Biochim Biophys Acta 1823(3):614–623
6. Panaretou B, Siligardi G, Meyer P, Maloney A, Sullivan JK, Singh S et al (2002) Activation of the ATPase activity of hsp90 by the stress-regulated cochaperone aha1. Mol Cell 10(6): 1307–1318
7. Retzlaff M, Hagn F, Mitschke L, Hessling M, Gugel F, Kessler H et al (2010) Asymmetric activation of the hsp90 dimer by its cochaperone aha1. Mol Cell 37(3):344–354
8. Sahasrabudhe P, Rohrberg J, Biebl MM, Rutz DA, Buchner J (2017) The plasticity of the Hsp90 co-chaperone system. Mol Cell 67(6): 947–61e5
9. Li W, Sahu D, Tsen F (2012) Secreted heat shock protein-90 (Hsp90) in wound healing and cancer. Biochim Biophys Acta 1823(3): 730–741
10. Li W, Tsen F, Sahu D, Bhatia A, Chen M, Multhoff G et al (2013) Extracellular Hsp90 (eHsp90) as the actual target in clinical trials: intentionally or unintentionally. Int Rev Cell Mol Biol 303:203–235
11. Poggio P, Sorge M, Secli L, Brancaccio M (2021) Extracellular HSP90 machineries build tumor microenvironment and boost cancer progression. Front Cell Dev Biol 9:735529
12. Wong DS, Jay DG (2016) Emerging roles of extracellular Hsp90 in cancer. Adv Cancer Res 129:141–163
13. Backe SJ, Votra SD, Stokes MP, Sebestyén E, Castelli M, Torielli L et al (2023) PhosY-secretome profiling combined with kinase-substrate interaction screening defines active c-Src-driven extracellular signaling. Cell Reports 42(6):112539. https://doi.org/10.1016/j.celrep.2023.112539
14. Bourboulia D, Stetler-Stevenson WG (2010) Matrix metalloproteinases (MMPs) and tissue inhibitors of metalloproteinases (TIMPs): positive and negative regulators in tumor cell adhesion. Semin Cancer Biol 20(3):161–168
15. Baker-Williams AJ, Hashmi F, Budzynski MA, Woodford MR, Gleicher S, Himanen SV et al (2019) Co-chaperones TIMP2 and AHA1 competitively regulate extracellular HSP90:client MMP2 activity and matrix proteolysis. Cell Rep 28(7):1894–1906. e6
16. Eustace BK, Sakurai T, Stewart JK, Yimlamai D, Unger C, Zehetmeier C et al (2004) Functional proteomic screens reveal an essential extracellular role for hsp90 alpha in cancer cell invasiveness. Nat Cell Biol 6(6): 507–514
17. Song X, Wang X, Zhuo W, Shi H, Feng D, Sun Y et al (2010) The regulatory mechanism of extracellular Hsp90{alpha} on matrix metalloproteinase-2 processing and tumor angiogenesis. J Biol Chem 285(51): 40039–40049
18. Stellas D, El Hamidieh A, Patsavoudi E (2010) Monoclonal antibody 4C5 prevents activation of MMP2 and MMP9 by disrupting their interaction with extracellular HSP90 and inhibits formation of metastatic breast cancer cell deposits. BMC Cell Biol 11:51
19. Sager RA, Khan F, Toneatto L, Votra SD, Backe SJ, Woodford MR et al (2022) Targeting extracellular Hsp90: a unique frontier against cancer. Front Mol Biosci 9:982593
20. Sánchez-Pozo J, Baker-Williams AJ, Woodford MR, Bullard R, Wei B, Mollapour M et al (2018) Extracellular phosphorylation of TIMP-2 by secreted c-Src tyrosine kinase controls MMP-2 activity. Iscience 1C:87–96
21. Ra HJ, Parks WC (2007) Control of matrix metalloproteinase catalytic activity. Matrix Biol 26(8):587–596
22. Corcoran ML, Emmert-Buck MR, McClanahan JL, Pelina-Parker M, Stetler-Stevenson WG (1996) TIMP-2 mediates cell surface binding of MMP-2. Adv Exp Med Biol 389: 295–304
23. Corcoran ML, Hewitt RE, Kleiner DE Jr, Stetler-Stevenson WG (1996) MMP-2: expression, activation and inhibition. Enzyme Protein 49(1–3):7–19
24. Itoh Y, Binner S, Nagase H (1995) Steps involved in activation of the complex of pro-matrix metalloproteinase 2 (progelatinase A) and tissue inhibitor of metalloproteinases (TIMP)-2 by 4-aminophenylmercuric acetate. Biochem J 308(Pt 2):645–651
25. Strongin AY, Collier I, Bannikov G, Marmer BL, Grant GA, Goldberg GI (1995) Mechanism of cell surface activation of 72-kDa type IV collagenase. Isolation of the activated form of the membrane metalloprotease. J Biol Chem 270(10):5331–5338
26. Kleiner DE, Stetler-Stevenson WG (1994) Quantitative zymography: detection of picogram quantities of gelatinases. Anal Biochem 218(2):325–329
27. Cortes S, Baker-Williams AJ, Mollapour M, Bourboulia D (1709) Detection and analysis of extracellular Hsp90 (eHsp90). Methods Mol Biol 2018:321–329

Chapter 18

Proteomic Profiling of the Extracellular Vesicle Chaperone in Cancer

Kisho Ono and Takanori Eguchi

Abstract

Molecular chaperones are widely distributed intracellular proteins that play essential roles in maintaining proteome function by assisting in the folding of client proteins. Molecular chaperones, such as heat shock proteins (HSPs), are found intracellularly and extracellularly. Extracellular vesicles (EVs), such as exosomes, contain HSPs and horizontally transfer the functional chaperones into various recipient cells. Besides, mass spectrometry has enabled a comprehensive analysis of exosomal and EV proteins, which is useful in basic biomedical research to clinical biomarker search. We have performed deep proteome analysis of EVs, including exosomes, from metastatic tongue and prostate cancers and detected >700 protein types, including cytoplasmic, ER, mitochondrial, small, and large HSPs. Here, we provide protocols for isolating exosomes/EVs and deep proteome analysis to detect the EV chaperone.

Key words Chaperone proteins, Heat shock protein, Exosome isolation methods, Extracellular vesicles, Proteome analysis, Mass spectrometry

1 Introduction

Molecular chaperones are proteins that are widely distributed throughout the cell and play an essential role in maintaining proteome function by accelerating or slowing the folding of target proteins, preventing them from folding improperly, aggregating, or remaining immature. The original definition of a chaperone was "a protein that helps another protein form a conformation but is not itself a final component [1]," but today the term is often used to refer to all the accessory proteins in the various processes that a protein goes through during its life [2]. Representative chaperones include chaperonins (GroEL-Hsp60, CCT, etc.) and heat shock proteins (HSPs) such as the Hsp70, Hsp90, and Hsp104 families [2–7]. All of them are involved in the structural formation and functional expression of many types of proteins using adenosine triphosphate (ATP). Although there are many different chaperones

Stuart K. Calderwood and Thomas L. Prince (eds.), *Chaperones: Methods and Protocols*, Methods in Molecular Biology, vol. 2693, https://doi.org/10.1007/978-1-0716-3342-7_18,

in the cell and their mechanisms of action are varied, their basic actions are common. In an attempt to expand our knowledge of chaperone mechanisms in cellular protein homeostasis, Finka et al. performed a meta-analysis of human chaperones using high-throughput proteomic data from 11 immortalized human cell lines [8]. The results showed that the chaperone polypeptides were about 10% of the total protein mass of human cells, half of which were Hsp90 and Hsp70 [8]. HSPs are ubiquitously expressed molecular chaperones that are abundant in cancer [9, 10]. Several research groups have shown that naturally occurring antitumor agents such as geldanamycin and radicol act by inhibiting the ATPase activity of Hsp90, and the development of various anticancer agents has attracted attention [11–13]. Various HSP-targeted therapies and biomarker developments are currently being evaluated in clinical trials [14].

Proteomic analysis using mass spectrometry is now a powerful analytical technique that can identify and quantify thousands to 10,000 different human proteins in about 2 h. Proteins extracted from cells and other sources are fragmented into peptides by digestive enzymes, and the peptides are sequentially subjected to mass spectrometry while being separated by nanoscale liquid chromatography. This is a quantum leap from the days of mass spectrometry, when proteins extracted from gels were analyzed one by one, and nowadays peptides can be sequenced at a rate of 20 to 100 peptides per second [15, 16]. Proteomic analysis at the single-cell level has become a reality due to improvements in chromatographic techniques and sample preparation methods, in addition to the improved performance of mass spectrometers [17].

Comprehensive analysis of extracellular vesicle (EV)-related proteins by mass spectrometry has recently gained momentum, and a wide range of research is being promoted, from basic research on EVs to disease biomarkers discovery for prevention, diagnosis, and treatment. The number of relevant papers has increased exponentially, and a keyword search on PubMed for "extracellular vesicle" and "proteomics" yielded a total of 2749 hits as of July 2022 (Fig. 1). On the other hand, although various biochemical isolation methods have been developed to recover EVs from cell supernatants and body fluids, isolating only EVs from diverse biomolecular populations, including abundant proteins (e.g., albumin and lipoproteins), is still a challenging task. In particular, blood is the most difficult body fluid to isolate EVs and identify EV proteins because of its high content of non-endoplasmic reticulum material, including free proteins and protein aggregates [18].

Several researchers have used mass spectrometry (MS) to determine and characterize the EV proteome in various cancer tissues and biofluids [19–21]. Hoshino et al. reported a database of the EV proteome of 426 human samples, including tissue explants and body fluids from controls and cancer samples [22]. However,

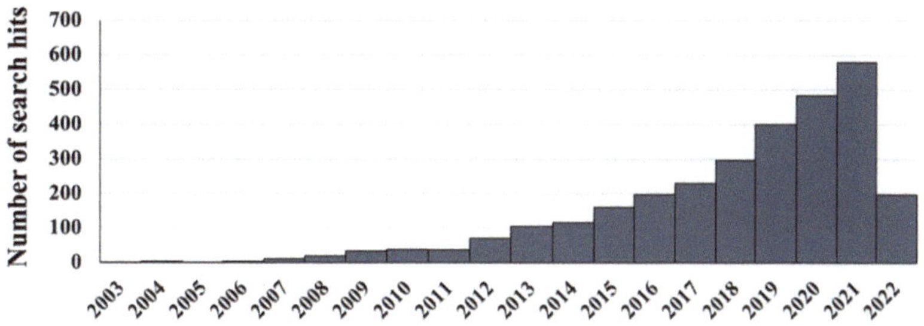

Fig. 1 A time-series graph of the PubMed search hits for "proteomics" and "extracellular vesicles". The number of hits on these keywords is increasing steadily every year

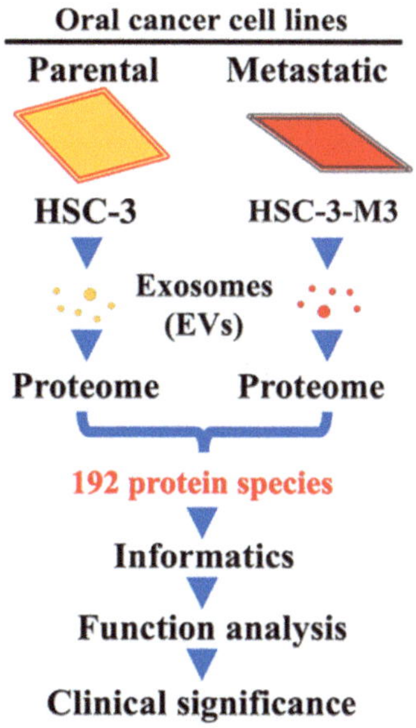

Fig. 2 Flow chart of comprehensive EV/exosome analysis methods [3]. We detected 192 protein species in EVs/exosomes derived from oral carcinoma cell lines, and among them, protein types related to cancer metastatic potential were informarized, functionally analyzed, and evaluated for clinical significance

there are few precedents for the analysis of HSP profiles of EVs derived from clinical samples of cancer patients. Albakova et al. examined in detail the expression of HSPs in EVs from different biological fluids from various MS studies and found that the HSP profiles of EVs differed by cancer type and body fluids [14]. In a previous study, we demonstrated that EVs secreted by highly metastatic oral cancer cells contain significantly more chaperone proteins, such as HSP90, by analyzing EVs using MS methods (Figs. 2 and 3) [3]. Based on the results, we reported that EV-HSP90 is significantly involved in malignant transformation events of oral cancer [4]. The discovery of the role of EVs in proteome and

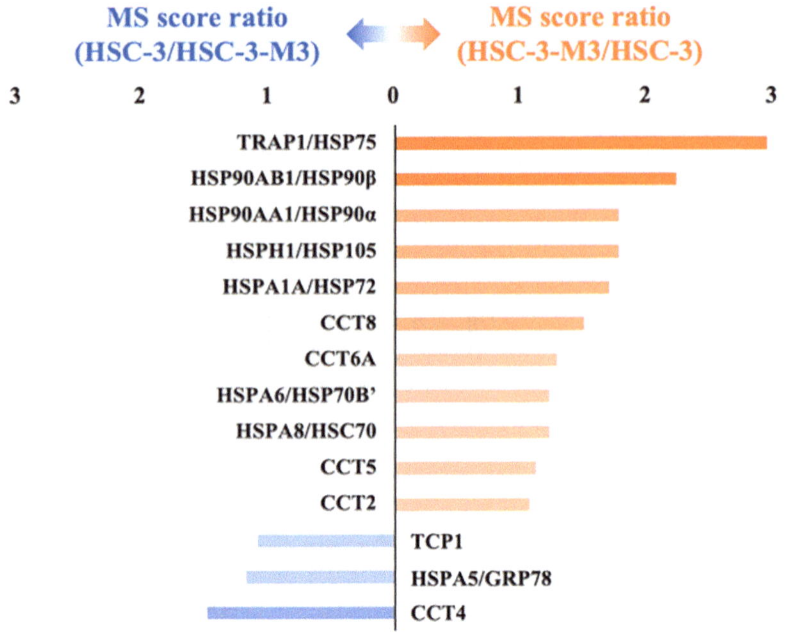

Fig. 3 EV chaperonomics of oral carcinoma cells. Among 14 EV chaperone types found in each cell line, 11 chaperone types (79%) were highly detected in HSC-3-M3 (highly metastatic cell subline), whereas only 1 chaperone type (21%) was highly detected in HSC-3 (parental oral carcinoma cell line). The original chart is published in ref. [3]

genetic information transfer and the identification of HSPs in EVs provide new opportunities and challenges for the determination of clinical biomarkers [3, 14, 23]. On the other hand, it should be noted that MS-generated data may have many missing values and may not reflect the actual HSP proteome composition of EVs, which may be caused by the difficulty of analyzing different types of liquid biopsies by MS-based proteomics and differences in EV isolation methods [24, 25].

We have also established protocols for EV proteome analysis [3, 5] using liquid chromatography-tandem mass spectrometry (LC-MS/MS), by which several hundreds of protein types are detectable, including various chaperone proteins (Figs. 2 and 3). This chapter provides protocols of exosomes/EVs sampling from 2D and 3D cultured cells and tissue specimens (Subheading 3.1), UF method (Subheading 3.2), PBP method (Subheading 3.3), UC method (Subheading 3.4), AP method (Subheading 3.5), SEC method (Subheading 3.6), preparation of non-EV/exosome fraction (Subheading 3.7), basic analysis methods of EVs/exosomes (Subheading 3.8), and proteome analysis for detecting chaperone proteins (Subheading 3.9).

2 Materials

2.1 Cell and Tissue Culture

2.1.1 Common Materials

1. Medium for the cell type of interest (DMEM, RPMI 1640, etc.).
2. PBS.
3. Centrifuge.
4. 50 mL tubes, 15 mL tubes.

2.1.2 2D or 3D Culture

1. Cells producing chaperone proteins (*see* **Note 1**).
2. Serum for maintaining cells, if required.
3. EV/exosome-free medium or EV/exosome-free serum (*see* **Note 2**).
4. Cell stress using a water bath that induces chaperone proteins (*see* **Note 3**).
5. 15 cm dishes (*see* **Note 4**).
6. 10 cm or 6-well ultralow attachment (ULA) plate or 6-well NanoCulture Plate (NCP) (*see* **Note 5**).
7. Trypsin–EDTA (0.25%).

2.1.3 Tissue-Exudative Extracellular Vesicles (Te-EVs)

1. Resected tissue specimen (e.g., a 2–3 mm cube) (*see* **Note 6**).
2. Microcentrifuge tubes (1.5 mL and 2 mL).
3. Rotator.

2.2 UF Method

1. Amicon Ultra-15 Centrifugal Filter Devices for M.W. 100 kD.
2. (Optional) Amicon Ultra-15 Centrifugal Filter Devices for M.W. 10 kD.
3. (Optional) 0.8 μm pore filter and syringe (*see* **Notes** 7 and **8**).
4. (Optional) 0.2 μm pore filter and syringe (*see* **Notes** 7 and **8**).

2.3 PBP Method

1. PEG polymer, e.g., Exosome Isolation Reagent (Thermo Fisher).
2. Centrifuge.
3. PBS (−).
4. Protease inhibitor cocktail (*see* **Note 9**).

2.4 UC Method

1. Ultracentrifuge.
2. Rotor (e.g., RP-42 rotor, Hitachi).
3. Tubes fitting to the rotor.
4. PBS (−).
5. Protease inhibitor cocktail (*see* **Note 9**).

2.5 AP Method (see Notes 10 and 11)

1. exoEasy Maxi Spin Columns.
2. Collection tubes (50 mL).
3. Buffer XBP.
4. Buffer XWP.
5. Buffer XE.
6. Refrigerated centrifuge with a swinging bucket rotor (*see* **Note 12**) and a fixed-angle rotor.

2.6 SEC Method

1. qEV Columns (*see* **Note 13**).
2. PBS (−) filtered by a 0.2 μm filter.

2.7 Basic EV Analyses

1. Vesicle/particle analyzer (*see* **Note 14**).
2. Cuvettes compatible with the vesicle/particle analyzer.
3. Transmission electron microscopy (TEM) (e.g., H-7650, Hitachi) (*see* **Note 15**).
4. 400-mesh copper grid coated with formvar/carbon films.
5. Parafilm.
6. 2% uranyl acetate.
7. 10× RIPA buffer: 0.5 M Tris-HCl pH 7.4, 1.5 M NaCl, 2.5% deoxycholic acid, 10% NP-40, 10 mM EDTA (*see* **Note 16**).
8. Micro BCA protein assay reagent (Thermo Fisher) (*see* **Notes 17** and **18**).
9. Microplate reader.

2.8 Proteome Analysis (LC-MS/MS)

1. Gloves, masks, lab head caps, a lab coat, and goggles (*see* **Note 19**).
2. 80% ethanol and clean bench (*see* **Note 19**).
3. 1.5 mL tubes.
4. Tris SDS buffer: 1% SDS and 2.5 mM Tris–HCl (2-carboxyethyl)phosphine hydrochloride.
5. Alkylating agent: 12.5 mM iodoacetamide.
6. Acetone, pre-chill at −20 °C for protein precipitation.
7. 100 mM ammonium bicarbonate.
8. Ultrasonicator, e.g., Bioruptor (Diagenode).
9. MS-grade trypsin.
10. A mass spectrometer, e.g., Agilent 6330 Ion Trap LC-MS System or QExactive (Thermo Fisher).
11. Nano-LC (AdvanceLC; Michrom Bioresources).
12. A column oven set.

13. Pre-column [L-column micro: 0.3 mm inner diameter, 5 mm length; Chemicals Evaluation and Research Institute (CERI)].
14. In-house-made 20 cm column (inner diameter 100 μm, 3 μL-column; CERI).
15. Proteomics software, e.g., 2D image-converted analysis of LC-MS/MS (2DICAL2, Mitsui Knowledge Industry) or Proteome Discoverer 1.4 (Thermo Fisher).
16. Mascot algorithm (ver. 2.6.0).
17. IPI human database (ver. 3.8.7).

3 Methods

3.1 Tissue/Cell Culture

3.1.1 2D Cell Culture

1. Seed cells in several 15 cm or 10 cm dishes (*see* **Note 20**), and culture until 70–80% confluency. For example, seed 3×10^6 cells per 15 cm dish or 1×10^6 cells per 10 cm dish.
2. For a medium change, aspirate the culture medium carefully. Do not aspirate cells.
3. Wash cells or tissues with PBS carefully.
4. Culture the cells in EV/exosome-free or serum-free medium for 1–2 days (*see* **Note 2**). Adding HSP-inducing cell stress such as heat shock stress (HSS) can increase the yield of HSP-EVs (*see* **Notes 1** and **3**).
5. Transfer the culture supernatant to 50 mL tubes and go to **step 3.2**.

3.1.2 Spheroid Culture and Tumoroid Culture

1. Seed cells in 10 cm ULA dishes for preparing tumoroids or spheroids. For example, seed 1×10^6 cells within 8 mL mTeSR1 medium (*see* **Note 2**). Per 10 cm ULA dish. Start with five dishes.
2. Culture the spheroids/tumoroids for 7–14 days to grow (*see* **Note 21**).
3. For a medium change, aspirate half of the culture medium carefully, and then add a half volume of fresh medium. Do not aspirate spheroids/tumoroids.
4. Culture the cells in EV/exosome-free or serum-free medium for 1–3 days (*see* **Note 22**). Adding HSP-inducing cell stress such as HSS can increase the yield of HSP-EVs (*see* **Notes 1** and **3**).
5. Transfer the culture supernatant to 50 mL tubes and go to **step 3.2**.

3.1.3 Te-EVs

1. Wash the resected tissue specimens with PBS.
2. Soak the specimen in a 1.5 mL serum-free medium in a 2 mL tube.
3. Rotate the specimen at 37 °C for 2–3 h.
4. Alternatively, shortly culture the specimen at 4 °C for 120 min.
5. Centrifuge at 2000 × *g* for 30 min at 4 °C to remove detached cells.
6. Transfer the supernatant to another tube.
7. Centrifuge at 10,000 × *g* for 30 min at 4 °C to remove cell debris (*see* **Note 23**).

3.2 UF Method (See Note 24)

The UF method can concentrate EVs and exosomes using an ultrafiltration filter by allowing the relatively higher molecular-weight exosomes or EVs to remain above the filter and the smaller molecular-weight vesicle-free proteins (e.g., less than 100 kD) to pass through the filter. Optionally, a 0.8 μm pore filter is used to remove L-EVs such as apoptotic bodies and cell debris and filtrate smaller EVs, while a 0.2 μm pore filter is used to remove mEVs and L-EVs and filtrate sEVs. mEVs and L-EVs can be collected by not using such pore filters and vice versa [5, 6].

1. **Option 1**: Removal of large EVs and cell debris. Filter the conditioned medium with a 0.8 μm syringe filter. Alternatively, centrifugate the conditioned medium at 3000 × *g* for 15 min at 4 °C.
2. **Option 2**: Removal of medium EVs and large EVs. Filter the supernatant with a 0.2 μm syringe filter (*see* **Notes** 7 and **8**).
3. For the AP method, move to Subheading 3.5.
4. For the PBP, UC, and SEC methods, samples should be concentrated. Apply the supernatant from Subheading 3.1 or the pass-through from Subheading 3.2, **steps 1** or **2** to an Amicon Ultra-15 Centrifugal Filter Devices for M.W. 100 kD.
5. Centrifuge at 5000 × *g* at 4 °C for concentrating the sample to less than 1 mL (*see* **Note 25**). The vesicle-free chaperone proteins (less than 100 kD) will pass through the filter.
6. Use the concentrate (remaining above the filter) for the PBP or UC methods.
7. Use the pass-through as a non-EV/exosome fraction that contains vesicle-free chaperone proteins in Subheading 3.7.

3.3 PBP Method

1. Transfer the concentrate above the filter (Subheading 3.2, **step 5**) into a low-adhesive microtube.
2. Suspend the sample in the same volume of polymer (e.g., total exosome isolation reagent), and incubate overnight at 4 °C.

3. Centrifuge the sample at 10,000 × *g* for 60 min at 4 °C.
4. Aspirate the supernatant and leave an EV/exosome pellet.
5. Suspend the EV/exosome pellet in 100 μL PBS (*see* **Notes 16**, **26**, and **27**).

3.4 UC Method

1. Centrifuge the samples (the concentrate remaining above the filter in Subheading 3.2, **step 5**) at 100,000 × *g* for 70 min (e.g., RP-42 rotor, Hitachi).
2. Aspirate the supernatant carefully (*see* **Note 28**).
3. Add PBS (−) and rinse the EVs/exosomes pellet.
4. Centrifuge the EV/exosome samples at 100,000 × *g* for 70 min.
5. Aspirate the supernatant carefully (*see* **Note 28**).
6. Suspend the EVs/exosomes pellet in 100 μL of PBS (−) (*see* **Notes 16**, **26**, and **27**).

3.5 AP Method

1. Add buffer XBP to the same volume of the samples, and invert the tube 5 times gently. Let the mixture warm up to RT.
2. Add up to 16 mL of the sample/XBP mixture onto the exoEasy spin column and centrifuge at 500 × *g* for 1 min.
3. (Optional) Keep the pass-through as non-affinity fractions (*see* **Note 29**).
4. Place the column back into the same collection tube. If the sample volume is larger than 8 mL, repeat **step 2** until the entire volume has been passed through the column.
5. (Optional) To remove residual liquid from the membrane, centrifuge at 5000 × *g* for 1 min.
6. Add 10 mL buffer XWP and centrifuge at 5000 × *g* for 5 min. Discard the flow-through together with the collection tube (see **Note 30**).
7. Transfer the spin column to a new collection tube.
8. Add 400 μL buffer XE (see **Note 31**) to the membrane and incubate for 1 min. Centrifuge at 500 × *g* for 5 min to collect the eluate.
9. Reapply the eluate to the spin column and incubate for 1 min. Centrifuge at 5000 × *g* for 5 min to collect the eluate (i.e., EVs), and transfer to an appropriate tube.

3.6 SEC Method

3.6.1 Concentration Step

1. Prepare the conditioned medium (> 50 mL) as described above.
2. Apply the conditioned medium to an Amicon Ultra-15 Centrifugal Filter Unit for M.W. 100 kD. The vesicle-free chaperone proteins (less than 100 kD) will pass through the filter.
3. Centrifuge at 5000 × *g* at 4 °C for concentrating the sample.

4. Repeat **steps 2** and **3** several times until concentrating the conditioned medium up to 500 μL (*see* **Note 32**). Use the concentrate for the SEC method.
5. (Optional) Keep the pass-through as a vesicle-free small factor fraction that contains vesicle-free chaperone proteins and concentrate in Subheading 3.7.

3.6.2 SEC Step

1. Carefully remove the top cap, and attach the column in an upright position to stand ready for use.
2. Remove the bottom cap and allow the buffer to run through the column.
3. Attach the buffer reservoir to the top of the column.
4. Flush the column with at least one column volume (15 mL) of PBS.
5. Continue to allow the buffer to run through the column. The column will stop flowing when all the buffer has entered the loading frit.
6. Load the concentrated sample (from Subheading 3.6.1, **step 4**) onto the loading frit (*see* **Note 33**).
7. Immediately start collecting the void volume (this includes the sample volume).
8. Allow the sample to run into the column. The column will stop flowing when all the sample has entered the loading frit.
9. Top up the column with PBS and collect the void fraction.
10. To collect accurate volumes, only load the required volume to the top of the column, wait for the volume to run through, and repeat (*see* **Note 34**).
11. The first 500 μL eluate is fraction #1, the second 500 μL eluate is fraction #2, and so on.
12. Collect fractions sequentially, e.g., Fr. #1, #2. . .#21.

3.6.3 Gathering Fraction Step

1. Gather every three fractions, e.g., Fr. #1–3, #4–6. . . .
2. Apply the gathered fraction to an Amicon Ultra-0.5 mL Centrifugal Filter for M.W. 100 kD.
3. Centrifuge at 14,000 × *g* at 4 °C for concentration.
4. Transfer the concentrate (remaining above the filter) into another tube.
5. Equalize the volumes of the gathered fractions by adding buffer.

3.7 Non-EV Fraction

1. Prepare the pass-through from Subheading 3.2, **step 5**. This fraction contains vesicle-free chaperone proteins.

2. Apply the pass-through to an Amicon Ultra-15 Centrifugal Filter Devices for M.W. 10 kD.
3. Centrifuge at 5000 × *g* at 4 °C for concentrating the non-EV sample.
4. Use the concentrate (remaining above the filter) as a non-EV/exosome fraction (*see* **Notes 35** and **36**).

3.8 Basic EV Analyses

Analyze the size and number of EVs after the isolation. The protein concentration of the EV fraction is also useful for downstream applications.

3.8.1 Measure the Numbers and Size Distribution of Vesicles/Particles (Video Drop, qNano, or NanoSight). Alternatively, Measure the Vesicle/Particle Size Distribution (Using Zetasizer) (see Note 37)

1. Start the Zetasizer and wait for 30 min for stability.
2. Choose the buffer type.
3. Prepare EV/exosome samples, PBS for sample dilution, and cuvettes.
4. Dilute the EV/exosome sample with PBS, e.g., 1:10.
5. Transfer 40 μL of the diluted EV/exosome liquid to a cuvette and measure.
6. Measure a few times to examine reproducibility and dispersion (*see* **Note 38**).

3.8.2 Visualize the Vesicles by Negative Staining and TEM

1. Hydrophilically treat a 400-mesh copper grid coated with formvar/carbon films.
2. Place the EV/exosome suspension (0.1–0.3 μg/μL in 5–10 μL) on Parafilm.
3. Float the grid on the EV liquid and leave for 15 min.
4. Negative staining with 2% uranyl acetate solution for 2 min.
5. Visualize EVs on the grid with a TEM.

3.8.3 Measure the Protein Concentration Using a Micro BCA Protein Assay

1. Mix the isolated EV/exosome sample, 10× RIPA buffer (*see* **Notes 16** and **26**), and PBS. Pipette a few times for mixing.
2. Incubate for 30 min at RT to dissolve the EV/exosome membrane.
3. Perform protein assay according to the manufacturer's protocol.

3.9 Proteome Analysis (LC-MS/MS)

There are two main methods for the enzymatic digestion of proteins. For the in-gel digestion method, (i) separate protein samples by electrophoresis shortly, (ii) cut out the stained gel (2–3 mm), and (iii) digest the proteins within the gel. However, keratin is often contaminated in these steps (*see* **Note 19**). Here we provide another method called in-solution digestion.

1. Incubate EVs in the presence of 1% SDS and 2.5 mM Tris (2-carboxyethyl) phosphine hydrochloride for 10 min at 85 °C.

2. For protein alkylation, add 12.5 mM iodoacetamide, and incubate for 15 min at RT.
3. Add more than a fourfold volume of chilled acetone and mix well for protein precipitation.
4. Incubate at −30 °C for 2 h (or overnight).
5. Centrifuge at 14,000 × *g* for 10 min.
6. Remove supernatant.
7. Evaporate the pellet by keeping the tube's lid open and incubating for 30 min.
8. Add 100 mM ammonium bicarbonate.
9. Ultrasonicate three times for 30 s with intervals of 30 s.
10. Add 0.1 mg/ml trypsin and incubate at 37 °C overnight.
11. Inject the resulting peptides to LC-MS/MS.
12. Analyze the acquired MS/MS spectra by software.
13. Search with the following parameters. Select "trypsin" as an enzyme used. Set the number of missed cleavages allowed as "3." Select "carbamidomethylation on Cys" as a fixed modification. Search "oxidized methionine" as variable modifications. Precursor mass tolerances: 10 ppm. Tolerance of MS/MS ions: 0.02 Da.
14. Search proteins/peptides with keywords: HSP, HSC, T-complex, and CCT (*see* Fig. 3).

4 Notes

1. Cells producing the high-level proteins or chaperone proteins of interest have an advantage for detection. Besides, stimulating cells with cell stress inducers such as HSS might also be effective [5].
2. The serum contains EVs/exosomes. Therefore, culture cells in serum-free or EV/exosome-free medium before EV/exosome isolation. EV/exosome-free serum is preparable by the exosome/EV isolation methods of this chapter. Use serum as a starting material in this case. Alternatively, commercially available serum-free medium such as mTeSR1 is useful for serum-free, EV-free culture [5, 6].
3. Cell stress such as HSS is useful for both serum-free and exosome-free cultures and especially useful for inducing HSPs [5]. A serum-free medium containing a set of growth factors required for a specific purpose, such as an mTeSR1 medium for stem cell culture, is useful [5, 6].
4. A bioreactor is also useful for large-scale culture.

5. Isolation of EVs/exosomes from tumoroids is reported in Refs. [5, 6]. More practical examples of 3D culture are available in Refs. [4, 26–30].
6. As patient-derived tissues are rare and precious, comprehensive analyses, such as RNA-seq or mass spectrometry, are recommended. On the other hand, western blotting requires too much sampling with little information output.
7. The 0.2 μm filter syringe is useful for separating large EVs and the selective purification of exosomes/sEVs (50–200 nm). Do not use a 0.2 μm filter syringe if you intend to analyze large EVs, such as microvesicles, stressome, and/or apoptotic bodies [5, 6].
8. The high pressure in the filtration step can destruct the structure of EVs/exosomes [31, 32].
9. Proteases in the EVs/exosomes can degrade EV/exosome proteins. Such proteolysis affects downstream applications such as western blotting. For inhibiting protease-dependent protein degradation, add a 10× protease inhibitor cocktail to the EV/exosome samples.
10. For affinity purification of EVs/exosomes, the MagCapture Exosome Isolation Kit PS (Fujifilm, Irvine Scientific) [33] and exoEasy Maxi Kit (Qiagen) are commercially available.
11. To use the exoEasy Maxi spin column, the maximum volume of serum or plasma is 4 mL, and the maximum volume of culture supernatant is 16 mL (strongly depends on cell type and culture conditions).
12. Centrifugation of the exoEasy Maxi spin columns is performed in a standard laboratory centrifuge with a swinging bucket rotor, preferably capable of up to 5000 × *g* (it is possible to reduce the steps performed at 5000 × *g* down to a minimum force of 3000 × *g* without performance loss).
13. The same qEV column can be reusable approximately five times. Wash the qEV column with 10–15 mL PBS after use for reuse.
14. Zetasizer (Malvern) is useful for vesicle/particle distribution analysis. For counting vesicles/particles and their distribution analysis, use Video Drop (Meiwa Fosis), qNano (Meiwa Fosis), or NanoSight (Malvern).
15. Cryogenic electron microscopy (Cryo-EM) is a state-of-the-art EV/exosome analysis technique.
16. Detergents such as NP-40 or Triton X-100 are needed to dissolve the EV membrane for protein assay and western blotting. Otherwise, proteins will be predominantly detected at

very high molecular-weight bands. See failure examples in supplemental figures in Ref. [4].

17. Exosome/EV protein concentration is often very low and under the detection limit of the regular BCA assay kit. Therefore, a micro BCA kit is necessary.
18. Body fluids such as serum and urine are rich in albumin and globulin bound with EVs/exosomes. In such cases, protein assay is useless.
19. Proteome analysis using mass spectrometry is very sensitive to the contaminations of Keratin derived from dust, saliva, sweat, and dandruff. Therefore, clean gloves, masks, lab head caps, a lab coat, and goggles must be worn. The sample should be handled on a clean bench or laboratory bench cleaned with 80% ethanol.
20. For the PBP method, 2–4 10 cm dishes are required. For the SEC method, 5–20 15 cm dishes are required.
21. Culturing cells for 7 days on the ULA dish often maximizes the size of tumoroids. See experimental examples [6].
22. Use 12 ml medium for a 15 cm dish and 4 ml medium for a 10 cm dish for concentrating EVs/exosomes in the culture supernatant.
23. Alternatively, centrifuge at maximum speed (e.g., 17,000 × *g*) for 30 min at 4 °C to remove cell debris.
24. The UF method is applicable both before and after the other EV/exosome isolation methods and for the re-purification of EVs/exosomes.
25. This concentration step might take a long time. Check the amount of the remaining concentrate every 10 min. If this step takes longer, the culture supernatant might contain much ECM complex. In this case, use Centrifugal Filter Devices for M.W. 500 kD instead of 100 kD.
26. Keep isolated EVs/exosomes at 4 °C for up to 1 week or at ≤ −20 °C for long-term storage.
27. According to the downstream application, choose a buffer type for EV/exosome suspension. For functional assay in vitro and in vivo, PBS is good enough. For protein assay and western blotting, solve the EV/exosome membrane with RIPA buffer.
28. EV/exosome pellet is invisible. Be careful not to aspirate the pellet.
29. This non-affinity fraction contains non-affinity EVs and proteins. Concentrate it in Subheading 3.6, or recollect EVs, e.g., using the PBP method, for further analyses.
30. Centrifugation speed can be reduced from 5000 × *g* down to a minimum force of 3000 × *g* without performance loss.

31. Eluates can be concentrated, e.g., using ultrafiltration. If an ultrafiltration step is performed, eluting in 1–2 mL buffer XE is recommended.
32. This concentration step might take a long time. Check the amount of the remaining concentrate every 10 min. If this step takes longer, the culture supernatant might contain much extracellular matrix (ECM) complex. In this case, use Centrifugal Filter Devices for M.W. 500 kD instead of 100 kD.
33. Load 500 μL concentrated samples in this protocol.
34. In our protocol, one fraction is 500 μL.
35. Growth factors [5] such as cytokines, chemokines [34], metalloproteinases [35], and HSPs [4, 5, 36] are found in both EV and non-EV fractions.
36. For a relative quantitative comparison of factors in EV and non-EV fractions, adjust the sample volumes according to the number of cells or protein concentrations of whole cell lysate.
37. The EV/exosome samples used for ZetaSizer are reusable for another experiment.
38. Sharply peaked graphs indicate a high concentration of EV/exosomes. Gentle-sloping graphs indicate lower concentrations of EVs, exosomes, particles, or protein aggregates [37].

Acknowledgments

K.O. was supported by JSPS Kakenhi (grant numbers 19 K24072 and 21 K17115) and the KAWASAKI Foundation for Medical Science and Medical Welfare. T.E. was supported by JSPS Kakenhi (grant numbers 17 K11642-TE, 20 K09904-CS, 19H03817-MT, 20H03888-HN, 20 K20611-MT, 20H03888-HN, 21H03119-TY, and 21 K08902-HY). The authors thank Junya Futagawa, Kazuko Kobayashi, Kuniaki Okamoto, Chiharu Sogawa, Yuka Okusha, Masaki Matsumoto, Seiji Tamaru, Koji Ueda, and Haruo Urata for useful information, discussion, materials, or experimentation.

References

1. Ellis J (1987) Proteins as molecular chaperones. Nature 328(6129):378–379. https://doi.org/10.1038/328378a0
2. Hartl FU, Bracher A, Hayer-Hartl M (2011) Molecular chaperones in protein folding and proteostasis. Nature 475(7356):324–332. https://doi.org/10.1038/nature10317
3. Ono K, Eguchi T, Sogawa C, Calderwood SK, Futagawa J, Kasai T et al (2018) HSP-enriched properties of extracellular vesicles involve survival of metastatic oral cancer cells. J Cell Biochem 119(9):7350–7362. https://doi.org/10.1002/jcb.27039
4. Ono K, Sogawa C, Kawai H, Tran MT, Taha EA, Lu Y et al (2020) Triple knockdown of CDC37, HSP90-alpha and HSP90-beta diminishes extracellular vesicles-driven malignancy events and macrophage M2 polarization

in oral cancer. J Extracell Vesicles. 9(1): 1769373. https://doi.org/10.1080/20013078.2020.1769373

5. Eguchi T, Sogawa C, Ono K, Matsumoto M, Tran MT, Okusha Y et al (2020) Cell stress induced stressome release including damaged membrane vesicles and extracellular HSP90 by prostate cancer cells. Cells 9(3):755. https://doi.org/10.3390/cells9030755
6. Taha EA, Ono K, Eguchi T (2019) Roles of extracellular HSPs as biomarkers in immune surveillance and immune evasion. Int J Mol Sci 20(18):4588. https://doi.org/10.3390/ijms20184588
7. Richter K, Haslbeck M, Buchner J (2010) The heat shock response: life on the verge of death. Mol Cell 40(2):253–266. https://doi.org/10.1016/j.molcel.2010.10.006
8. Finka A, Goloubinoff P (2013) Proteomic data from human cell cultures refine mechanisms of chaperone-mediated protein homeostasis. Cell Stress Chaperones 18(5):591–605. https://doi.org/10.1007/s12192-013-0413-3
9. Calderwood SK, Khaleque MA, Sawyer DB, Ciocca DR (2006) Heat shock proteins in cancer: chaperones of tumorigenesis. Trends Biochem Sci 31(3):164–172. https://doi.org/10.1016/j.tibs.2006.01.006
10. Ciocca DR, Calderwood SK (2005) Heat shock proteins in cancer: diagnostic, prognostic, predictive, and treatment implications. Cell Stress Chaperones 10(2):86–103. https://doi.org/10.1379/csc-99r.1
11. Whitesell L, Mimnaugh EG, De Costa B, Myers CE, Neckers LM (1994) Inhibition of heat shock protein HSP90-pp60v-src heteroprotein complex formation by benzoquinone ansamycins: essential role for stress proteins in oncogenic transformation. Proc Natl Acad Sci USA 91(18):8324–8328. https://doi.org/10.1073/pnas.91.18.8324
12. Schulte TW, Akinaga S, Soga S, Sullivan W, Stensgard B, Toft D et al (1998) Antibiotic radicicol binds to the N-terminal domain of Hsp90 and shares important biologic activities with geldanamycin. Cell Stress Chaperones 3(2):100–108. https://doi.org/10.1379/1466-1268(1998)003<0100:arbttn>2.3.co;2
13. Blachere NE, Li Z, Chandawarkar RY, Suto R, Jaikaria NS, Basu S et al (1997) Heat shock protein-peptide complexes, reconstituted in vitro, elicit peptide-specific cytotoxic T lymphocyte response and tumor immunity. J Exp Med 186(8):1315–1322. https://doi.org/10.1084/jem.186.8.1315
14. Albakova Z, Siam MKS, Sacitharan PK, Ziganshin RH, Ryazantsev DY, Sapozhnikov AM (2021) Extracellular heat shock proteins and cancer: New perspectives. Transl Oncol 14(2): 100995. https://doi.org/10.1016/j.tranon.2020.100995
15. Shishkova E, Hebert AS, Coon JJ (2016) Now, more than ever, proteomics needs better chromatography. Cell Syst 3(4):321–324. https://doi.org/10.1016/j.cels.2016.10.007
16. Meier F, Brunner AD, Koch S, Koch H, Lubeck M, Krause M et al (2018) Online parallel accumulation-serial fragmentation (PASEF) with a novel trapped ion mobility mass spectrometer. Mol Cell Proteomics 17(12):2534–2545. https://doi.org/10.1074/mcp.TIR118.000900
17. Kelly RT (2020) Single-cell proteomics: progress and prospects. Mol Cell Proteomics 19(11):1739–1748. https://doi.org/10.1074/mcp.R120.002234
18. Tian Y, Gong M, Hu Y, Liu H, Zhang W, Zhang M et al (2020) Quality and efficiency assessment of six extracellular vesicle isolation methods by nano-flow cytometry. J Extracell Vesicles 9(1):1697028. https://doi.org/10.1080/20013078.2019.1697028
19. Hallal S, Russell BP, Wei H, Lee MYT, Toon CW, Sy J et al (2019) Extracellular Vesicles from Neurosurgical Aspirates Identifies Chaperonin Containing TCP1 Subunit 6A as a Potential Glioblastoma Biomarker with Prognostic Significance. Proteomics 19(1–2): e1800157. https://doi.org/10.1002/pmic.201800157
20. Crescitelli R, Lässer C, Jang SC, Cvjetkovic A, Malmhäll C, Karimi N et al (2020) Subpopulations of extracellular vesicles from human metastatic melanoma tissue identified by quantitative proteomics after optimized isolation. J Extracell Vesicles. 9(1):1722433. https://doi.org/10.1080/20013078.2020.1722433
21. Dhondt B, Geeurickx E, Tulkens J, Van Deun J, Vergauwen G, Lippens L et al (2020) Unravelling the proteomic landscape of extracellular vesicles in prostate cancer by density-based fractionation of urine. J Extracell Vesicles. 9(1):1736935. https://doi.org/10.1080/20013078.2020.1736935
22. Hoshino A, Kim HS, Bojmar L, Gyan KE, Cioffi M, Hernandez J et al (2020) Extracellular Vesicle and Particle Biomarkers Define Multiple Human Cancers. Cell 182(4):1044–61.e18. https://doi.org/10.1016/j.cell.2020.07.009
23. Kalluri R, LeBleu VS (2020) The biology, function, and biomedical applications of exosomes. Science 367(6478):eaau6977. https://doi.org/10.1126/science.aau6977
24. Patel GK, Khan MA, Zubair H, Srivastava SK, Khushman M, Singh S et al (2019) Comparative analysis of exosome isolation methods

using culture supernatant for optimum yield, purity and downstream applications. Sci Rep 9(1):5335. https://doi.org/10.1038/s41598-019-41800-2

25. Pietrowska M, Wlosowicz A, Gawin M, Widlak P (2019) MS-Based proteomic analysis of serum and plasma: problem of high abundant components and lights and shadows of albumin removal. Adv Exp Med Biol 1073:57–76. https://doi.org/10.1007/978-3-030-12298-0_3
26. Sogawa C, Eguchi T, Namba Y, Okusha Y, Aoyama E, Ohyama K et al (2021) Gel-Free 3D tumoroids with stem cell properties modeling drug resistance to cisplatin and imatinib in metastatic colorectal cancer. Cells 10(2):344. https://doi.org/10.3390/cells10020344
27. Sogawa C, Eguchi T, Tran MT, Ishige M, Trin K, Okusha Y et al (2020) Antiparkinson drug benztropine suppresses tumor growth, circulating tumor cells, and metastasis by acting on SLC6A3/DAT and reducing STAT3. Cancers (Basel) 12(2):1–22
28. Sogawa C, Eguchi T, Okusha Y, Ono K, Ohyama K, Iizuka M et al (2019) A reporter system evaluates tumorigenesis, metastasis, beta-catenin/MMP regulation, and druggability. Tissue Eng Part A 25(19–20):1413–1425. https://doi.org/10.1089/ten.TEA.2018.0348
29. Namba Y, Sogawa C, Okusha Y, Kawai H, Itagaki M, Ono K et al (2018) Depletion of lipid efflux pump ABCG1 triggers the intracellular accumulation of extracellular vesicles and reduces aggregation and tumorigenesis of metastatic cancer cells. Front Oncol 8(376):1–16. https://doi.org/10.3389/fonc.2018.00376
30. Ono K, Sato K, Nakamura T, Yoshida Y, Murata S, Yoshida K et al (2022) Reproduction of the Antitumor Effect of Cisplatin and Cetuximab Using a Three-dimensional Spheroid Model in Oral Cancer. Int J Med Sci 19(8): 1320–1333. https://doi.org/10.7150/ijms.74109
31. Thery C, Witwer KW, Aikawa E, Alcaraz MJ, Anderson JD, Andriantsitohaina R et al (2018) Minimal information for studies of extracellular vesicles 2018 (MISEV2018): a position statement of the International Society for Extracellular Vesicles and update of the MISEV2014 guidelines. J Extracell Vesicles. 7(1):1535750. https://doi.org/10.1080/20013078.2018.1535750
32. Yang D, Zhang W, Zhang H, Zhang F, Chen L, Ma L et al (2020) Progress, opportunity, and perspective on exosome isolation - efforts for efficient exosome-based theranostics. Theranostics 10(8):3684–3707. https://doi.org/10.7150/thno.41580
33. Nakai W, Yoshida T, Diez D, Miyatake Y, Nishibu T, Imawaka N et al (2016) A novel affinity-based method for the isolation of highly purified extracellular vesicles. Sci Rep 6:33935. https://doi.org/10.1038/srep33935
34. Lima LG, Ham S, Shin H, Chai EPZ, Lek ESH, Lobb RJ et al (2021) Tumor microenvironmental cytokines bound to cancer exosomes determine uptake by cytokine receptor-expressing cells and biodistribution. Nat Commun 12(1):3543. https://doi.org/10.1038/s41467-021-23946-8
35. Taha EA, Sogawa C, Okusha Y, Kawai H, Oo MW, Elseoudi A et al (2020) Knockout of MMP3 weakens solid tumor organoids and cancer extracellular vesicles. Cancers (Basel) 12(5):1260. https://doi.org/10.3390/cancers12051260
36. Eguchi T, Ono K, Kawata K, Okamoto K, Calderwood SK. Regulatory roles of HSP90-rich extracellular vesicles. In: Asea AAA, Kaur P, editors. Heat shock protein 90 in human diseases and disorders. Heat shock proteins. Cham: Springer Nature; 2019. p. 3–17
37. Lu Y, Eguchi T, Sogawa C, Taha EA, Tran MT, Nara T et al (2021) Exosome-based molecular transfer activity of macrophage-like cells involves viability of oral carcinoma cells: size exclusion chromatography and concentration filter method. Cells 10(6):1328. https://doi.org/10.3390/cells10061328

Chapter 19

A Modified Differential Centrifugation Protocol for Isolation and Quantitation of Extracellular Heat Shock Protein 90 (eHsp90)

Cheng Chang, Xin Tang, and Wei Li

Abstract

Studies of the past 15 years have revealed a critical role for extracellular heat shock protein 90alpha (eHsp90α) in the development of several human disorders, including wound healing, cachexia (muscle wasting), inflammatory diseases, and cancers. The two established functions of highly purified eHsp90α protein are to promote cell survival and to stimulate cell migration. However, the mechanism of secretion and the method of isolation of eHsp90α remained to be standardized. Among the half a dozen reported methodologies, differential centrifugation is considered the "gold standard" largely for its quantitative recovery of eHsp90α from a conditioned medium of cultured cells. Herein, we describe a revised protocol that isolates three fractions of extracellular vesicles with distinct ranges of diameters and the leftover vesicle-free supernatant for biochemical analyses, especially eHsp90α, from tumor cell-conditioned media. Quantitation of the relative amount of eHsp90α can be carried out with known amounts of recombinant Hsp90α protein on the same SDS-PAGE. We believe that this modified methodology will prove to be a useful tool for studying eHsp90α in cultured cells and beyond.

Key words Extracellular vesicles, Exosomes, Differential centrifugation, Extracellular heat shock protein 90

1 Introduction

Extracellular vesicles (EVs) are cell-derived lipid bilayer vesicles ranging from 30 nm to 2000 nm in diameter according to their origin [1]. EVs are secreted by donor cells into the extracellular space where they transport various cargos, including nucleic acids, lipids, and proteins to recipient cells for cell-to-cell communication either within the same or between different microenvironments [2]. In recent years, EVs have become therapeutic agents for various human diseases [1]. Based on the size, morphology, markers, contents, and origins, EVs can be classified into three main categories, including (1) microvesicles, (2) exosomes, and (3) apoptotic

Stuart K. Calderwood and Thomas L. Prince (eds.), *Chaperones: Methods and Protocols*, Methods in Molecular Biology, vol. 2693, https://doi.org/10.1007/978-1-0716-3342-7_19,

bodies [1–4]. Microvesicles are medium size vesicles whose diameters range from 100 nm to 1000 nm. Enclosed and containing cytoplasmic cargos, microvesicles bud outward directly from the plasma membrane [5]. Exosomes are the smallest vesicles whose diameters range from 30 nm to 150 nm [1, 6–8]. Different from the origin of microvesicles, exosomes undergo the maturation process inside the cell. Firstly, the plasma membrane undergoes invagination to generate early endosomes. Then, the maturing endosomes continue to bud inward and generate intraluminal vesicles (ILVs). The endosome-containing ILVs are also known as multivesicular bodies (MVBs). Some of the MVBs fuse with lysosomes and be degraded with their cargos, whereas other MVBs fuse with the plasma membrane and release ILVs. After this point, the secreted ILVs are called exosomes [7, 8]. Apoptotic bodies are the largest vesicles whose diameters range from 50 nm to 2000 nm. During apoptosis, the dying cell begins to condensate its nuclear chromatin and disintegrated the cellular content into distinct membrane-enclosed vesicles, so-called apoptotic bodies [9]. All three types of EVs can be taken up by recipient cells via different mechanisms, including (1) directly fusing with the plasma membrane and delivering its cargos into the cytosol, (2) attaching to the cell surface via ligand–receptor interaction, or (3) being taken up through endocytosis [1, 2]. Either way, the EVs play a crucial role in regulating the biological process of recipient cells. Taking tumor-derived EVs as an example, numerous studies have implicated the important roles of EVs in promoting cancer metastasis [10–13], stimulating cancer proliferation [13–15], inducing angiogenesis [14–16], driving the formation of pre-metastatic niche [17, 18], and promoting immune escape by modulating T cell activities [19–21].

One major challenge for isolating EVs from the cell-conditioned medium is the lack of a standardized methodology, especially when it comes to quantitative analyses of their contents [22]. There were three commonly used methods for EV isolation, including (1) differential centrifugation, (2) ultrafiltration (UF), and (3) immunoaffinity isolation. Differential centrifugation is the most commonly used method for EV isolation. Based on the different sizes and densities of EVs, differential centrifugation uses a sequence of different centrifugal forces to pellet and separate the larger and denser vesicles from the smaller and less-dense vesicles [3]. Although differential centrifugation has been considered a "golden standard" for the quantitation of EV isolation, there is no universally accepted protocol [3]. Only the three steps of sequential centrifugations reached consensus in different laboratories (Fig. 1). The disadvantage of differential centrifugation is the complexity of the operation process and the strict requirements for the equipment. For example, the acceleration, type of rotor, radius of rotation, sedimentation path length, and sample viscosity,

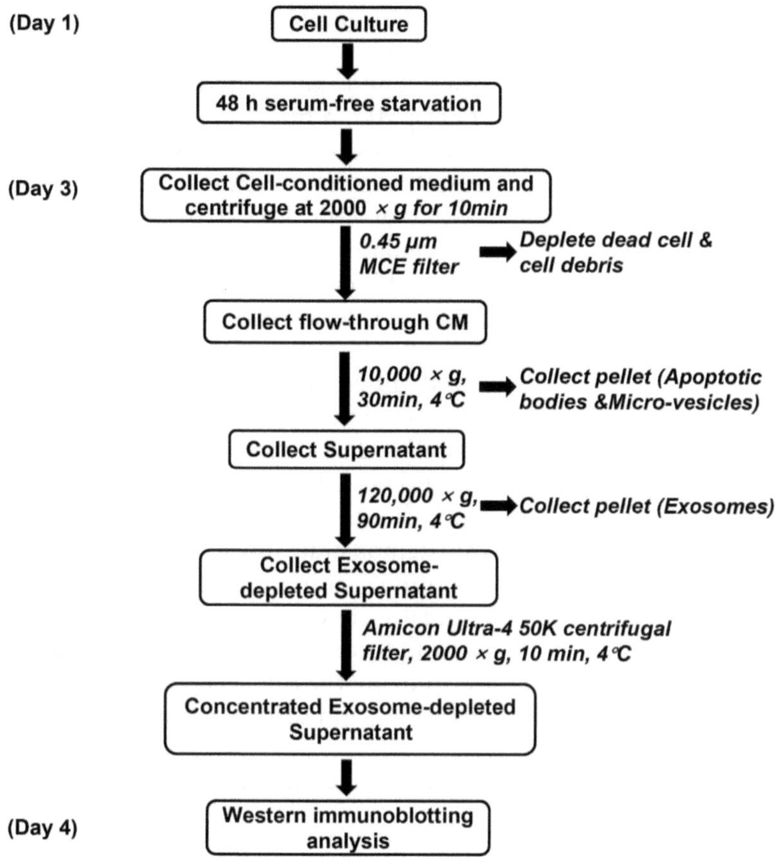

Fig. 1 Protocol for isolating three extracellular fractions from MDA-MB-231 cell-conditioned medium for the presence of eHsp90α. Each step was established based on the reproducible results of multiple experiments in our laboratory. The primary emphasis was to avoid contamination of the small number of dead cells

determine the final efficiency of EV isolation [23–26]. Ultrafiltration uses membranes with different pore sizes to trap and separate vesicles from other smaller components or molecules [27, 28]. The main disadvantages of UF are the reduced EVs recovery rate and the low purity of isolated EVs due to the membrane plugging [29, 30]. Immunoaffinity isolation is the best method to isolate specific EVs. Based on specific proteins, polysaccharides, and lipids that EVs express on the surface, this method uses antibodies, lectins, or lipid-binding proteins coated affinity capture beads to pull down specific EV populations [26, 31]. The disadvantage of this method is the low yield of EVs due to the limited binding capacity of affinity beads [26, 32].

One major protein found on the external surface of exosomes is heat shock protein 90 alpha (Hsp90α) [33]. The discovery of extracellular Hsp90α (eHsp90α) can be traced back to the late 1970s [34], while the function of eHsp90α remained to be elucidated at that time. Two papers in 2004 and 2007 lay the foundation for the functions of eHsp90α in promoting tumor cell invasion and skin wound healing [35, 36]. Furthermore, the level of eHsp90α in human plasma is also shown to be positively correlated with cancer progression [37–39]. Given the importance of eHsp90α in tumorigenesis, however, there is no standardized protocol to describe how to isolate and quantify the eHsp90α from the cell-conditioned medium. Herein, taking advantage of the conventional differential centrifugation protocol [22], we demonstrate an optimized differential centrifugation protocol to separate and quantify eHsp90α from the cell-conditioned medium. The optimized differential centrifugation protocol allows us to separate three fractions from the cell-conditioned medium that contained eHsp90α. These three fractions are the large vesicles (microvesicles and apoptotic bodies), exosomes, and extracellular vesicle-depleted supernatant. Besides isolating and separating three fractions, we also demonstrate how to detect and quantify eHsp90α in each fraction by using immunoblotting analysis. Different from the original differential centrifugation protocol that only focuses on isolating exosomes, this optimized method provides new insights for studying the cell-secreted Hsp90α in terms of its existing forms, quantities, and secretion mechanisms.

2 Materials

2.1 For Cell Culture

1. Human triple-negative breast cancer cell line MDA-MB-231 (from the laboratory of Dr. Michael Press, University of Southern California, Los Angeles, CA).
2. Dulbecco's Modified Eagle's Medium (DMEM) with high glucose supplemented with 10% fetal bovine serum (FBS) (Thermo Scientific, MA) and 1% penicillin–streptomycin (P/S) (Thermo Scientific, MA).
3. Dulbecco's Phosphate Buffered Saline (DPBS) (Thermo Scientific, MA).
4. Trypsin (2.5%) without phenol red (Thermo Scientific, MA).

2.2 For Centrifugations

1. Avanti®J-E centrifuge with JLA-10.500 rotor (Beckman Coulter, CA).
2. Optima™ L-100 XP Ultracentrifuge with SW-41Ti rotor (Beckman Coulter, CA).
3. Polypropylene Centrifuge Tubes (Backman Coulter, CA).
4. Eppendorf Centrifuge 5424 (Hamburg, Germany).
5. 0.45 μm MCE membrane (Cat. #: HAWP04700 Millipore Sigma, MA).

6. Amicon® Ultra-4 Centrifugal filters, Ultracel-50 K (Cat. #: UFC805024 Millipore Sigma, MA).
7. 50 ml Polypropylene Conical Tube (Corning, NY).
8. Phosphate Buffered Saline (PBS) (Thermo Scientific, MA).

2.3 For Immunol (Western) Blotting Analysis

1. Cell lysis buffer: 20 mM Tris-base, 50 mM NaCl, 50 mM napyrophosphate, 30 mM NaF, 100 μM naorthovanadate, 5 μM zinc chloride, 2 mM Iodoacetic acid, 1.0% Triton-X-100, pH adjusted to 7.4 with 1 N HCl. Add ddH_2O to the final volume of 1 L and filter with 0.45 μm MEC membrane.
2. 4 × Tris-HCl-SDS pH 8.8: To 300 ml ddH_2O, add 91.0 g Tris-base and 2.0 g SDS. Adjust pH to 8.8 with 1 N HCl, add ddH_2O to the final volume of 500 ml, and filter with 0.45 μm MEC membrane.
3. 4 × Tris-HCl-SDS pH 6.8: To 200 ml ddH_2O, add 30.25 g Tris-base and 2.0 g SDS. Adjust pH to 6.8 with 1 N HCl, add ddH_2O to the final volume of 500 ml, and filter with 0.45 μm MEC membrane.
4. SDS-PAGE: 10% resolving gel: 15 ml H_2O, 12 ml 30% acrylamide/Bis solution 37.5:1, 9 ml 4 × Tris-HCl-SDS pH 8.8, 10% APS 150 μl, TEMED 15 μl. 8% stacking gel: 6.25 ml H_2O, 1.2 ml 30% acrylamide/Bis solution 37.5:1, 2.5 ml 4 × Tris-HCl-SDS pH 6.8, 100 μl 10% ammonium persulfate (APS), 10 μl TEMED.
5. 4 × SDS-PAGE sample buffer: Tris-base 3.04 g, glycerol 40 ml, SDS 4 g, β-mercaptoethanol 4 ml, ddH_2O 40 ml. Mix well, and adjust pH to 6.8 with 1 N HCl. Add 2 mg bromophenol blue, add ddH_2O to the final volume of 100 ml, and filter with 0.45 μm MEC membrane.
6. 10 × running buffer: 60.4 g Tris-base, glycine 288 g, SDS 20 g. Add ddH_2O to the final volume of 2 L.
7. 1 × transfer buffer: 6.04 g Tris-base, glycine 28.8 g, 200 ml methanol, SDS 2 g. Add ddH_2O to the final volume of 2 L.
8. 10 × TBS: 100 mM Tris-base, 1.5 M NaCl. Adjust pH to 7.4 with 1 N HCl, and add ddH_2O to the final volume of 1 L.
9. 10 × TTBS: 10 × TBS with 0.5% Triton-X-100.
10. Rabbit anti-Hsp90 alpha antibody (Cat. #: NB120–2928, Novus Biologicals, CO).
11. Rabbit anti-β-actin mAb (Cat. #: CA038, Transduction Laboratories, CA).
12. Rabbit anti-CD9 mAb (Cat. #: 13403S, Cell Signaling, MA).
13. m-IgG_k BP-HPR (Cat. # sc-516,102, Santa Cruz, TX).

14. Goat Anti-Rabbit IgG Antibody, (H + L) HRP conjugate (Cat. #: AP307P MilliporeSigma, MA).
15. Nonfat dry milk (Cell Signaling, MA).
16. Nitrocellulose membrane 0.45 μm (Bio-Rad, CA).
17. Amersham™ Ecl™ Western Blotting Detection Reagents (Cytiva, MA).
18. Premium Clear Blue X-Ray Film (Bioland, CA).

3 Methods

3.1 Cell Culture

1. MDA-MB-231 cells are cultured in the desired number of 15 cm cell culture dishes at 37 °C in a humidified incubator with 5% CO_2 in DMEM culture medium with high glucose supplemented with 10% FBS and 1% P/S for at least 48 h to reach approximately 80% cell confluence.
2. Cells then are subcultured into the desired number of 15 cm cell culture dishes for preparation and collection of the cell-conditioned medium (CM).
3. (Day 1) When the cell confluence reaches 80%, aspirate the serum-containing medium, and wash the cells gently three times with 10 ml 37 °C pre-warmed DPBS per dish.
5. Add 12 ml serum-free DMEM to each plate and incubated for an additional 48 h (*see* **Note 1**).

3.2 Separation of Cell-Conditioned Medium into Three Fractions

1. (Day 3) After 48 h of cell culture, collect CM (*see* **Note 2**) and centrifuge at 2000 *g* at 4 °C for 10 min to remove floating cells.
2. Transfer the CM into a clean 50 ml conical tube (*see* **Note 3**).
3. Filter the CM through an 0.45 μm MCE membrane by vacuum aspiration to further remove any debrides and save the flow through CM (*see* **Note 4**).
4. Transfer the CM to a clean 50 ml conical tube, and centrifuge at 4 °C for 30 min at a speed of 10,000 *g* to collect the pellet of apoptotic bodies and microvesicles. Transfer the supernatant into a new tube, and make the tube as "exosome-contained supernatant." Transfer the supernatant, 6 ml each, into six 10 ml Beckman ultracentrifuge tubes (*see* **Note 5**).
5. Invert the pellet- (apoptotic bodies and microvesicles) containing 50 ml conical tube on paper towels to remove residual medium as completely as possible, use 1 ml cold PBS to resuspend the pellet, and transfer to a clean 1.5 ml Eppendorf (EP) tube (*see* **Note 6**).
6. Centrifuge the EP tube at 4 °C for 30 min at a speed of 10,000 *g* to re-pellet the apoptotic bodies and microvesicles.

7. Gently discard the supernatant, and add 100 μl cold PBS to resuspend apoptotic bodies and microvesicle pellet. One could either immediately add 4 × sample buffer (with 10% β-mercaptoethanol) and boil it for Western immunoblotting analysis or store it at −80 °C for later use.
8. Pre-cool the ultracentrifugation rotor inside the machine to 4 °C. Ultracentrifuge the exosome-contained supernatant at 120,000 *g* for 90 min at 4 °C.
9. Collect the supernatant, mark it as exosome-depleted supernatant, and keep it on ice.
10. Add 10 ml cold PBS to resuspend exosome pellets from different Beckman ultracentrifuge tubes to the same one and re-ultracentrifuge (120,000 g for 90 min at 4 °C) to obtain purified exosomes (*see* **Note** 7).
11. Gently discard the supernatant and invert the Beckman ultracentrifuge tube on paper towels to remove residual PBS completely. A semitransparent pellet on the bottom of the ultracentrifugation tube should become visible.
12. Add 100 μl cold PBS to resuspend the exosome pellet, and either immediately add 4 × sample buffer (with 10% β-mercaptoethanol) or boil it for Western immunoblotting analysis right away or store at −80 °C.
13. Use Amicon Ultra-4 50 K centrifugal filter to concentrate exosome-depleted supernatant by using 2000 g centrifugation for 10 min at 4 °C.
14. Discard flow through, and repeat step 13 until the final volume of exosome-depleted supernatant is concentrated to <200 μl (*see* **Note 8**).
15. After the concentration, pipette the filter membrane several times and transfer the 200 μl concentrated exosome-depleted supernatant to a clean EP tube, store at either 4 °C for short term (within 2 days), or immediately add 4 × sample buffer (with 10% β-mercaptoethanol) and boil it for Western immunoblotting analysis (*see* **Note 9**).

3.3 Immunoblotting Analysis

1. (Day 4) Add 10 μl 4 × sample buffer (with 10% β-mercaptoethanol) to apoptotic bodies/microvesicle and purified exosome samples.
2. Keep the EP tube lid open, and boil samples until the final volume of samples decreases to 40 μl (*see* **Note 10**).
3. The apoptotic bodies/microvesicles and purified exosome samples are ready for Western immunoblotting analysis. Freeze down the samples at −20 °C if not immediately used.

4. (Day 4) Add 4 × sample buffer (with 10% β-mercaptoethanol) to the exosome-depleted supernatant in a 1:3 ratio.
5. Boil samples for 5 min. The exosome-depleted supernatant sample is ready for Western immunoblotting analysis. Freeze down the samples at −20 °C if not immediately used.
6. Briefly centrifuge three samples (apoptotic bodies/microvesicles, exosomes, exosome-depleted supernatant).
7. Load samples on 10% SDS-PAGE gel with final 1 × running buffer.
8. Electrophoresis at 180 V for 50 min (*see* **Note 11**).
9. Prepare fresh transfer buffer. Transfer the gel onto the nitrocellulose membrane at 80 V for 1 h, and keep it on ice.
10. Block membrane with 5% nonfat dry milk in 1 × TBS for 1 h.
11. Wash the membrane with 1 × TBS for 5 min followed by 1 × TTBS three times, each wash lasting 5 min.
12. Use anti-Hsp90α antibody at 1:1000 dilution, anti-β-actin antibody at 1:1000 dilution, and anti-CD9 at 1:1000 dilution; shake and incubate with the membrane at 4 °C overnight.
13. Wash the membrane with 1 × TBS for 5 min followed by 1 × TTBS three times, each wash lasting 5 min.
14. During step 13, prepare secondary antibody: (1) goat anti-rabbit and (2) goat anti-mouse antibody at 1:10000 dilution with 1 × TBS 5% nonfat dry milk; shake and incubate at room temperature for 1 h.
15. Wash the membrane with 1 × TBS for 5 min followed by 1 × TTBS three times, each wash lasting 5 min.
16. Apply ECL solution to the membrane for 30 s. Gently remove excess ECL solution by dipping the membrane on a paper towel.
17. Place membrane on an X-ray film developing cassette.
18. Place X-ray film to the membrane for different times of length (*see* **Note 12**) in a dark room.
19. Detect Hsp90α protein at 90 kDa. Detect β-actin protein at 42 kDa. Detect CD9 protein at 23-27 kDa (Fig. 2).

4 Notes

1. Gently add 12 ml DMEM along the edge of the cell culture dish to avoid cell detachment. In addition, 10 ml DMEM may not be enough to cover a 15 cm culture dish, due to unleveled surface and medium evaporation.

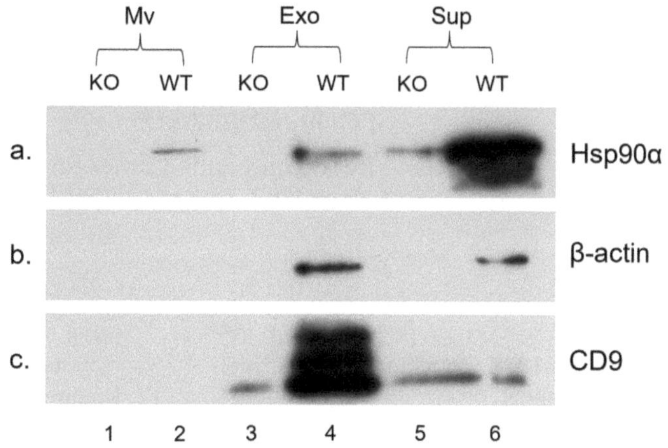

Fig. 2 A representative data of immunoblotting (Western) analysis of the indicated extracellular fractions with the indicated antibodies. The microvesicles and apoptotic body fraction (Mv), the exosome fraction (Exo), and the EV-depleted supernatant fraction (Sup) from parental (WT) or Hsp90α-knock-out (KO) MDA-MB-231 cells were immunoblotted with anti-Hsp90α, anti-β-actin, and anti-CD9 antibodies

2. To collect CM, avoid touching and disrupting the attached cell monolayer.
3. Avoid freezing down the CM, since freezing and thawing could disrupt the integrity of the extracellular vesicles.
4. Ideally operate this step in a 4 °C cold room; otherwise, operate this step quickly on ice.
5. Since the maximum volume of each of the Beckman ultracentrifugation tubes is 10 ml, four will be required for 36 ml CM.
6. Respin down microvesicles and apoptotic bodies after suspending the pellets in 1 ml cold PBS to remove containment particles from the CM.
7. Respin down exosomes after suspending the pellets in 10 ml cold PBS to remove containment particles from the supernatant.
8. The final volume of concentrated exosome-depleted supernatant varies depending on the different initial volume of CM (e.g., if the initial volume of CM is 100 ml, the final volume of concentrated supernatant will be adjusted to 500 μl).
9. Do not freeze down the sample at this step since the repeated freezing and thawing could cause protein degradation. Prepare the samples with sample buffer (with 10% β-mercaptoethanol) immediately.
10. This step aims to increase the protein concentration in extracellular vesicle samples. At least three plates of cultured cells are

needed to visualize the vesicle markers by Western immunoblotting analysis.

11. Do not run off the 15 kDa maker, since the molecular weight of CD9 is 23-27 kDa.
12. The recommended times for ECL film development are 5 min, 1 min, 30 s, and 15 s.

References

1. El Andaloussi S, Mäger I, Breakefield XO, Wood MJ (2013) Extracellular vesicles: biology and emerging therapeutic opportunities. Nat Rev Drug Discov 12:347–357
2. Zaborowski MP, Balaj L, Breakefield XO, Lai CP (2015) Extracellular vesicles: composition, biological relevance, and methods of study. Bioscience 65:783–797
3. Van der Pol E, Böing AN, Harrison P, Sturk A, Nieuwland R (2012) Classification, functions, and clinical relevance of extracellular vesicles. Pharmacol Rev 64:676–705
4. Tkach M, Théry C (2016) Communication by extracellular vesicles: where we are and where we need to go. Cell 164:1226–1232
5. Heijnen HF, Schiel AE, Fijnheer R, Geuze HJ, Sixma JJ (1999) Activated platelets release two types of membrane vesicles: microvesicles by surface shedding and exosomes derived from exocytosis of multivesicular bodies and α-granules. Blood J Am Soc Hematol 94: 3791–3799
6. Cocucci E, Meldolesi J (2015) Ectosomes and exosomes: shedding the confusion between extracellular vesicles. Trends Cell Biol 25: 364–372
7. Théry C, Zitvogel L, Amigorena S (2002) Exosomes: composition, biogenesis and function. Nat Rev Immunol 2:569–579
8. Kalluri R (2016) The biology and function of exosomes in cancer. J Clin Invest 126:1208–1215
9. Akers JC, Gonda D, Kim R, Carter BS, Chen CC (2013) Biogenesis of extracellular vesicles (EV): exosomes, microvesicles, retrovirus-like vesicles, and apoptotic bodies. J Neuro-Oncol 113:1–11
10. Becker A, Thakur BK, Weiss JM, Kim HS, Peinado H, Lyden D (2016) Extracellular vesicles in cancer: cell-to-cell mediators of metastasis. Cancer Cell 30:836–848
11. Nishida-Aoki N, Tominaga N, Takeshita F, Sonoda H, Yoshioka Y, Ochiya T (2017) Disruption of circulating extracellular vesicles as a novel therapeutic strategy against cancer metastasis. Mol Ther 25:181–191
12. Kogure A, Yoshioka Y, Ochiya T (2020) Extracellular vesicles in cancer metastasis: potential as therapeutic targets and materials. Int J Mol Sci 21:4463
13. Xu R, Rai A, Chen M, Suwakulsiri W, Greening DW, Simpson RJ (2018) Extracellular vesicles in cancer—implications for future improvements in cancer care. Nat Rev Clin Oncol 15: 617–638
14. Al-Nedawi K, Meehan B, Micallef J, Lhotak V, May L, Guha A et al (2008) Intercellular transfer of the oncogenic receptor EGFRvIII by microvesicles derived from tumour cells. Nat Cell Biol 10:619–624
15. Skog J, Würdinger T, Van Rijn S, Meijer DH, Gainche L, Curry WT et al (2008) Glioblastoma microvesicles transport RNA and proteins that promote tumour growth and provide diagnostic biomarkers. Nat Cell Biol 10:1470–1476
16. Al-Nedawi K, Meehan B, Kerbel RS, Allison AC, Rak J (2009) Endothelial expression of autocrine VEGF upon the uptake of tumor-derived microvesicles containing oncogenic EGFR. Proc Natl Acad Sci 106:3794–3799
17. Gatti S, Bruno S, Deregibus MC, Sordi A, Cantaluppi V, Tetta C et al (2011) Microvesicles derived from human adult mesenchymal stem cells protect against ischaemia–reperfusion-induced acute and chronic kidney injury. Nephrol Dial Transplan 26:1474–1483
18. Rak J, Guha A (2012) Extracellular vesicles–vehicles that spread cancer genes. BioEssays 34: 489–497
19. Wieckowski EU, Visus C, Szajnik M, Szczepanski MJ, Storkus WJ, Whiteside TL (2009) Tumor-derived microvesicles promote regulatory T cell expansion and induce apoptosis in tumor-reactive activated CD8+ T lymphocytes. J Immunol 183:3720–3730
20. Kim JW, Wieckowski E, Taylor DD, Reichert TE, Watkins S, Whiteside TL (2005) Fas ligand–positive membranous vesicles isolated from sera of patients with oral cancer induce

apoptosis of activated T lymphocytes. Clin Cancer Res 11:1010–1020

21. Cai Z, Yang F, Yu L, Yu Z, Jiang L, Wang Q et al (2012) Activated T cell exosomes promote tumor invasion via Fas signaling pathway. J Immunol 188:5954–5961
22. Théry C, Amigorena S, Raposo G, Clayton A (2006) Isolation and characterization of exosomes from cell culture supernatants and biological fluids. Curr Protoc Cell Biol 30:3–22
23. Momen-Heravi F, Balaj L, Alian S, Trachtenberg AJ, Hochberg FH, Skog J et al (2012) Impact of biofluid viscosity on size and sedimentation efficiency of the isolated microvesicles. Front Physiol 3:162
24. Momen-Heravi F, Balaj L, Alian S, Mantel PY, Halleck AE, Trachtenberg AJ et al (2013) Current methods for the isolation of extracellular vesicles. Biol Chem 394:1253–1262
25. Cvjetkovic A, Lötvall J, Lässer C (2014) The influence of rotor type and centrifugation time on the yield and purity of extracellular vesicles. J Extracell Vesicles 3:23111
26. Clos-Sansalvador M, Monguió-Tortajada M, Roura S, Franquesa M, Borràs FE (2022) Commonly used methods for extracellular vesicles' enrichment: Implications in downstream analyses and use. Eur J Cell Biol 101:151227
27. Nordin JZ, Lee Y, Vader P, Mäger I, Johansson HJ, Heusermann W et al (2015) Ultrafiltration with size-exclusion liquid chromatography for high yield isolation of extracellular vesicles preserving intact biophysical and functional properties. Nanomedicine 11:879–883
28. Benedikter BJ, Bouwman FG, Vajen T, Heinzmann AC, Grauls G, Mariman EC et al (2017) Ultrafiltration combined with size exclusion chromatography efficiently isolates extracellular vesicles from cell culture media for compositional and functional studies. Sci Rep 7:1–13
29. Cheruvanky A, Zhou H, Pisitkun T, Kopp JB, Knepper MA, Yuen PS et al (2007) Rapid isolation of urinary exosomal biomarkers using a nanomembrane ultrafiltration concentrator. Am J Physiol Renal Physiol 292:1657–F1661
30. Lobb RJ, Becker M, Wen Wen S, Wong CS, Wiegmans AP, Leimgruber A et al (2015) Optimized exosome isolation protocol for cell culture supernatant and human plasma. J Extracell Vesicles 4:27031
31. Konoshenko MY, Lekchnov EA, Vlassov AV (2018) Laktionov PP (2018) isolation of extracellular vesicles: general methodologies and latest trends. Biomed Res Int 2018:1
32. Tauro BJ, Greening DW, Mathias RA, Ji H, Mathivanan S, Scott AM et al (2012) Comparison of ultracentrifugation, density gradient separation, and immunoaffinity capture methods for isolating human colon cancer cell line LIM1863-derived exosomes. Methods 56: 293–304
33. Tang X, Chang C, Guo J, Lincoln V, Liang C, Chen M et al (2019) Tumour-secreted Hsp90α on external surface of exosomes mediates tumour-stromal cell communication via autocrine and paracrine mechanisms. Sci Rep 9:1–13
34. Stone KR, Smith RE, Joklik WK (1974) Changes in membrane polypeptides that occur when chick embryo fibroblasts and NRK cells are transformed with avian sarcoma viruses. Virology 58:86–100
35. Eustace BK, Sakurai T, Stewart JK, Yimlamai D, Unger C, Zehetmeier C et al (2004) Functional proteomic screens reveal an essential extracellular role for hsp90α in cancer cell invasiveness. Nat Cell Biol 6:507–514
36. Li W, LiY GS, Fan J, Cheng CF, Bright AM et al (2007) Extracellular heat shock protein-90α: linking hypoxia to skin cell motility and wound healing. EMBO J 26:1221–1233
37. Wang X, Song X, Zhuo W, Fu Y, Shi H, Liang Y et al (2009) The regulatory mechanism of Hsp90α secretion and its function in tumor malignancy. Proc Natl Acad Sci 106:21288–21293
38. Burgess EF, Ham AJL, Tabb DL, Billheimer D, Roth BJ, Chang SS et al (2008) Prostate cancer serum biomarker discovery through proteomic analysis of alpha-2 macroglobulin protein complexes. PROTEOMICS Clin Appl 2:1223–1233
39. Wei W, Liu M, Ning S, Wei J, Zhong J, Li J et al (2020) Diagnostic value of plasma HSP90α levels for detection of hepatocellular carcinoma. BMC Cancer 20:1–9

Chapter 20

Immunohistochemistry of Human Hsp60 in Health and Disease: Recent Advances in Immunomorphology and Methods for Assessing the Chaperonin in Extracellular Vesicles

Celeste Caruso Bavisotto, Francesco Cappello, Everly Conway de Macario, Alberto J. L. Macario, and Francesca Rappa

Abstract

The subject matter of this chapter is defined by the title of its two previous editions, "Immunohistochemistry of human Hsp60 in health and disease: From autoimmunity to cancer," the latest of which appeared in 2018. Since then, considerable advances have been made in the fields of autoimmunity and cancer and some of them are closely linked to progress in the understanding of the chaperone system (CS). This is a physiological system composed of molecular chaperones, co-chaperones, chaperone cofactors, and chaperone interactors and receptors. The molecular chaperones are the chief members of the CS, and here we focus on one of them, Hsp60. Since extracellular vesicles (EVs) have also emerged as key factors in the functioning of the CS and in carcinogenesis, we have incorporated a detailed section about them. This chapter explains how to assess Hsp60 in tissues and in EVs for application in diagnosis, prognostication, and patient monitoring and, eventually, for developing methods using them as therapeutic targets and tools. We describe immunohistochemical techniques, immunofluorescence and double immunofluorescence-confocal microscopy, and methods for collecting and isolating EVs from blood plasma and for assessing their contents in Hsp60 and related microRNAs (miRNAs). All these procedures have proven to be reliable and useful in the study and management of various types of cancer and inflammatory and autoimmune conditions.

Key words Hsp60, Chaperone system, Extracellular vesicles, microRNA, Immunohistochemistry, Immunomorphology, Immunofluorescence, Double immunofluorescence, Confocal microscopy, Transmission electron microscopy, Cancer

1 Introduction

Hsp60 is a member of the chaperone system (CS) which, like other members of the CS, has canonical and noncanonical functions, the former to maintain protein homeostasis inside mitochondria and

Stuart K. Calderwood and Thomas L. Prince (eds.), *Chaperones: Methods and Protocols*, Methods in Molecular Biology, vol. 2693, https://doi.org/10.1007/978-1-0716-3342-7_20,

the latter displayed in various intra- and extracellular locations and related to inflammatory and autoimmune processes and carcinogenesis [1–4]. Critical reviews of many aspects of Hsp60 are available in the two previous versions of this chapter [5]. Since the publication of the latest version in 2018, our group continued to work with Hsp60 as reported [6–10]. Likewise, other groups have contributed to the field as illustrated in recent publications [11–13]. Consequently, there is a plethora of information on Hsp60 evolution and structure and functions in health and disease already available that can easily be consulted, so to avoid duplication, we will not expand on this topic here.

This chapter follows essentially the same structure as its predecessors and presents methods for immunohistochemistry and immunocytochemistry of human and experimental-model tissue and cell samples to determine the presence and levels of Hsp60. This chapter also includes the methods for immunofluorescence and double immunofluorescence confocal microscopy, the latter to detect the colocalization of Hsp60 with other molecules of interest. The current version of the chapter includes an entirely new section, pertaining to extracellular vesicles (EVs), and explains how to collect them from plasma and how to purify and study them, focusing on Hsp60 and related microRNAs (miRNAs). EVs-Hsp60 have the potential as biomarkers useful for differential diagnosis and patient follow-up in certain diseases, such as some types of cancer, that are characterized by the release of EVs into the blood. Information about this interesting topic can be found in various publications [14–16].

2 Immunomorphological Assessment of Hsp60

2.1 Materials

Ten staining jars.

One staining humid tray.

Hydrophobic marker pen (PAP pen).

Super frost slides (characterized by a special treatment process that electrostatically adheres tissue sections and cells to the glass without the need for adhesives or protein coatings) or chamber slides for cells.

Coverslips for slides.

Xylene (stored at room temperature, i.e., 22–24 °C).

Ethanol absolute (stored at room temperature).

Ethanol 96% (stored at room temperature).

Distilled water (stored at room temperature).

Sodium citrate buffer (pH 6).

Acetone.

Absorbent paper.

Dulbecco's phosphate-buffered saline (PBS) (pH 7.4) (stored at room temperature).

Aqueous 3% H_2O_2 (peroxidase solution) (stored at 4 °C).

3% bovine serum albumin (BSA) or ready-to-use protein-blocking solution, serum-free (both stored at 4 °C).

Primary antibody for human Hsp60 (stored as indicated by the supplier) to be diluted in the diluent provided with the antibody, in a similar commercially available alternative or in PBS (stored at 4 °C). Make a fresh dilution, following the directions in the antibody instruction datasheet; this dilution is stable at 4 °C for variable periods of times (a few hours to a few days), depending on the antibody.

Ready-to-use universal biotinylated secondary antibody diluted in PBS (stored at 4 °C).

Ready-to-use streptavidin–peroxidase complex diluted in PBS (stored at 4 °C).

Ready-to-use chromogen, i.e., aminoethilcarbazole (AEC) or diaminobenzidine (DAB) (stored at 4 °C, in the dark). These reagents must be used with caution since they are highly toxic and must always be manipulated wearing protective gloves.

Ready-to-use aqueous hematoxylin (stored at 4 °C).

Ready-to-use aqueous mounting medium or nonaqueous mounting medium (stored at 23 °C).

Primary antibody for another molecule which we want to assess in double immunofluorescence.

Fluorochrome-conjugated secondary antibody.

4',6-diamidino-2-phenylindole (DAPI).

Aqueous mounting medium for fluorescence microscopy.

2.2 Methods

2.2.1 Immunohistochemistry

A detailed protocol of the immunohistochemical procedure on formalin-fixed and paraffin-embedded histological sections (*see* **Note 1**).

For this procedure superfrost slides are necessary to prevent detachment of the tissue sections. Sections with a thickness of 4–5 μm are obtained from paraffin blocks of biopsies with a cutting microtome (*see* **Note 2**) [17–19].

1. To remove the paraffin on the slides and to rehydrate the sections:
   - Put the sections in an oven for 30 min at 60 °C.
   - Immerse the sections in xylene for 30 min at 60 °C, twice.
   - Immerse the sections in absolute ethanol for 10 min.

- Immerse the sections in a descending scale of ethanol, at room temperature, from ethanol 96% through 80%/70%/50%/30%, one time for 5 min for each alcohol.
- Place the sections in distilled water for 5 min to remove excess alcohol (*see* **Note 3**).

2. To perform the antigen unmasking, immerse the sections in sodium citrate buffer (pH 6) at 85–95 °C for 8 min, and, later, immerse them in acetone at −20 °C for 8 min, to prevent detachment of the section from the slide. Many antibodies need antigen retrieval to improve the demonstration of antigens. It consists of a pretreatment with an antigen retrieval solution that breaks the protein cross-links formed by formalin fixation and thereby uncovers hidden antigenic sites. Some primary antibodies do not need antigen retrieval before staining. It must be remembered that antigenic unmasking should only be done when required because it is a procedure that could alter the integrity of the histological section.
3. Wash the sections twice with PBS (phosphate-buffered saline pH 7.4), each for 5 min (*see* **Note 4**).
4. Blot away (with absorbent paper) excess water from the slides: first allow water to drain, and then gently dry the slide with absorbent paper.
5. Delimit the histological section on the slide with the PAP pen to make it visible throughout the steps that follow and to reduce the number of solutions and antibodies that will be used (*see* **Note 5**).
6. Apply on the top of the section a drop of aqueous 3% H_2O_2 (peroxidase solution) for 10 min, to inhibit any endogenous peroxidase activity. Then wash the sections twice again with PBS for 5 min, drain excess PBS, and dry the slides.
7. Apply on the top of the section a drop of 3% bovine serum albumin (BSA) or ready-to-use protein-blocking solution, for 10 min to block nonspecific antigenic sites.
8. Remove the excess blocking solution without rinsing the slides with PBS, and apply directly diluted primary anti-human Hsp60 antibody on the section (20–100 μL of diluted antibody solution for each section, depending on its size). For antibody dilution follow the indications in the antibody instruction datasheet. The calculation of the amount of diluted primary antibody must be done before starting the reaction in relation to the extent of the histological sections. Simultaneously, apply only PBS on the negative control section in the negative control slide. Likewise, in the positive control slide, apply a primary antibody known to give a positive reaction in the tissue under examination (e.g., antibody anti-cytokeratins will give a

positive reaction in epithelial specimens). Keep all the reagents on the slides for 1 h, at room temperature, in a humid tray now closed with a cover, so it becomes a humid chamber (evaporation must be avoided in order not to dry out the sections). Even better, the sections can be incubated with the primary antibody even overnight at 4 °C.

9. Rinse the slides twice with PBS, each for 5 min. Drain PBS and dry the slides.
10. Incubate the sections with a universal biotinylated secondary antibody diluted in PBS in the humid tray closed with a cover, for 10 min at room temperature.
11. Rinse the slides twice with PBS, each for 5 min. Drain PBS and dry the slides.
12. Incubate the sections with streptavidin–peroxidase complex diluted in PBS in the humid tray closed with a cover, for 10 min at room temperature.
13. Rinse the slides twice with PBS, each for 5 min. Drain PBS and dry the slides.
14. Apply on the top of the section the chromogen in the dark for 5–10 min (as indicated in the instruction datasheet provided by the supplier).
15. Rinse the slides twice with fresh water, each for 5 min. Drain water and dry the slides.
16. Apply hematoxylin for nuclear counterstaining, for 1–2 min.
17. Rinse the slides twice with fresh water, each for 5 min. Drain water and dry the slides.
18. Mount coverslips with an aqueous mounting medium. It is possible to mount coverslips with a nonaqueous mounting medium after dehydration and clearing of the tissue sections (in an ascending scale of ethanol from 30%, through 50%/70%/80% and absolute ethanol and then in xylene). The coverslip should be placed on the counter, and a small drop of the mounting medium should be placed on it. Immediately afterward (do not allow the mount to dry!), put the immunostained slide on the drop, wait a few seconds to spread the mounting, and press the coverslip onto the slide to prevent the formation of bubbles. The excess mounting medium must be removed with absorbent paper (*see* **Note 6**).
19. The observation of the sections is performed with an optical microscope; the immunopositivity appears brown (if DAB chromogen is used) or red (if AEC chromogen is used). The optical microscope must be equipped with a set of objectives (5x, 10x, 20x, 40x, and 63x) that, in combination with 10x eyepieces, results in a magnification (50x, 100x, 200x, 400x,

630x) useful for a complete evaluation of the stained slides. The lower magnification is useful for a panoramic view of the histological specimen and to evaluate the quality of the immunostaining. The pathologist's experience plays a deciding role in the evaluation of the quality of the stain. One must learn how to detect false positives, if the negative control slide is positive, it is very likely that the reaction in the experimental slides produced false positives. In these cases, the test must be repeated! On the contrary, if the positive control slide does not show the expected positivity in terms of intensity and distribution, the reaction in the experimental slides will likely show false negativity. Also, in this case the test must be repeated! At higher magnifications (100x and 200x) it is possible to better localize the immunopositivity (where the positivity is localized, in the epithelial, or connective, or inflammatory cells) and evaluate the specificity of the stain. At higher magnification (400x or 630x), it is possible to distinguish in which cell compartment the positivity is located (e.g., in the nucleus, in the cytoplasm, or on the cell membrane). The pathologist's experience plays a deciding role in the evaluation of the quality of the stain. The evaluations and quantifications of immunopositivity are made at a magnification of 400x (Fig. 1). Usually, the percentage of immunopositivity is calculated in a 400x high-power field (HPF) and repeated for 10 HFP. The arithmetic means of counts is used when a statistical analysis is performed. All observations must be performed by two independent observers. Reading and scoring must be done masking (blinding) to the reader the identity of the sections to prevent bias.

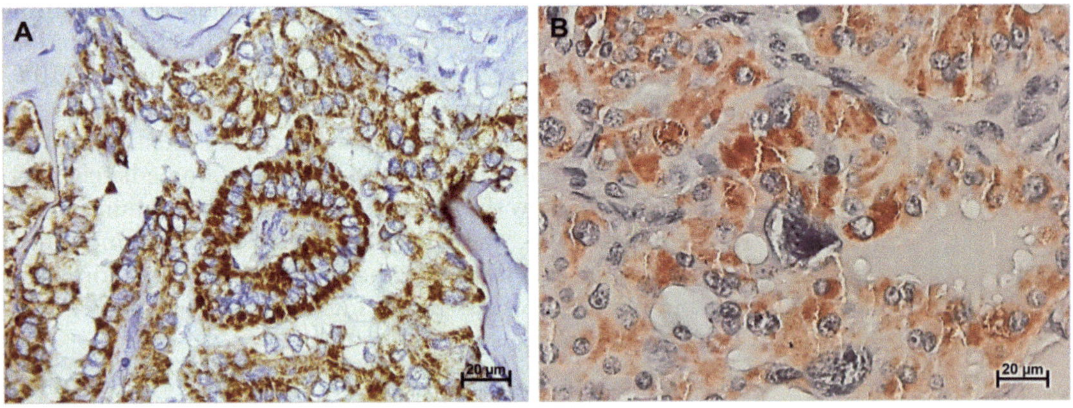

Fig. 1 (**a**) Representative image of Hsp60 revealed by immunohistochemistry on thyroid papillary carcinoma. Magnification 400x, scale bar 20 μm. (**b**) Representative image of Hsp60 revealed by immunohistochemistry on thyroid Hurtle cell carcinoma. Magnification 400x, scale bar 20 μm

2.2.2 Immunofluorescence and Confocal Microscopy

For this procedure, sections with a thickness of 4–5 μm are obtained from paraffin blocks of biopsies with a cutting microtome and are placed in the super frost slides that are necessary to prevent detachment of the same sections [9, 20–22] (*see* **Notes 7** and **8**).

1. To remove the paraffin on the slides and to rehydrate the sections:
   - Put the sections in an oven for 30 min at 60 °C.
   - Immerse the sections in xylene for 30 min at 60 °C, twice.
   - Immerse the sections in absolute ethanol for 10 min.
   - Immerse the sections in a descending scale of ethanol, at room temperature (i.e., 22–24 °C), from ethanol 96% through 80%/70%/50% and 30%, one time for 5 min each alcohol.
   - Place the sections in distiller water for 5 min to remove excess alcohol (*see* **Note 3**).
2. Perform antigen retrieval to break the protein cross-links formed by formalin fixation and paraffin inclusion: immerse the sections in sodium citrate buffer (pH 6) at 75 °C for 8 min, and, later, immerse them in acetone at −20 °C for 8 min, to prevent detachment of the section from the slide.
3. Wash the sections twice with PBS, each for 5 min (*see* **Note 4**).
4. Blot away (with absorbent paper) excess water from the slides: first allow water to drain, and then gently dry the slide with absorbent paper.
5. Delimit the histological section on the slide with the PAP pen to make it visible throughout the steps that follow and to reduce the number of solutions and antibodies that will be used.
6. Apply on the top of the section a drop of bovine serum albumin (BSA) or ready-to-use protein-blocking solution, for 30 min to block nonspecific antigenic sites.
7. Incubate the slides in a humid chamber with primary human Hsp60 antibody overnight at 4 °C.
8. Wash the sections twice with PBS (phosphate-buffered saline pH 7.4), each for 5 min.
9. Incubate the sections whit a fluorochrome-conjugated secondary antibody for 1 h at room temperature, in the dark. Use a secondary antibody that is reactive to your primary antibody host species!
10. Wash the sections twice with PBS, each for 5 min.
11. Incubate the sections whit dihydrochloride (DAPI) for 5 min at room temperature, in the dark, to counterstain the nuclei.
12. Wash the sections twice with PBS, each for 5 min.

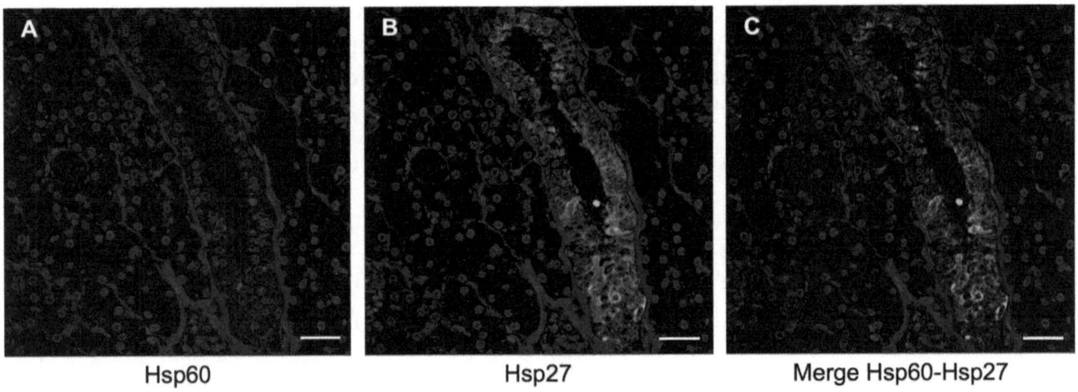

Fig. 2 Representative images of double immunofluorescence on salivary gland by confocal microscopy. (**a**) Immunofluorescence for Hsp60 (red). (**b**) Immunofluorescence for Hsp27 (green). (**c**) Merged images of double immunofluorescence for Hsp60 and Hsp27. Nuclei stained with DAPI are shown in blue. Magnification 400x, scale bar 25 μm

13. Mount coverslips with an aqueous mounting medium for fluorescence microscopy or with some drops of PBS (*see* **Note 8**).
14. The observation and evaluations of the sections are performed with a confocal microscope equipped with several objectives (10x; 20x; 40x; 63x). The staining intensity is expressed as the mean pixel intensity (PI) normalized to the CSA (cross-sectional area expressed in pixels) using the confocal microscopy software (*see* **Note 9**) (Fig. 2). The immunofluorescence test (Fig. 3) can be done on frozen sections, and, on this occasion, the procedure begins at **step 3**, because the dewaxing and the antigen retrieval are not necessary. On cell culture it is possible to perform immunofluorescence starting the protocol from **step 6**. In this case the cells are placed in chamber slides.

3 Assessment of Hsp60 and microRNAs in EVs from Plasma

This protocol describes the isolation of EVs from plasma, using differential ultracentrifugation, which is nowadays the "gold standard" method for the isolation of EVs [16, 23–26].

3.1 Materials

3.1.1 Sample Collection

Refrigerated centrifuge.

EDTA collection tubes.

Cryovials.

3.1.2 EV Isolation

Refrigerated centrifuge.

Ultracentrifuge and fixed-angle or swinging-bucket rotor.

15 mL tubes.

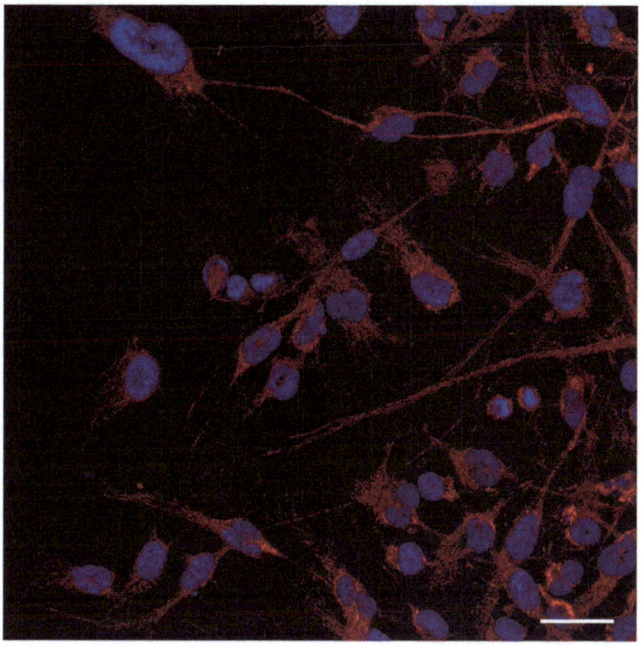

Fig. 3 Representative image of immunofluorescence for Hsp60 (red) in a Glioblastoma cell line. Nuclei stained with DAPI appear blue. Magnification 400x, scale bar 25 μm

Syringe 0.2 μm filter.

Ultracentrifuge tubes.

PBS, phosphate-buffered saline.

RIPA (radioimmunoprecipitation assay) lysis buffer (0.3 M NaCl, 0.1% SDS, 25 mm HEPES pH 7.5, 1.5 mm $MgCl_2$, 0.2 mm EDTA, 1% Triton X-100, 0.5 mm DTT, 0.5% sodium deoxycholate).

3.1.3 EV Morphological Characterization

2.5% glutaraldehyde.

1% uranyl acetate.

Reynolds' solution.

Laemmli buffer.

Formvar carbon-coated EM grids.

Parafilm.

Forceps.

Glass dish.

Filter paper.

Grid storage boxes.

Transmission electron microscope (TEM).

3.1.4 Protein Isolation and Assessment of Hsp60

Bradford solution.

10X Laemmli buffer.

1.5 M Tris-HCl, pH 8.8.

0.5 M Tris-HCl, pH 6.8.

Acrylamide/Bis acrylamide solution.

Ammonium persulfate.

TEMED (tetramethylethylenediamine).

SDS-PAGE running buffer.

Nitrocellulose membranes.

Transfer buffer.

Tris-buffered saline (TBS; 10).

TBS containing 0.05% Tween 20.

BSA (bovine serum albumin) blocking buffer.

Western blot equipment and supplies.

(*see* **Note 10**).

3.1.5 microRNA Isolation and Hsp60-Related microRNA Analysis

3.1.5.1 RNA Extraction

Chloroform.

Ethanol (70% and 96–100%).

Sterile, RNase-free pipet tips.

Sterile, RNase-free 1.5 mL or 2 mL microcentrifuge tubes.

Microcentrifuge(s) (with rotor for 2 mL tubes) for centrifugation at 4 °C and room temperature (i.e., 22–24 °C).

3.1.5.2 microRNA Analysis

Pipettes and filtered RNase-free pipette tips.

Reagent for cDNA synthesis and miRNA analysis.

Real-time PCR machine.

3.2 Methods

3.2.1 Sample Collection

1. Draw and collect blood (4 mL) into EDTA collection tubes (the EDTA concentration per mL of blood may vary depending on the manufacturer).
2. Invert the tubes carefully 10 times to mix blood and anticoagulant, and store them at room temperature until centrifugation (*see* **Note 11**).
3. Centrifuge at 1800× *g* for 30 min at room temperature.
4. After centrifugation are obtained three layers: at the top of the plasma, in the intermediate layer leucocytes (buffy coat), and in the bottom erythrocytes.

5. Carefully aspirate, aliquot in cryovials, and store the plasma at −80 °C until use (*see* **Note 12**).

3.2.2 EV Isolation

1. Centrifuge plasma at 2000× *g* for 30 min at 4 °C.
2. Carefully transfer supernatant to 15 mL tubes, and centrifuge at 11,000× *g* for 30 min at 4 °C to remove cell debris.
3. Dilute supernatant with an equal volume of PBS and filter through a 0.2 μm filter.
4. Carefully transfer supernatant to ultracentrifuge tubes and centrifuge at 110,000× *g* for 2 h at 4 °C to pellet the EVs.
5. Resuspend EV pellet in 1 mL of PBS, and centrifuge at 110,000× *g* for 2 h at 4 °C (*see* **Note 13**).
6. Resuspend EV pellet in 50–100 μL of PBS (for morphology analyses) or RIPA buffer (for Western blotting).
7. Store at −80 °C.

3.2.3 EV Morphological Characterization by Transmission Electron Microscopy (TEM)

The EVs isolated can be biochemically characterized for the presence of typical markers by Western blot (*See* Subheading 3.2.4). The biophysical features of EVs can be studied by Nanoparticle Tracking Analysis, e.g., Dynamic Light Scattering (DLS), Atomic Force Microscopy (AFM), and Electron Microscopy (EM). In our laboratory evaluation of the size and morphology of EVs by TEM has been standardized (Fig. 4), according to the following protocol:

1. Add to the EV pellet resuspended in PBS 50 μL freshly made fixative (2.5% glutaraldehyde in PBS) for 30 min at room temperature.
2. Place formvar carbon-coated EM grids in filter paper using clean forceps.
3. Place by layering 10 μL drops of exosome preparations on formvar carbon-coated EM grids on filter paper.
4. Leave to dry for 10 min at room temperature.
5. Cover with parafilm a glass dish, and carefully lay the grid-mounted preparations.
6. Prepare the grid-mounted preparations for contrast staining by treating them with uranyl acetate (1%) for 5 min at room temperature.
7. Treat the grid-mounted preparations with Reynolds' solution for 5 min at room temperature.
8. Rinse them eight times in distilled water for 2 min.
9. Store in a grid storage box at room temperature until TEM observation.

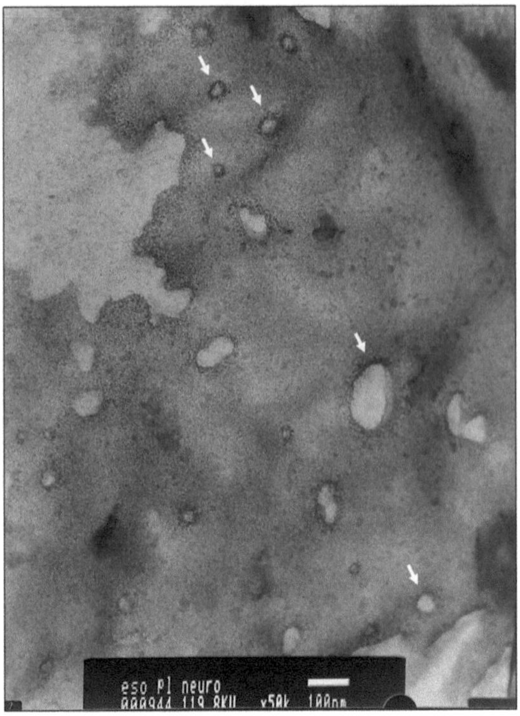

Fig. 4 Electron micrograph (image obtained with TEM) of EVs purified from plasma showing typical characteristics. The negative stain of EV suspension shows spherical structures enclosed by a phospholipid bilayer, some of them shown by the arrows. The diameter of the EVs isolated using the procedure described here typically ranges from 30 to 150 nm (scale bar 100 nm)

3.2.4 Protein Isolation and Assessment of Hsp60

1. Measure the protein content of EVs using the Bradford assay.
2. Add 10X Laemmli buffer to 50 μg of EVs proteins resuspended in RIPA buffer and heated for 5 min at 95 °C.
3. Follow the standard Western blotting procedure using a 12% polyacrylamide gel (SDS-PAGE).
4. Transfer to a nylon membrane and analyze by immunoblotting with anti-Hsp60 antibody (Fig. 5).

3.2.5 microRNA Isolation and Hsp60-Related microRNA Analysis

To analyze Hsp60-related microRNAs, there are several protocols. Studies from our lab provided good results using the miRNeasy Mini Kit (QIAGEN) for total RNA purification from EVs, through the enrichment of miRNAs and other small RNA (less than 200 nt). For miRNA expression analysis in our lab, we optimized the primer-based real-time PCR procedure, using miScript SYBR Green PCR Kit (Qiagen) [12].

3.2.5.1 RNA Extraction

1. Mix the 50 μL of EVs pellet resuspended in PBS with 700 μL of QIAzol Lysis Reagent by pipetting several times to disrupt membranes (*see* **Note 14**).

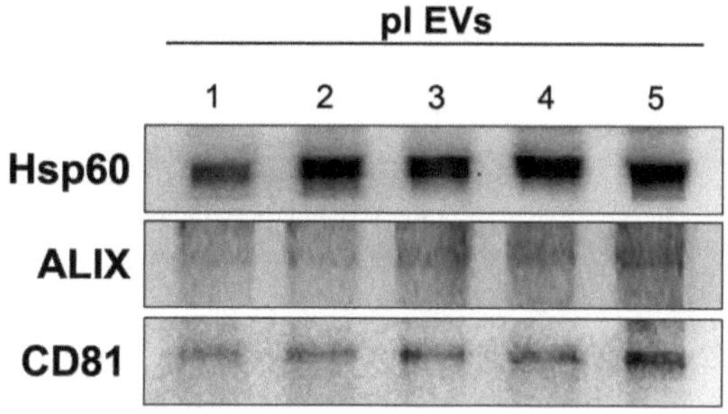

Fig. 5 Western blots of five different preparations of EVs isolated from different plasmas, showing the respective levels of Hsp60 (mouse anti-Hsp60, LK1 clone, Sigma, St. Louis, MO, USA), and of the EV markers ALIX (mouse anti-Alix, 1A12 clone, Santa Cruz Biotechnology, Inc., Dallas, TX, USA), and CD81 (mouse anti-CD81, B-11 clone, Santa Cruz Biotechnology, Inc.)

2. Place the tube for 5 min at room temperature.
3. Add 140 μL chloroform to the tube containing the homogenate and cap it securely. Shake the tube vigorously for 15 s.
4. Place the tube for 5 min at room temperature.
5. Centrifuge for 15 min at 12,000× *g* at 4 °C. After centrifugation the homogenate is separated into three phases: an upper, colorless, aqueous phase containing RNA; a white interphase; and a lower, red, organic phase. The volume of the aqueous phase should be approximately 350 μL.
6. Carefully transfer the upper aqueous phase to a new tube (*see* **Note 15**).
7. Add 350 μL of 70% ethanol and mix thoroughly by vortexing.
8. Pipet the sample into the RNeasy Mini spin column placed in a 2 mL collection tube, and centrifuge at 10,000× *g* for 15 s at room temperature.
9. Add 450 μL of 100% ethanol to the flow-through and mix thoroughly by vortexing.
10. Place the sample into the RNeasy MinElute spin column placed in a 2 mL collection tube, and centrifuge for 15 s at 10,000× *g* at room temperature.
11. Discard the flow-through.
12. Pipet 500 μL Buffer RPE into the RNeasy MinElute spin column, and centrifuge for 15 s at 10,000× *g*.
13. Discard the flow-through.
14. Add 500 μL of 80% ethanol to the RNeasy MinElute spin column and centrifuge for 2 min at 10,000× *g* to dry the spin column membrane.
15. Discard the flow-through and the collection tube.

16. Place the RNeasy MinElute spin column into a new 2 mL collection tube, and centrifuge for 5 min at 10,000× *g*.
17. Place the RNeasy MinElute spin column into a 1.5 mL collection tube, and pipet 14 μL RNase-free water onto the spin column membrane. Close the lid gently and centrifuge for 1 min at 10,000× *g* to elute the miRNA-enriched fraction.

3.2.5.2 microRNA Analysis

1. Prepare cDNA using the miScript PCR system starting from 200 ng per sample.
2. Incubate for 60 min at 37 °C.
3. Incubate for 5 min at 95 °C to inactivate miScript Reverse Transcriptase Mix, and place on ice.
4. Store at −20 °C or proceed immediately with real-time PCR.
5. Perform real-time PCR using the miScript SYBR Green PCR Kit.
6. Prepare the reaction mix containing 2x QuantiTect SYBR Green PCR Master Mix, 10x miScript Universal Primer, 10x miScript Primer Assay, and RNase-free water.
7. Add the appropriate volume of reaction mix in template cDNA.
8. Centrifuge for 1 min at 1000× *g* at room temperature to remove bubbles (*see* **Note 16**).
9. Program the real-time cycler according to the PCR cycle conditions: 95 °C for 15 min, 45 cycles of 94 °C for 10 s, 55 °C for 30 s, and 70 °C for 30 s.
10. Perform dissociation curve analysis of the PCR products, using the software of real-time cyclers and following the instructions provided by the supplier.

4 Notes

1. For an accurate evaluation, the immunostaining must be repeated at least twice, preferably three times on a set of new slides.
2. One slide for the positive control and one for the negative control should be used in the slide set. If there are two histological sections per slide, one can be used for controls.
3. The xylene and the alcohol must be kept under the hood in tightly closed containers to avoid personal exposure to xylene and alcohol vapors. In addition, the alcohols must be changed often so as not to be altered by evaporation and hydration. These measures are necessary to obtain an optimal dewaxing of the sections.

4. The PBS solution must be prepared fresh on the same day of use or the day before.
5. Make sure that each reagent (i.e., aqueous 3% H_2O_2, primary antibody dilution, biotinylated secondary antibody, streptavidin, and chromogen) added on the PAP pen delimitation of the slide covers the histological section completely. The section must never be allowed to dry!
6. Immunohistochemistry can be performed on frozen sections. In this case, **steps 1** and **2** are not carried out, and after fixing the sections with acetone, we proceed with **step 3**. Dewaxing and antigen retrieval are not necessary.

   Immunostaining can be carried out on cells (immunocytochemistry) from cultures (cells are placed on chamber slides). Also in this case, **steps 1** and **2** are not done, and the procedure, after cell fixation, begins with serum blocking (**step** 7).
7. For an accurate evaluation, the immunofluorescence must be repeated at least twice, preferably three times on a set of new slides.
8. One slide for the negative control should be used in the slide set. If there are two histological sections per slide, one can be used for control. In this case it is necessary to apply drops of PBS instead of primary antibodies and always respect the timing of the procedure. The negative control is important for setting up the confocal microscope before acquiring the images necessary for the quantification of immunopositivity expressed in pixel intensity (PI).
9. The immunofluorescence can be used also for assessing the colocalization of Hsp60 with other molecules. In this case, the technique is named double immunofluorescence. The procedure of double immunofluorescence is identical but two primary antibodies must be incubated (anti-Hsp60 and anti-the-other molecule that we want to investigate). The first six steps are identical; in **step** 7 we proceed with the incubation of the first primary antibody (usually the primary antibody obtained from mice is chosen first) overnight at 4 °C and, on the following day, after washing with PBS twice for 5 min, proceed with the incubation with the second primary antibody (obtained from a source different from mice!) overnight at 4 °C. Then the sections are washed with PBS, twice for 5 min, followed by incubation with the first fluorochrome-conjugated secondary antibody, which reacts with the first primary antibody used, for 1 h at room temperature, in the dark. After washing with PBS twice for 5 min, continue with the incubation with the secondary antibody conjugated to a different fluorochrome, which reacts with the second primary antibody used, for 1 h at room temperature, in the dark. It is

important to use two different fluorochromes! After these steps the procedure is concluded with **steps 10–14** described above. The colocalization ratio (%) analysis is performed to test the colocalization of Hsp60 and other molecules by confocal microscopy software (Fig. 2).

10. Prepare all solutions using ultrapure water.
11. Samples should undergo centrifugation immediately using a centrifuge without a brake to stop.
12. Aspirating plasma must be done very carefully to avoid breaking cells and the transfer of any cells of the buffy coat layer. If this occurs, the plasma appears turbid, and it should be centrifuged and aspirated again to remove the remaining insoluble matter.
13. To improve the yield, after the first ultracentrifugation step, it is possible to resuspend with 1 mL of PBS and store at 4 °C overnight and then proceed with the subsequent ultracentrifugation step.
14. Homogenized EVs lysates can be stored at −80 °C for 3–4 months.
15. It is important to use sterile, RNase-free tips and tubes.
16. This step is not necessary for reactions set up in rotor discs.

Acknowledgments

A.J.L.M. and E.C. de M. were partially supported by IMET and IEMEST.

References

1. Macario AJL, Conway de Macario E, Cappello F (2013) The Chaperonopathies. Diseases with defective molecular chaperones. Springer, Dordrecht/Heidelberg/New York/London
2. Bross P, Fernandez-Guerra P (2016) Front Mol Biosci 3:49. https://doi.org/10.3389/fmolb.2016.00049
3. Gomez-Llorente Y, Jebara F, Patra M et al (2020) Structural basis for active single and double ring complexes in human mitochondrial Hsp60-Hsp10 chaperonin. Nat Commun 11:1916. https://doi.org/10.1038/s41467-020-15698-8
4. Macario AJL, Conway de Macario E (2023) The chaperone system in autoimmunity, inflammation, and virus-induced diseases: role of chaperonins. In: Fink G (ed) Stress: immunology and inflammation. Handbook of stress series, vol 5. Chapter 13. Elsevier/Academic, San Diego, USA
5. Cappello F, Conway de Macario E, Zummo G, Macario AJL (2011) Immunohistochemistry of human Hsp60 in health and disease: from autoimmunity to cancer. Methods Mol Biol 787:245–254. https://doi.org/10.1007/978-1-61779-295-3_18
6. Cappello F, Conway de Macario E, Rappa F et al (2018) Immunohistochemistry of human Hsp60 in health and disease: from autoimmunity to cancer. Methods Mol Biol 1709:293–305. https://doi.org/10.1007/978-1-4939-7477-1_21
7. Caruso Bavisotto C, Alberti G, Vitale AM et al (2020) Hsp60 post-translational modifications: functional and pathological consequences. Front Mol Biosci 7:95. https://doi.org/10.3389/fmolb.2020.00095

8. Vitale AM, Conway de Macario E, Alessandro R et al (2020) Missense mutations of human Hsp60: a computational analysis to unveil their pathological significance. Front Genet 11:969. https://doi.org/10.3389/fgene.2020.00969. Section Genetics of Common and Rare Diseases
9. Basset CA, Rappa F, Lentini VL et al (2021) Hsp27 and Hsp60 in human submandibular salivary gland: quantitative patterns in healthy and cancerous tissues with potential implications for differential diagnosis and carcinogenesis. Acta Histochem 123:151771. https://doi.org/10.1016/j.acthis.2021.151771
10. Alberti G, Vergilio G, Paladino L et al (2022) The chaperone system in breast cancer: roles and therapeutic prospects of the molecular chaperones Hsp27, Hsp60, Hsp70, and Hsp90. Int J Mol Sci 23:7792
11. Sun B, Li G, Yu Q et al (2021) HSP60 in cancer: a promising biomarker for diagnosis and a potentially useful target for treatment. J Drug Target 30:31–45. https://doi.org/10.1080/1061186X.2021.1920025
12. Tang Y, Zhou Y, Fan S et al (2022) The multiple roles and therapeutic potential of HSP60 in cancer. Biochem Pharmacol 201:115096. https://doi.org/10.1016/j.bcp.2022.115096
13. Tikhomirova TS, Matyunin MA, Lobanov MY et al (2022) In-depth analysis of amino acid and nucleotide sequences of Hsp60: how conserved is this protein? Proteins 90:1119–1141. https://doi.org/10.1002/prot.26294
14. Caruso Bavisotto C, Scalia F, Marino Gammazza A et al (2020) Extracellular vesicle-mediated cell-cell communication in the nervous system: focus on neurological diseases. Int J Mol Sci 20:pii: E434. https://doi.org/10.3390/ijms20020434
15. David S, Vitale AM, Fucarino F et al (2021) The challenging riddle about the Janus-type role of Hsp60 and related extracellular vesicles and miRNAs in carcinogenesis and the promises of its solution. Appl Sci 11:1175. https://doi.org/10.3390/app11031175
16. Graziano F, Iacopino DG, Cammarata G, Scalia G, Campanella C, Giannone AG et al (2021) The triad Hsp60-miRNAs-extracellular vesicles in brain tumors: assessing its components for understanding tumorigenesis and monitoring patients. Appl Sci 11:2867. https://doi.org/10.3390/app11062867
17. Rappa F, Pitruzzella A, Marino Gammazza A et al (2016) Quantitative patterns of Hsps in tubular adenoma compared with normal and tumor tissues reveal the value of Hsp10 and Hsp60 in early diagnosis of large bowel cancer. Cell Stress Chaperones 21:927–933. https://doi.org/10.1007/s12192-016-0721-5
18. McKee C, Sigala B, Soeda J et al (2015) Amphiregulin activates human hepatic stellate cells and is upregulated in non-alcoholic steatohepatitis. Sci Rep 5:8812. https://doi.org/10.1038/srep08812
19. Rappa F, Sciume C, Bello M et al (2014) Comparative analysis of hsp10 and hsp90 expression in healthy mucosa and adenocarcinoma of the large bowel. Anticancer Res 34:4153–4159
20. Alberti G, Campanella C, Paladino L et al (2022) The chaperone system in glioblastoma multiforme and derived cell lines: diagnostic and mechanistic implications. Front Biosci (Landmark Ed) 27. https://doi.org/10.31083/j.fbl2703097
21. Mishra A, Mirzaei H, Guidi N et al (2021) Fasting-mimicking diet prevents high-fat diet effect on cardiometabolic risk and lifespan. Nat Metab 3:1342–1356. https://doi.org/10.1038/s42255-021-00469-6
22. Barone R, Marino Gammazza A, Paladino L et al (2021) Morphological alterations and stress protein variations in lung biopsies obtained from autopsies of COVID-19 subjects. Cell 10:3136. https://doi.org/10.3390/cells10113136
23. Théry C, Witwer KW, Aikawa E et al (2018) Minimal information for studies of extracellular vesicles 2018 (MISEV2018): a position statement of the International Society for Extracellular Vesicles and update of the MISEV2014 guidelines. J Extracell Vesicles 7:1535750. https://doi.org/10.1080/20013078.2018.1535750
24. Théry C, Clayton A, Amigorena S et al (2006) Isolation and characterization of exosomes from cell culture supernatants. Chapter 3. Unit 3.22. https://doi.org/10.1002/0471143030.cb0322s30
25. Campanella C, Rappa F, Sciumè C et al (2015) Heat shock protein 60 levels in tissue and circulating exosomes in human large bowel cancer before and after ablative surgery. Cancer 121: 3230–3239. https://doi.org/10.1002/cncr.29499
26. Caruso Bavisotto C, Cipolla C, Graceffa G et al (2019) Immunomorphological pattern of molecular chaperones in normal and pathological thyroid tissues and circulating exosomes: potential use in clinics. Int J Mol Sci 20:4496. https://doi.org/10.3390/ijms20184496

Chapter 21

Multiplex Immunostaining Method to Distinguish HSP Isoforms in Cancer Tissue Specimens

Hotaka Kawai, Kisho Ono, and Takanori Eguchi

Abstract

Heat shock proteins (HSPs) are often expressed in all nucleated cells, but their expression profiles differ. In particular, HSP90α and HSP90β have high sequence identity and have not been fully examined for their individual and compensatory functions as molecular chaperones, differences in client proteins, and extracellular distributions with exosomes. Immunohistochemical staining is a technique to visualize the presence and localization of target antigens using specific antibodies, of which the multiplex immunostaining method can reveal differences in protein expression in the same tumor tissue and the localization of proteins of interest within tumor tissue or single cells. The common multiplex immunostaining method uses multiple secondary antibodies of different reacting animal species to identify and detect different antigens, thus requiring different animals to be immunized with each primary antibody. Furthermore, the fluorescent-antibody method is the predominant multiplex staining method but has the critical disadvantage that permanent specimens cannot be prepared. Here, we outline a multiplex staining method for HSP90α and HSP90β based on the enzyme–antibody method that allows permanent specimens to be prepared without the restriction of immunized animal species.

Key words HSP90 isoforms, Tumor microenvironment, Protein localization, Immunohistochemistry, Multiple staining, Enzyme, antibody method

1 Introduction

HSP90 is an ATP-dependent molecular chaperone belonging to the heat shock protein (HSP) family of stress-inducible proteins. HSP90 is composed of four paralogues: HSP90α and HSP90β in the cytoplasm, glucose-regulated protein 94 (GRP94) in the endoplasmic reticulum, and TNF receptor-associated protein 1 (TRAP1) in mitochondria [1]. Hundreds of client proteins, including growth factors, signaling kinases, transcription factors, and cell surface receptors, depend on members of the HSP90 family for their conformational maturation, stabilization, and proper subcellular localization [2, 3]. However, this chaperone function is used in cancer cells to protect the dysregulated proteome, stabilize

Stuart K. Calderwood and Thomas L. Prince (eds.), *Chaperones: Methods and Protocols*, Methods in Molecular Biology, vol. 2693, https://doi.org/10.1007/978-1-0716-3342-7_21,

cancer proteins that would otherwise be degraded, and promote oncogenesis, making HSP90 an attractive therapeutic target [4, 5]. In other words, despite its cytoprotective role, HSP90 is involved in pathological processes such as cancer progression and the promotion of metastasis [6].

HSP90α and HSP90β, sharing 86% amino acids, are expressed in all nucleated cells, but their expression profiles differ [7]. HSP90β is a constitutively expressed paralog, while HSP90α is stress-inducible and often overexpressed in cancer [8–10]. Recent reports have shown that HSP90α and HSP90β are secreted by cancer cells and cells exposed to various stress conditions such as hypoxia, heat shock, DNA damage, and growth factor stimulation [9–12]. The extent to which the two cytoplasmic isoforms of HSP90 differ functionally has not yet been fully explored. HSP90α and HSP90β are also implicated in many cancers and neurodegenerative diseases [1, 10, 13–16]. On the other hand, the high sequence identity between HSP90α and HSP90β further complicates the rational design of inhibitors that selectively target only one paralog [17]. In terms of their functions as molecular chaperones, whether HSP90α and HSP90β bind to various target molecules or only one of them has not been examined, and this remains to be explored in the future.

Immunohistochemical (IHC) staining is a technique that uses specific antibodies to visualize the presence and localization of target antigens and is widely used in pathological diagnosis and drug discovery research. IHC is classified into enzyme–antibody methods (EAM) using enzyme-labeled antibodies and specific substrates (e.g., peroxidase-labeled antibody and diaminobenzidine [DAB]) and fluorescent-antibody method (FAM) using fluorescent-labeled antibodies. The EAM is the most widely used because it can be observed with a standard optical microscope, and specimens are easy to handle. However, compared to FAM, EAM has disadvantages such as frequent nonspecific reactions and is unsuitable for multiple staining. Conversely, FAM has attracted attention as a way to overcome the disadvantages of EAM, but it is often problematic because tissue autofluorescence can interfere with the detection of the target antigen. Based on their mutual advantages and disadvantages, we must properly use different IHC methods for different research purposes. In particular, tumor tissue comprises cancer cells, immune cells, fibroblasts, vascular endothelial cells, and various other cells that make up the tumor microenvironment (TME) [18, 19]. It is known that the complex interactions of these component cells in the TME regulate tumor growth, invasion, and resistance to therapy [18, 19]. Therefore, immunostaining imaging analysis using tissue samples in which the TME is maintained is ideal for understanding the dynamics in tumor tissue.

IHC has undergone many technical improvements, and there are now a wide variety of methods and applications. Among them, the multiple immunostaining method is one of the applications that enable the simultaneous detection of multiple antigen localizations on the same section and the same cell to observe and demonstrate their interrelationship. Furthermore, it is a useful method for comparing differences in protein expression in the same tumor tissue. The greatest advantage is that it is not limited by the molecular weight of the protein being compared and can reveal the localization of the molecule of interest in the tumor tissue or in a single cell. We have performed IHC staining on tumor specimens obtained from patients with oral squamous cell carcinoma to investigate the expression and tissue localization of HSP90α and HSP90β in the TME [14]. The results showed that the rate of HSP90α-positive cancer cells in advanced stages was significantly higher than that in early stages; conversely, the rate of HSP90β-positive cancer cells in early stages was significantly higher than that in advanced stages [14]. On the other hand, however, the HSP90β-positive cell rate in the advanced-stage oral cancer group was significantly higher in stromal cells than in tumor cells [14].

In recent years, the FAM has been widely used in biological research, especially for the specific differentiation of two or three different substances in a single specimen. In the case of multiple staining using the FAM, multiple secondary antibodies of different reacting animal species are used to distinguish and detect different antigens; thus, the animals immunized with each primary antibody must be different. For this reason, it has not been possible to double-stain antibodies from the same immunized animal. In many cases, the same immunized animal is used for the same proven antibodies, and the labor required to change new antibodies is immeasurable. Furthermore, specimens prepared by the FAM, for which much effort has been expended, cannot be permanent. In other words, it is difficult to preserve many tissue specimens for a long time, making them unsuitable for retrospective studies.

This chapter outlines a method for performing multiple staining of EAM that allows permanent specimens to be prepared without restrictions on immune animal species. The application of this research method will not only enable us to analyze the localization of HSP90 but also elucidate its interaction with the client molecule as a chaperone function. We provide step-by-step protocols from sample preparation to immunoreaction, multiple staining, and mounting (Subheadings 3.1, 3.2, 3.3, 3.4, 3.5, 3.6, 3.7, 3.8, and 3.9).

2 Materials

All chemicals are of reagent grade. Use ultrapure water for all solutions.

2.1 Sample Preparation

1. Microtome.
2. Water bath.
3. Flattening table for paraffin-embedded sections.
4. Charged glass slides (*see* **Note 1**).

2.2 Deparaffinization/Rehydration

1. Xylene.
2. Graded ethanol (100%, 90%, 80%, 70%, and 50%).
3. Methanol.
4. 3% hydrogen peroxide (H_2O_2) in methanol as a blocking solution for endogenous peroxidase (*see* **Note 2**).
5. Glass staining jars.
6. Slotted stainless steel slide rack fitting into the glass staining jar.

2.3 Antigen Retrieval

1. Antigen retrieving buffer (e.g., Target Retrieval Solution, pH 6.1, DAKO).
2. Microwavable vertical staining jars (*see* **Note 3**).
3. 20 mM Tris-buffered saline (TBS), PH 7.4.
4. Microwave oven.

2.4 Immunoreaction

1. Primary antibodies specific to the targets, such as HSP90α and HSP90β (*see* **Note 4**).
2. Blocking solution and antibody diluent. As a blocking reagent, select a suitable one for the immunodetection system (*see* **Note 5**).
3. 20 mM TBS, pH 7.4.
4. Wash buffer. 20 mM TBS, pH 7.4.
5. Humid chambers.
6. Shaker.

2.5 Immunostaining Components

1. Avidin–biotin complex immunohistochemical detection system (*see* **Note 6**).
2. Substrate for the detection system: 3,3′-diaminobenzidine (DAB) or 3-amino-9-ethyl carbazole (AEC) (*see* **Note 7**).
3. Counterstaining solution: methyl green or hematoxylin (optional).

3 Methods

The main steps of this protocol are sample preparation, deparaffinization/rehydration, antigen retrieval, immunoreaction, immunostaining, and dehydration/mounting. Do not allow the sample to slide dry until it is mounted on the coverslip.

3.1 Sample Preparation

1. Prepare paraffin-embedded tissues of interest, and store them until sectioning (*see* **Note 8**).
2. Use a microtome to cut the embedded samples into 5-µm-thick sections, and then float them in a 45 °C water bath containing distilled water.
3. Mount the sections on charged glass slides and allow them to dry at 42 °C overnight (*see* **Note 9**).

3.2 Deparaffinization/Rehydration

Perform this procedure according to Table 1.

1. To remove the paraffin, place the slides sequentially into three staining jars of xylene for 5 min each time (*see* **Note 10**).

Table 1
Rehydration procedure. Deparaffinization and dehydration procedure, including the blocking reaction for endogenous peroxidase

Solution	Repetitions	Incubation time (min)	
Xylene	Three changes	5 5 5	Deparaffinization
100% ethanol	Three changes	5 5 5	To rinse off the xylene and start rehydration
100% methanol	One time	5	Blocking endogenous peroxidase activity
5% H_2O_2-methanol	One time	30	
90% ethanol	One time	5	
80% ethanol	One time	5	
70% ethanol	One time	5	
50% ethanol	One time	5	
TBS	Three changes	5 5 5	To rinse off the ethanol

2. For rehydration, place the slides sequentially into three jars of 100% ethanol for 5 min each time and then into one jar of 100% methanol for 5 min.
3. Quench endogenous peroxidase activity by placing and keeping the slides in 3% H_2O_2-methanol for 30 min.
4. Place the slides into a graded ethanol series for 5 min for each concentration. Start from 90% and proceed downward to 50% ethanol.
5. To complete the rehydration process, wash the slides three times in TBS buffer for 5 min each time.

3.3 Antigen Retrieval

PFA fixation forms protein cross-links and masks the epitope sites of target proteins in the specimen, which can cause weak staining or false negatives in the immunohistochemical detection of certain proteins.

1. Add 50 ml of 1 × concentrated target retrieval solution in a microwavable vertical staining jar.
2. Insert the sample slides into the slots of the jar.
3. Heat the sections twice by microwaving until the buffer starts to boil, usually requiring 3 min each time (500 W microwave oven) (*see* **Note 11**).
4. Cool the slides at room temperature for 1 h.
5. Wash the slides in the TBS buffer twice for 5 min each time.

3.4 Immunoreaction (First Cycle)

All incubation steps should be carried out in a humid chamber to avoid drying of the specimens.

1. Block each section with 100–200 μl of blocking solution for 20 min at room temperature in the humid chamber.
2. Aspirate off the blocking buffer.
3. Apply 100 μl of primary antibody diluted in blocking solution onto each section.
4. As a negative control, if necessary, apply 100 μl of normal rabbit IgG in blocking solution onto another slide of the same sample. The same concentration of IgG as that of the primary antibody should be used.
5. Incubate the slide at 4 °C overnight in the humid chamber.
6. Wash the slides three times with TBS wash buffer for 5 min each, with gentle shaking.
7. Apply one drop of immunodetection reagent onto each section.
8. Incubate the slide for 30 min at room temperature in the humid chamber.

9. Wash the slide three times with TBS for 5 min each, with gentle shaking.
10. Wash the slide with TBS for 5 min before starting the DAB chromogen reaction.

3.5 Immunostaining (First Cycle: Brown Color)

This protocol is for chromogenic staining with DAB (brown color).

1. Prepare DAB reagent or suitable substrate according to the manufacturer's instructions immediately before use.
2. Apply 100 μl of DAB or suitable substrate solution to each section, and monitor its staining carefully.
3. As soon as the brown reaction product develops, immerse the slide in deionized water.
4. Counterstain the section with methyl green or hematoxylin for 20 s, if necessary.
5. Wash the slide in tap water until the water becomes clear.

3.6 Detachment of First Cycle Antibodies from Specimen

In this process, the primary antibody is detached from the antigen by heat treatment to create an environment in which the antibodies of the next cycle can react easily.

1. Add 50 ml of 1 × concentrated target retrieval solution in a microwavable vertical staining jar.
2. Insert the sample slides into the slots of the jar.
3. Heat the sections twice by microwaving until the buffer starts to boil, usually requiring 3 min each time (500-W microwave oven) (*see* **Note 11**).
4. Cool the slides at room temperature for 1 h.
5. Wash the slides in the TBS buffer twice for 5 min each time.

3.7 Immunoreaction (Second Cycle)

All incubation steps should be carried out in a humid chamber to avoid drying of the specimens.

1. Apply 100 μl of primary antibody diluted in blocking solution onto each section.
2. As a negative control, if necessary, apply 100 μl of normal rabbit IgG in blocking solution onto another slide of the same sample. The same concentration of IgG as that of the primary antibody should be used.
3. Incubate the slide at 4 °C overnight or at room temperature for 2 h in the humid chamber (*see* **Note 12**).
4. Wash the slides three times with TBS buffer for 5 min each, with gentle shaking.
5. Apply one drop of immunodetection reagent onto each section.

6. Incubate the slide for 30 min at room temperature in the humid chamber.
7. Wash the slide three times with TBS for 5 min each, with gentle shaking.
8. Wash the slide with TBS for 5 min before starting the AEC reaction.

3.8 Immunostaining (Second Cycle: Red Color)

This protocol is for chromogenic staining with AEC (red color).

1. Apply 100 μl of AEC or suitable substrate solution to each section, and monitor its staining carefully.
2. As soon as the red reaction product develops, immerse the slide in deionized water.
3. Counterstain the section with methyl green or hematoxylin for 20 s, if necessary.
4. Wash the slide in tap water until the water becomes clear.

3.9 Dehydration and Mounting

1. Dehydrate slides according to Table 2.
2. When all the necessary wash steps have been completed, mount the sections with coverslips using suitable mounting media, being careful to avoid introducing air bubbles.
3. Examine the sections under a microscope and capture images as required.

Table 2
Rehydration procedure

Solution	Repetitions	Incubation time (min)
70% ethanol	One time	5
80% ethanol	One time	5
90% ethanol	One time	5
100% ethanol	Three changes	5 5 5
Xylene	Three changes	5 5 5

4 Notes

1. Charged glass slides stabilize the adhesion of sections. Using silane-treated or otherwise charged slides effectively prevents possible losses during microwave processing.
2. Depending on the type of antibody used, a series of blocking steps may be required to minimize nonspecific reactions of the antibody and endogenous enzymes. Some cell and tissue types contain endogenous peroxidase. This protocol may result in high levels of nonspecific background staining when using the HRP conjugate immunodetection system. To block this endogenous peroxidase activity, 5% H_2O_2-methanol is used. Some cell surface proteins are sensitive to 5% H_2O_2-methanol quenching.
3. Vertical staining jars such as a Coplin jar hold slides vertically for easy manipulation. A microwavable one is necessary. We commonly use microwavable jars made of polymethylpentene (PMP).
4. Select antibodies that do not cross-react with the other HSP family members. We have verified GTX10975 (GeneTex) for HSP90α and GTX101448 (GeneTex) for HSP90β. The dilution ratio for both primary antibodies is 1:100.
5. We recommend that you use the blocking specified in the staining kit. We use the avidin–biotin complex method (Vector Laboratories, Burlingame, CA), and the blocking is choused from the materials included in the kit.
6. We routinely use commercially available, sensitive, ready-to-use polymer-based immunodetection reagents. In this chapter, we use the avidin–biotin complex method (Vector Laboratories, Burlingame, CA), which includes reagents for human, mouse, and rat tissues. The principle of the ABC detection kit is based on the affinity of avidin (or streptavidin) for biotin.
7. To make the two antibodies colored by each, we stained the two antibodies with two different colors, DAB (brown) and AEC (red).
8. We prepare tissue samples fixed with freshly made 4% PFA-PBS.
9. Drying slides at low temperatures will reduce the adhesive strength of the tissue sections. Once slides are completely dry, they should be stored at 4 °C before staining. It is a better practice that sections be freshly cut before use.
10. Incomplete deparaffinization results in spotty staining, background staining, and decreased immune response.
11. Since equipment differs, the heating protocol must be optimized for the make and model of microwave used in one's

laboratory, whose optimization is performed by making a trial run using blank slides. The vibration of the water due to boiling may cause the sections to come off the glass. In such cases, the boiling time should be adjusted to be shorter.

12. To shorten the reaction time, incubate the primary antibody at room temperature for the second cycle to shorten the reaction time.

References

1. Taha EA, Ono K, Eguchi T (2019) Roles of Extracellular HSPs as Biomarkers in Immune Surveillance and Immune Evasion. Int J Mol Sci 20(18):4588. https://doi.org/10.3390/ijms20184588
2. Wayne N, Mishra P, Bolon DN (2011) Hsp90 and client protein maturation. Methods Mol Biol 787:33–44. https://doi.org/10.1007/978-1-61779-295-3_3
3. Marzec M, Eletto D, Argon Y (2012) GRP94: an HSP90-like protein specialized for protein folding and quality control in the endoplasmic reticulum. Biochim Biophys Acta 1823(3): 774–787. https://doi.org/10.1016/j.bbamcr.2011.10.013
4. Whitesell L, Lindquist SL (2005) HSP90 and the chaperoning of cancer. Nat Rev Cancer 5(10):761–772. https://doi.org/10.1038/nrc1716
5. Wu BX, Hong F, Zhang Y, Ansa-Addo E, Li Z (2016) GRP94/gp96 in cancer: biology, structure, immunology, and drug development. Adv Cancer Res 129:165–190. https://doi.org/10.1016/bs.acr.2015.09.001
6. Bellaye PS, Burgy O, Causse S, Garrido C, Bonniaud P (2014) Heat shock proteins in fibrosis and wound healing: good or evil? Pharmacol Ther 143(2):119–132. https://doi.org/10.1016/j.pharmthera.2014.02.009
7. Hickey E, Brandon SE, Smale G, Lloyd D, Weber LA (1989) Sequence and regulation of a gene encoding a human 89-kilodalton heat shock protein. Mol Cell Biol 9(6):2615–2626. https://doi.org/10.1128/mcb.9.6.2615-2626.1989
8. Sreedhar AS, Kalmár E, Csermely P, Shen YF (2004) Hsp90 isoforms: functions, expression and clinical importance. FEBS Lett 562(1–3): 11–15. https://doi.org/10.1016/s0014-5793(04)00229-7
9. Ono K, Eguchi T, Sogawa C, Calderwood SK, Futagawa J, Kasai T et al (2018) HSP-enriched properties of extracellular vesicles involve survival of metastatic oral cancer cells. J Cell Biochem 119(9):7350–7362. https://doi.org/10.1002/jcb.27039
10. Eguchi T, Sogawa C, Ono K, Matsumoto M, Tran MT, Okusha Y et al (2020) Cell Stress Induced Stressome Release Including Damaged Membrane Vesicles and Extracellular HSP90 by Prostate Cancer Cells. Cells 9(3). https://doi.org/10.3390/cells9030755
11. Li W, Sahu D, Tsen F (2012) Secreted heat shock protein-90 (Hsp90) in wound healing and cancer. Biochim Biophys Acta 1823(3): 730–741. https://doi.org/10.1016/j.bbamcr.2011.09.009
12. Hance MW, Dole K, Gopal U, Bohonowych JE, Jezierska-Drutel A, Neumann CA et al (2012) Secreted Hsp90 is a novel regulator of the epithelial to mesenchymal transition (EMT) in prostate cancer. J Biol Chem 287(45):37732–37744. https://doi.org/10.1074/jbc.M112.389015
13. Calderwood SK, Borges TJ, Eguchi T, Lang BJ, Murshid A, Okusha Y et al (2021) Extracellular Hsp90 and protection of neuronal cells through Nrf2. Biochem Soc Trans 49(5): 2299–2306. https://doi.org/10.1042/bst20210370
14. Ono K, Sogawa C, Kawai H, Tran MT, Taha EA, Lu Y et al (2020) Triple knockdown of CDC37, HSP90-alpha and HSP90-beta diminishes extracellular vesicles-driven malignancy events and macrophage M2 polarization in oral cancer. J Extracell Vesicles 9(1): 1769373. https://doi.org/10.1080/20013078.2020.1769373
15. Karagöz GE, Duarte AM, Akoury E, Ippel H, Biernat J, Morán Luengo T et al (2014) Hsp90-tau complex reveals molecular basis for specificity in chaperone action. Cell 156(5): 963–974. https://doi.org/10.1016/j.cell.2014.01.037
16. Luo W, Sun W, Taldone T, Rodina A, Chiosis G (2010) Heat shock protein 90 in neurodegenerative diseases. Mol Neurodegener 5:24. https://doi.org/10.1186/1750-1326-5-24

17. Huck JD, Que NLS, Sharma S, Taldone T, Chiosis G, Gewirth DT (2019) Structures of Hsp90α and Hsp90β bound to a purine-scaffold inhibitor reveal an exploitable residue for drug selectivity. Proteins 87(10):869–877. https://doi.org/10.1002/prot.25750
18. Eguchi T, Sheta M, Fujii M, Calderwood SK (2022) Cancer extracellular vesicles, tumoroid models, and tumor microenvironment. Semin Cancer Biol 86:112–126. https://doi.org/10.1016/j.semcancer.2022.01.003
19. Oo MW, Kawai H, Takabatake K, Tomida S, Eguchi T, Ono K et al (2022) Resident stroma-secreted chemokine CCL2 governs myeloid-derived suppressor cells in the tumor microenvironment. JCI Insight 7(1):e148960. https://doi.org/10.1172/jci.insight.148960

Chapter 22

Large-Scale Databases and Portals on Cancer Genome to Analyze Chaperone Genes Correlated to Patient Prognosis

Kisho Ono and Takanori Eguchi

Abstract

Molecular chaperones, such as heat shock proteins (HSPs), have attracted attention as molecules involved in malignant events in cancers and are potential therapeutic targets and biomarkers for tumor therapy. Furthermore, mutations in chaperones can significantly impact cancer risk and prognosis. Bioinformatics is a particularly useful method for developing biomarkers as a practical consideration for the immediate clinical application of data. Many large-scale databases and portals on cancer genome are nowadays publicly available, including the International Cancer Genome Consortium (ICGC); The Cancer Genome Atlas (TCGA), renamed as Genomic Data Commons (GDC); Catalogue of Somatic Mutations in Cancer (COSMIC); and Cancer Cell Line Encyclopedia (CCLE). Referring to these databases, advanced web portals are publicized, including cBioPortal, Human Protein Atlas (HPA), Kaplan–Meier (KM) plotter, Gene Expression Profiling Interactive Analysis 2 (GEPIA2), Genomics of Drug Sensitivity in Cancer (GDSC), and Dependency Map (DepMap). Here, we assemble these databases and portals to clarify what is available and useful for current cancer research and provide protocols to utilize the HPA, KM plotter, and GEPIA2 for studies on chaperone genes in cancer patients. Utilizing these portals will reveal the correlation between tumor subtype-specific high expression of chaperone genes and patient prognosis. Our protocols are useful to increase systematic awareness of chaperones and find new biomarkers for diagnosis and prognosis and new targets for anticancer drugs.

Key words Chaperones, Heat shock proteins, Gene expression, Protein expression, Prognostic value, Large-scale database, Portals

1 Introduction

"Chaperone" is a general term for a protein that helps other protein molecules fold correctly and acquire function. Chaperones are also called molecular chaperones or protein chaperones. A chaperone is a protein usually coded by a single gene; e.g., human heat shock protein 90 alpha (HSP90α) is coded by the human *HSP90AA1* gene. Many public large-scale databases, portals, and platforms on cancer genome are nowadays available on the web for free, including the International Cancer Genome Consortium (ICGC); The

Stuart K. Calderwood and Thomas L. Prince (eds.), *Chaperones: Methods and Protocols*, Methods in Molecular Biology, vol. 2693, https://doi.org/10.1007/978-1-0716-3342-7_22,

Cancer Genome Atlas (TCGA), renamed as Genomic Data Commons (GDC); cBioPortal; Catalogue Of Somatic Mutations In Cancer (COSMIC); and Cancer Cell Line Encyclopedia (CCLE). Referring to these large-scale databases, advanced web portals are available for more specific purposes, including Human Protein Atlas (HPA), Kaplan–Meier plotter (KM plotter), Gene Expression Profiling Interactive Analysis 2 (GEPIA2), Genomics of Drug Sensitivity in Cancer (GDSC), and Dependency Map (DepMap). Here, we assemble these large-scale databases and portals to clarify what is available and useful for researchers and provide protocols to utilize the HPA, KM plotter, and GEPIA2 for studies of chaperone genes.

Heat shock proteins (HSPs) and other chaperones are genetically highly conserved proteins that play a key role in maintaining cellular homeostasis [1, 2]. However, some HSPs act as malignant chaperones in cancer development, progression, metastasis, and drug resistance [3–6]. Therefore, HSPs have been proposed as a potential therapeutic target for tumor therapy [3–5, 7]. Several studies suggest HSPs are potential diagnostic and prognostic biomarkers and therapeutic targets for tumors [4, 8]. Furthermore, it has been noted that mutations in HSPs may profoundly impact cancer risk and prognosis [9]. We recently demonstrated that highly metastatic head and neck squamous cell carcinoma (HNSC) and prostate carcinoma secrete HSP-rich extracellular vesicles (EVs), and overexpression of HSPs correlates with poor prognosis in HNSC and prostate cancer [8, 10]. In addition, various co-chaperones are important factors in tumor progression [4, 10]. In particular, CDC37, a co-chaperone of HSP90, is overexpressed in some tumor types and mediates carcinogenesis and maintenance of malignant phenotype by stabilizing the impaired conformation of mutant and/or overexpressed oncogenic kinases [4, 10, 11]. Chaperones, including HSPs, other chaperones, co-chaperones, and chaperonins, are considered prognostic factors that play key roles in tumor formation, progression, and resistance.

Many public large-scale databases, portals, and cancer genomes platforms are available on the web for free (Table 1). The International Cancer Genome Consortium (ICGC) coordinates large-scale cancer genome studies from 50 cancer types/subtypes. Systematic studies of more than 25,000 cancer genomes at the genomic, epigenomic, and transcriptomic levels reveal the repertoire of oncogenic mutations, uncover traces of the mutagenic influences, define clinically relevant subtypes for prognosis and therapeutic management, and enable the development of new cancer therapies. TCGA was renamed as Genomic Data Commons (GDC) Data Portal driven by the NIH Cancer Genome Project. Data for various cancer types were aggregated on genomes, epigenomes, transcriptomes, and mutation information. COSMIC is provided by Sanger Institute. Expert curation data are manually entered from papers to obtain information on known oncogenes, whereas systematic

Table 1
Databases and portals

Database	Outline
ICGC dcc.icgc.org	The International Cancer Genome Consortium (ICGC) coordinates large-scale cancer genome studies from 50 cancer types/subtypes. Systematic studies of more than 25,000 cancer genomes at the genomic, epigenomic, and transcriptomic levels reveal the **repertoire of oncogenic mutations**, uncover traces of the mutagenic influences, define clinically relevant subtypes for prognosis and therapeutic management, and enable the development of new cancer therapies
GDC (**TCGA**) portal.gdc.cancer.gov	Genomic Data Commons (GDC) Data Portal is driven by the NIH Cancer Genome Project. Data for various cancer types were aggregated on **genomes**, **epigenomes**, **transcriptomes**, and **mutation** information
cBioPortal cbioportal.org	cBioPortal is an **analyzer** provided by Sloan Kettering Cancer Institute, which **integrates** public data such as TCGA, ICGC, and CCLE and allows various analyses focusing on a gene of interest
COSMIC cancer.sanger.ac.uk/cosmic	Sanger Institute provides a **Catalogue Of Somatic Mutations In Cancer (COSMIC)**. Expert curation data are manually entered from papers to obtain information on known oncogenes, whereas systematic screen data are imported from databases to discover **new oncogenes** that are unknown
CCLE sites.broadinstitute.org/ccle	**Cancer Cell Line Encyclopedia** (CCLE) is a database of human cancer cell lines by the Broad Institute. It compiles data on gene expression, **mutations**, RNA **splicing**, DNA methylation, histone H3 modifications, microRNA expression, and protein arrays
Human Protein Atlas (HPA) proteinatlas.org	**HPA** is a **protein expression** and **localization** database in various human organs, tissues, and cells. Protein information is obtained by fluorescent antibody imaging, MS-based proteomics, transcriptomics, and systems biology methods
Kaplan–Meier plotter (KM plotter) kmplot.com	The KM plotter can assess the correlation between the expression of all genes (mRNA, miRNA, protein) and survival in 30 k + samples from 21 tumor types for a **meta-analysis**-based discovery and validation of **survival biomarkers**. Sources include GEO, EGA, and TCGA
GEPIA2 gepia2.cancer-pku.cn	**Gene Expression Profiling Interactive Analysis 2** (GEPIA2) features 198,619 **isoforms** and 84 cancer subtypes, extends gene expression quantification, and supports analysis of cancer subtypes. GEPIA2 adopts gene signature quantification inspired by **single-cell sequencing** studies and provides customized analysis where users **upload RNA-seq data** and compare it with TCGA and GTEx samples

(continued)

Table 1 (continued)

Database	Outline
Genomics of Drug Sensitivity in Cancer cancerrxgene.org	Genomics of Drug Sensitivity in Cancer (GDSC) is a database of **drug sensitivity of cancer cell lines** by the Sanger Institute. **Drug response** data and **genomic markers of sensitivity** are available on 1000 human cancer cell lines and screened with 100 s of compounds
DepMap depmap.org/portal	The **Dependency Map** (DepMap) portal empowers researchers to make discoveries related to cancer **vulnerabilities**. DepMap provides data and **screening results** of various cancer cells for RNAi and CRISPR screening to find synthetic lethal combinations

screen data are imported from databases to discover new oncogenes that are unknown. CCLE is a database of human cancer cell lines by the Broad Institute. It compiles data on gene expression, mutations, RNA splicing, DNA methylation, histone H3 modifications, microRNA expression, and protein arrays.

Advanced web portals are available for more specific purposes (Table 1). cBioPortal is an analyzer provided by Sloan Kettering Cancer Institute, which integrates public data such as TCGA and ICGC and allows various analyses focusing on a gene of interest. Human Protein Atlas (HPA) is a protein expression database and localization in various types of human organs, tissues, and cells. Protein information is obtained by fluorescent antibody imaging, MS-based proteomics, transcriptomics, and systems biology methods. Kaplan–Meier plotter can assess the correlation between the expression of all genes (mRNA, miRNA, protein) and survival in 30 k + samples from 21 tumor types for a meta-analysis-based discovery and validation of survival biomarkers. GEPIA2 features 198,619 isoforms and 84 cancer subtypes, extends gene expression quantification, and supports analysis of cancer subtypes. GEPIA2 adopts gene signature quantification inspired by single-cell sequencing studies and provides customized analysis where users upload RNA-seq data and compare it with TCGA and GTEx samples. Genomics of Drug Sensitivity in Cancer is a database of drug sensitivity of cancer cell lines by the Sanger Institute. The Dependency Map (DepMap) portal empowers researchers to make discoveries related to cancer vulnerabilities. DepMap provides data and screening results of various cancer cells for RNAi and CRISPR screening to find synthetic lethal combinations.

Basic research using cell lines and animal experiments is useful for elucidating the mechanism of action of cancer-related molecules. Besides, to promote research optimizing precision medicine for individual cancer patients, verifying whether the results and hypotheses obtained in each research are occurring in human bodies using clinical specimens is essential. Experiments with clinical specimens provide insight into what is happening in patients. 3D spheroid/organoid models and patient-derived xenograft (PDX) models have also received attention and been widely operationalized [12–16]. However, experiments using clinical specimens are sometimes faced with problems that do not allow for optimism, such as the possibility of unnecessary invasion of patients, complexity of experimental techniques and analysis methods, high costs, and difficulties in research planning and operation at each facility. In this context, comprehensive analyses of genetic abnormalities associated with metastasis, recurrence, and response and resistance to drugs have been conducted in various countries around the world using clinical specimens from various malignant tumors. For a great example, the analyzed data of RNA sequencing on tumors have been compiled into databases useful for further research [17]. Moreover, bioinformatics methods are particularly useful for establishing biomarkers. The collected data have been publicized and useful for practical considerations for immediate clinical application [18]. Analysis of these data sets is becoming more common to better understand diseases and discover new drugs and biomarkers.

Recent open data initiatives in basic and clinical research have dramatically increased the types of biomedical data available to the public [19]. This chapter introduces three public databases with which the authors have a particular experience. Furthermore, the practicality of the database is demonstrated using "HNSC," or most notable carcinomas, as examples, along with the search for some chaperone genes [8]. These databases reveal tumor-specific high expression and prognostic correlations for various cancer-related proteins, which have been identified as valid therapeutic targets and disease biomarkers. This chapter aims to analyze chaperone profiles in cancer and clarify the utility of database analysis to increase systematic awareness of molecular chaperones. Finding diagnostic and prognostic biomarkers and targets for anticancer drugs provide important clues for further mechanistic studies.

2 Materials

1. *Accessibility to a high-speed internet connection*
2. A personal computer or computing device with an Internet browser (*see* **Note 1**).
3. Software, e.g., Microsoft Office software (PowerPoint, Excel, and Word).

4. Chaperone gene name or ID of interest.
5. Databases.

5A. Human Protein Atlas (HPA)

HPA (https://www.proteinatlas.org/) is capable of identifying the mRNA expression of chaperone proteins in both cancerous cells and healthy cells [20]. We can compare specific chaperone protein expression levels and protein localization in normal and tumor tissues by using the "TISSUE" and "PATHOLOGY" modules [21]. Values derived from the RNA expression summary graph on the website should be used to determine what would be deemed as high, medium, and low expression (*see* **Notes 2** and **3**).

5B. Kaplan–Meier (KM) Plotter

KM plotter (http://kmplot.com/analysis) is capable of assessing the correlation between the expression of some genes (encoding mRNA, miRNA, or protein) and survival in 30 k + samples from 21 tumor types [22]. Sources for the databases include Gene Expression Omnibus (GEO), the European Genome-phenome Archive (EGA), and TCGA. The primary purpose of the tool is a meta-analysis-based discovery and validation of survival biomarkers. To analyze the prognostic value of specific genes, the enrolled patient sample will be divided into two groups according to various quartile expressions of the proposed biomarkers. The two patient cohorts will be compared on a Kaplan–Meier survival plot, and hazard ratios with 95% confidence intervals and log-rank P-values will be calculated (*see* **Note 4**).

5C. Gene Expression Profiling Interactive Analysis 2 (GEPIA2).

GEPIA2 (http://gepia2.cancer-pku.cn/) is capable of using as a valuable and highly cited resource for gene expression analysis based on 9736 tumors and 8587 normal samples from the TCGA and Genotype-Tissue Expression (GTEx) databases [23] (*see* **Note 5**). GEPIA2 provides customizable functions such as tumor/normal differential expression analysis, profiling according to cancer types or pathological stages, patient survival analysis, similar gene detection, correlation analysis, and dimensionality reduction analysis.

3 Methods

3.1 HPA

1. Open the URL: https://www.proteinatlas.org/.
2. Type the chaperone gene name (Fig. 1a).
3. Select the chaperone gene you are interested in from the list (*see* **Note 6**).

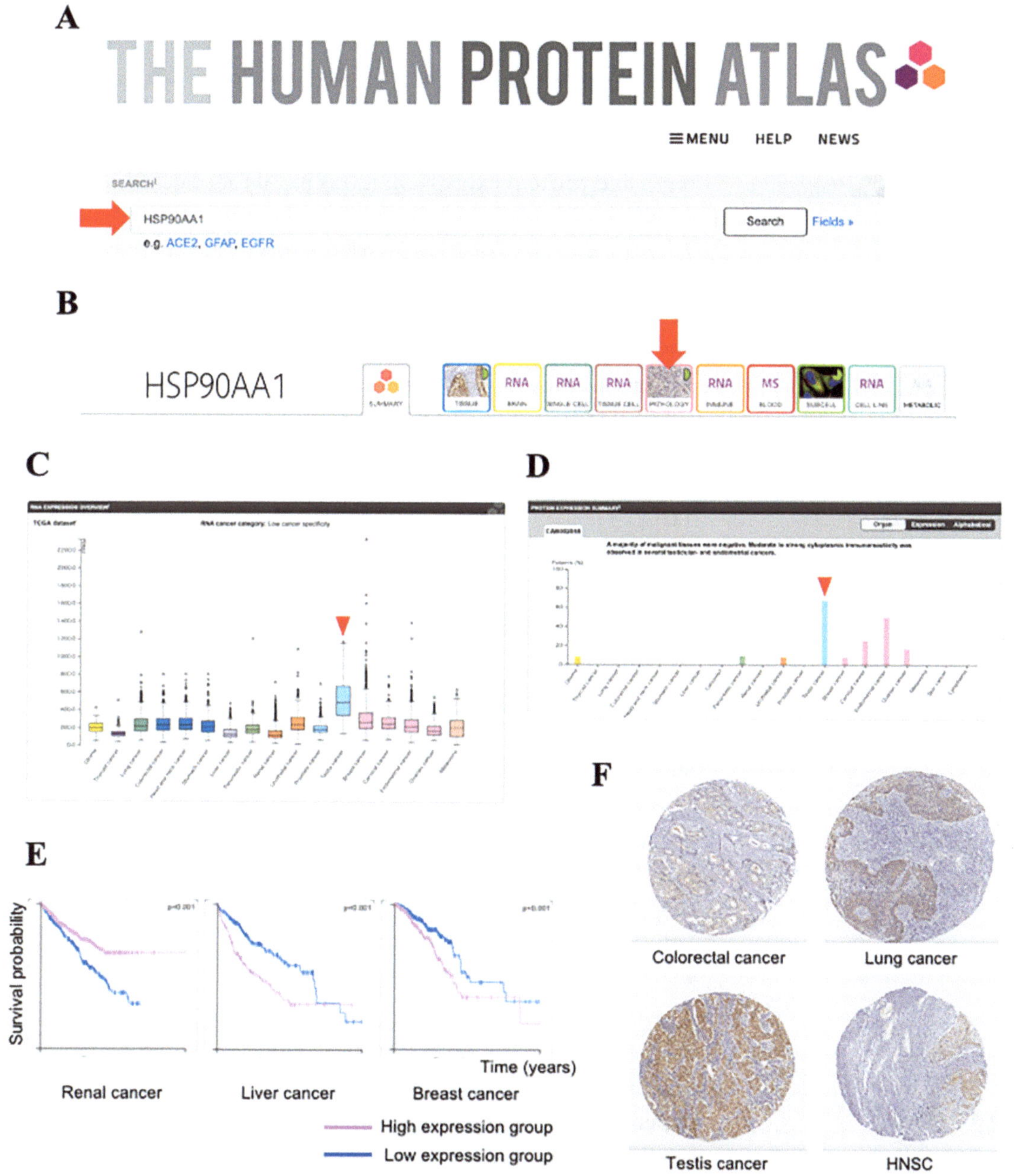

Fig. 1 Overview of pages in the Human Protein Atlas. (**a**) On top of the page, type the chaperone gene name. (**b**) Select the module you want to examine about the chaperone gene of interest. Select "PATHOLOGY" if you want to examine the position of chaperone genes in cancer (red arrow). (**c**) Comparison of mRNA expression levels of HSP90AA1 gene in 17 cancer types. The red arrows indicate that HSP90AA1 mRNA expression is most prominent in testis cancer. (**d**) Comparison of the protein expression levels of the HSP90AA1 gene in 20 cancer types. The red arrows indicate that HSP90AA1 protein expression is also most prominent in testis cancer. (**e**) Representative Kaplan–Meier survival plots for some cancers show a significant prognostic relationship between HSP90AA1 mRNA expression and patient survival. HSP90AA1, in particular, is a potential prognostic marker in renal cancer (favorable), liver cancer (unfavorable), and breast cancer (unfavorable). (**f**) Representative examples of tissue microarray stained images for HSP90AA1 from various cancers in the HPA database

4. An interactive overview of RNA and protein expression data will appear (*see* **Note 7**).
5. Select the "PATHOLOGY" module to investigate the role of selected genes in each cancer (Fig. 1b) (*see* **Note 8**).
6. Take an overview of the RNA expression levels in 17 cancer types and protein expression levels in 20 for selected chaperone genes (Fig. 1c and d).
7. Take an overview of Kaplan–Meier survival plots for representative cancers that show a significant prognostic relationship between mRNA expression of selected protein-coding genes and patient survival (Fig. 1e).
8. Select the cancer type of interest and generate Kaplan–Meier survival plots (*see* **Note 9**).
9. (Optional) Take an overview of images of tissue samples stained for selected chaperones in each cancer type (Fig. 1f) (*see* **Note 10**) [24].

3.2 KM Plotter

1. Open the URL: http://kmplot.com/analysis.
2. Select the cancer type and objective gene type (mRNA, miRNA, protein) (Fig. 2a) (*see* **Note 11**).
3. Type the chaperone gene name (Fig. 2a) (*see* **Note 12**).
4. In the "Survival" column, select survival factors such as overall survival (OS), relapse-free survival (RFS), etc., depending on the condition (Fig. 2a).
5. (Optional) If you choose "pan-cancer," select the cancer type of interest (Fig. 2a) (*see* **Note 13**).
6. (Optional) Restrict analysis to subtypes (stage, gender, etc.) or cell types (Fig. 2a) (*see* **Notes 13** and **14**).
7. Click on the "Draw Kaplan–Meier plot" module (Fig. 2a).
8. Take an overview of the Kaplan–Meier graph with a hazard ratio (HR) and log-rank P-values (Fig. 2b) (*see* **Note 15**).

3.3 GEPIA 2

1. Open the URL: http://gepia2.cancer-pku.cn/.
2. (Optional) As a quick start, enter a chaperone gene/isoform symbol or ID in the "Enter gene name" field, and click the "GoPIA!" button to search for the gene of interest (Fig. 3a, b) (*see* **Note 16**).
3. Select the "Expression DIY" module to profile the tissue-wise expression or cancer stage-wise expression of one or multigene in different cancer types using a dot or box plot (Fig. 3c, d) (*see* **Notes 17** and **18**).
4. Select the "Survival Analysis" module to perform survival analysis based on the expression status of one chaperone gene or a

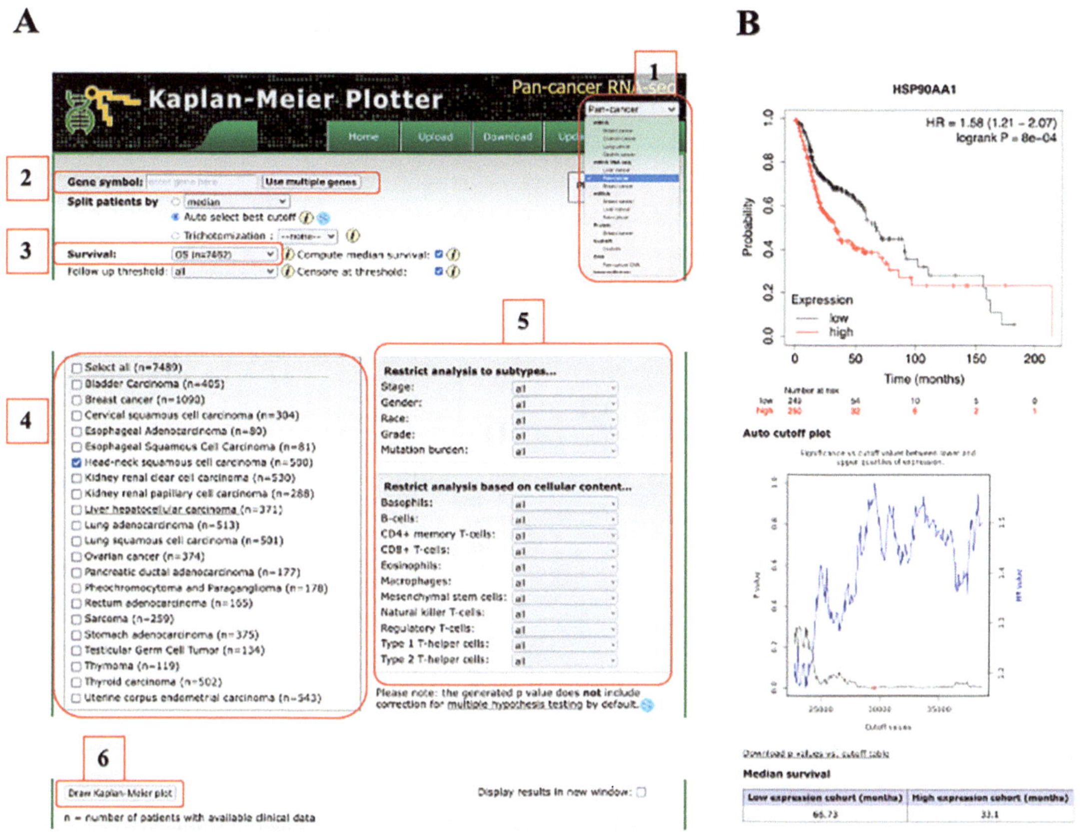

Fig. 2 Overview of pages in the Kaplan–Meier plotter. (**a**) On the first page, (1) select the cancer type and objective gene type; (2) type the chaperone gene name; (3) select survival factors; (4) select the cancer type of interest; (5) restrict the analysis to subtypes or cell types; (6) click on "Draw Kaplan–Meier plot" module. (**b**) The Kaplan–Meier graph shows the prognostic association between HSP90AA1 expression and HNSC survival rates with HR and log-rank P-values. Auto cutoff plot and median survival are also displayed

multigene signature, and plot a Kaplan–Meier curve (Fig. 3e) (*see* **Note 19**).

5. Select the "Correlation Analysis" module to compute the correlation of two genes or two signatures in multiple cancer types and tissues (*see* **Note 20**).

4 Notes

1. We recommend using the following browsers: Google Chrome, Firefox 3.0 or higher, Safari, Internet Explorer 9.0 or higher.
2. We determine any value from 3.0 FPKM (fragments per kilobase million, unit for RNA expression by two reads per fragment), and greater is considered high, 2.2–2.9 FPKM is

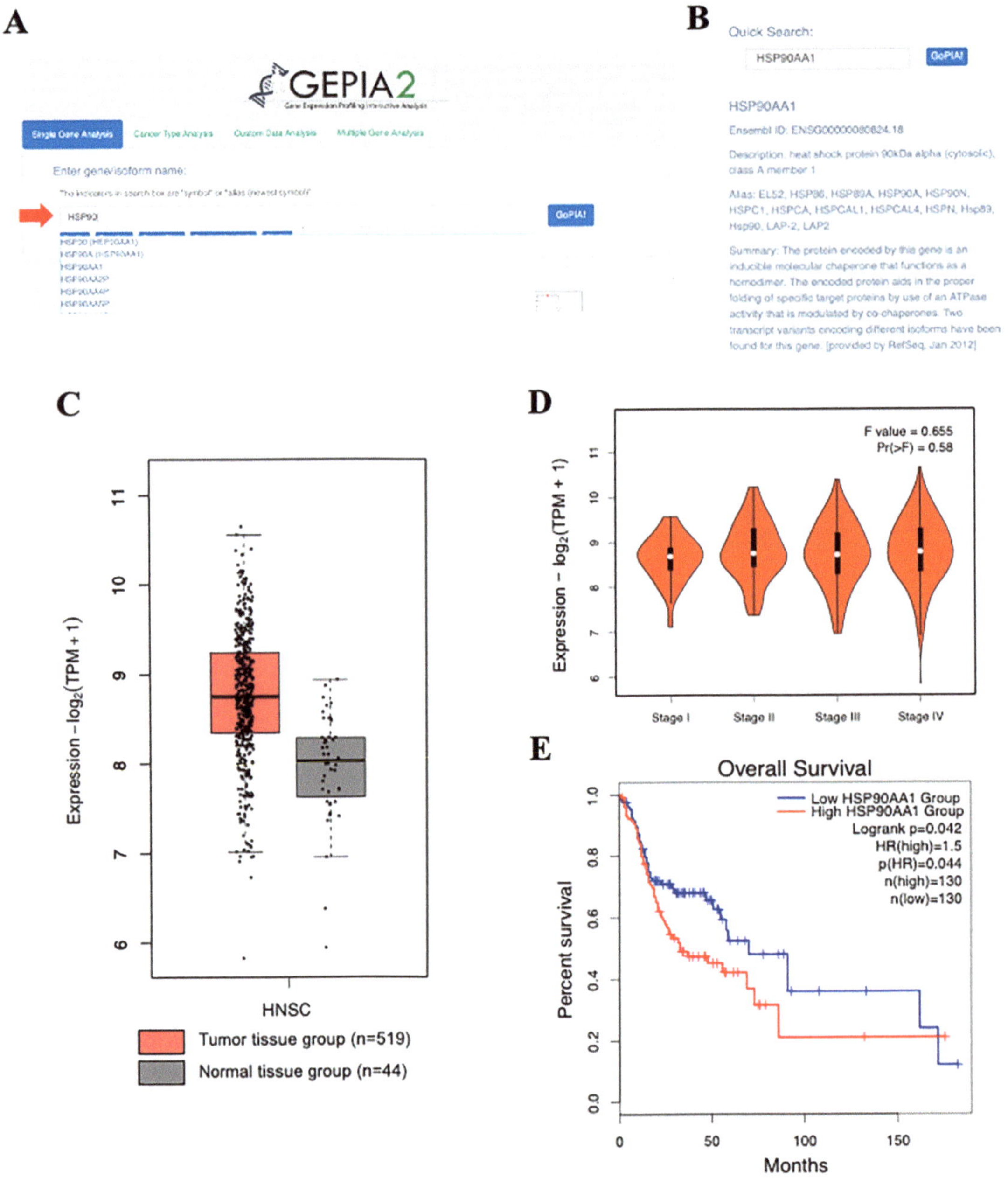

Fig. 3 Overview of pages in the GEPIA2. (a) On top of the page, type the chaperone gene name. (b) Summary of HSP90AA1 (selected gene). (**c**) Comparison of HSP90AA1 expression in normal and HNSC tissues using a box plot. (**d**) Comparison of HSP90AA1 expression by HNSC stages. (**e**) Survival analysis in HNSCs by differential expression of HSP90AA1 using Kaplan–Meier graph

considered medium, and 1–2.1 FPKM is considered low expression [25].

3. It contains immunohistochemical expression data for nearly 20 common kinds of cancers. Using immunohistochemistry images, we can compare the protein expression levels of different chaperones between normal and HNSC tissues [26]. The protein expression scores were based on immunohistochemical data, including staining intensity (not detected, low, medium, or high).
4. The "auto select best cutoff" model can be chosen, which means that all possible cutoff values were computed, and the best-performing threshold was used as the cutoff [27].
5. The "Survival Analysis" module is capable of analyzing the relationship between chaperone genes and overall survival (OS) and disease-free survival (DFS) in HNSC patients. Not only that, users can compare chaperone expression levels by HNSC stage.
6. The show/hide column function on the top left is useful for adding columns to the search result list if you need more information.
7. This chapter describes only the flow of the "PATHOLOGY" section specifically to clarify the position of molecular chaperones in cancer, but expression by normal human tissue organs, or each cell line, subcellular localization, etc., can be revealed by selecting other sections.
8. The pathology/cancer section consists of two different parts: (1) information on the association between genome-wide RNA expression levels and survival of cancer patients (for nearly 8000 cancer patients representing 17 major types of cancer) and (2) examples of protein expression patterns in cancer tissues (for 216 tumors representing the 20 most common forms of human cancer) [20].
9. Kaplan–Meier survival plots and a scatter plot showing the correlation between mRNA expression of the gene and patient survival of a specific cancer type are shown on a cancer type-specific gene summary page. The page is interactive, and users can immediately select a subgroup of patients based on, for example, tumor stage I, and generate specific plots for the selected subgroup on the website. The user can also use any specific expression cutoff (FPKM value) to produce different Kaplan–Meier and scatter plots.
10. To exemplify protein expression patterns both within one cancer type and between different types of cancer, a multitude of immunohistochemistry (IHC) images for more than 17,000 protein-coding genes in 20 human cancer types are provided in

this section [24]. An example of an IHC image for the HSP90AA1 gene from an HNSC patient is shown (Fig. 1f).

11. If no cancer name is of interest, select "pan-cancer" (Fig. 2a).
12. Multiple gene names can be entered by clicking on the "Use multiple genes" module (Fig. 2a). A maximum of 65 genes are allowed.
13. Multiple selections are possible.
14. Relevant factors include the status of hormone receptors (estrogen receptor, ER; progesterone receptor, PgR), and human epidermal growth factor receptor type 2 (HER2) for breast cancer, smoking status for lung cancer, alcohol consumption for liver cancer, etc. can also be selected.
15. There are four options: (1) utilize the scalable PDF provided in the results; (2) adjust "Settings" to generate a hi-res TIFF file; (3) use our PowerPoint template to change the font and the text size; (4) adjust "Settings" to export plot data as text, and format it in any other software.
16. The first page shows the selected gene summary (Fig. 3b), the gene expression profile across all tumor samples and paired normal tissues, the isoforms list of the selected gene, and the list of the most similar genes.
17. For an example of the expression analysis in box plot and stage plot, users can easily get the result that HSP90AA1 is overexpressed in HNSC compared with the normal tissue, while HSP90AA1 expression does not change with HNSC stage progression (Fig. 3c and d).
18. Users also can compare the expression of one gene in multiple cancers by box plot or compare multiple genes by a matrix plot in multiple gene comparison.
19. For example, in the survival analysis in the Kaplan–Meier curve, given the high expression of HSP90AA1, the patients in HNSC had a worse prognostic outcome (Fig. 3e).
20. Users can take a pair-wise gene correlation analysis of a given set of TCGA and/or GTEx expression data. Normalization is optional and customizable.

Acknowledgments

K.O. was supported by JSPS Kakenhi (grant numbers 19 K24072 and 21 K17115) and the KAWASAKI Foundation for Medical Science and Medical Welfare. T.E. was supported by JSPS Kakenhi, grant numbers 17 K11642-TE, 20 K09904-CS, 19H03817-MT, 20H03888-HN, 20 K20611-MT, 20H03888-HN, 21H03119-TY, and 21 K08902-HY.

References

1. Lang BJ, Guerrero ME, Prince TL, Okusha Y, Bonorino C, Calderwood SK (2021) The functions and regulation of heat shock proteins; key orchestrators of proteostasis and the heat shock response. Arch Toxicol. 95(6):1943–1970. https://doi.org/10.1007/s00204-021-03070-8
2. Murshid A, Eguchi T, Calderwood SK (2013) Stress proteins in aging and life span. Int J Hyperthermia. 29(5):442–447. https://doi.org/10.3109/02656736.2013.798873
3. Lang BJ, Prince TL, Okusha Y, Bunch H, Calderwood SK (2022) Heat shock proteins in cell signaling and cancer. Biochim Biophys Acta Mol Cell Res. 1869(3):119187. https://doi.org/10.1016/j.bbamcr.2021.119187
4. Ono K, Sogawa C, Kawai H, Tran MT, Taha EA, Lu Y et al (2020) Triple knockdown of CDC37, HSP90-alpha and HSP90-beta diminishes extracellular vesicles-driven malignancy events and macrophage M2 polarization in oral cancer. J Extracell Vesicles. 9(1): 1769373. https://doi.org/10.1080/20013078.2020.1769373
5. Prince TL, Lang BJ, Guerrero-Gimenez ME, Fernandez-Muñoz JM, Ackerman A, Calderwood SK (2020) HSF1: Primary Factor in Molecular Chaperone Expression and a Major Contributor to Cancer Morbidity. Cells. 9(4). https://doi.org/10.3390/cells9041046
6. Eguchi T, Calderwood SK, Takigawa M, Kubota S, Kozaki KI (2017) Intracellular MMP3 Promotes HSP Gene Expression in Collaboration With Chromobox Proteins. J Cell Biochem. 118(1):43–51. https://doi.org/10.1002/jcb.25607
7. Nitika BJS, Knighton LE, Takakuwa JE, Calderwood SK, Truman AW (2020) Chemogenomic screening identifies the Hsp70 co-chaperone DNAJA1 as a hub for anticancer drug resistance. Sci Rep. 10(1):13831. https://doi.org/10.1038/s41598-020-70764-x
8. Ono K, Eguchi T, Sogawa C, Calderwood SK, Futagawa J, Kasai T et al (2018) HSP-enriched properties of extracellular vesicles involve survival of metastatic oral cancer cells. J Cell Biochem. 119(9):7350–7362. https://doi.org/10.1002/jcb.27039
9. Taha EA, Ono K, Eguchi T (2019) Roles of Extracellular HSPs as Biomarkers in Immune Surveillance and Immune Evasion. Int J Mol Sci. 20(18). https://doi.org/10.3390/ijms20184588
10. Eguchi T, Sogawa C, Ono K, Matsumoto M, Tran MT, Okusha Y et al (2020) Cell Stress Induced Stressome Release Including Damaged Membrane Vesicles and Extracellular HSP90 by Prostate Cancer Cells. Cells. 9(3). https://doi.org/10.3390/cells9030755
11. Eguchi T, Prince TL, Tran MT, Sogawa C, Lang BJ, Calderwood SK (2019) MZF1 and SCAND1 Reciprocally Regulate CDC37 Gene Expression in Prostate Cancer. Cancers (Basel). 11(6). https://doi.org/10.3390/cancers11060792
12. Eguchi T, Sheta M, Fujii M, Calderwood SK (2022) Cancer extracellular vesicles, tumoroid models, and tumor microenvironment. Semin Cancer Biol. 86:112–126. https://doi.org/10.1016/j.semcancer.2022.01.003
13. Eguchi T, Sogawa C, Okusha Y, Uchibe K, Iinuma R, Ono K et al (2018) Organoids with cancer stem cell-like properties secrete exosomes and HSP90 in a 3D nanoenvironment. PLoS One. 13(2):e0191109. https://doi.org/10.1371/journal.pone.0191109
14. Sogawa C, Eguchi T, Namba Y, Okusha Y, Aoyama E, Ohyama K et al (2021) Gel-Free 3D Tumoroids with Stem Cell Properties Modeling Drug Resistance to Cisplatin and Imatinib in Metastatic Colorectal Cancer. Cells. 10(2). https://doi.org/10.3390/cells10020344
15. Ono K, Sato K, Nakamura T, Yoshida Y, Murata S, Yoshida K et al (2022) Reproduction of the Antitumor Effect of Cisplatin and Cetuximab Using a Three-dimensional Spheroid Model in Oral Cancer. International Journal of Medical Sciences. 19(8):1320–1333. https://doi.org/10.7150/ijms.74109
16. Taha EA, Sogawa C, Okusha Y, Kawai H, Oo MW, Elseoudi A et al (2020) Knockout of MMP3 Weakens Solid Tumor Organoids and Cancer Extracellular Vesicles. Cancers (Basel). 12(5). https://doi.org/10.3390/cancers12051260
17. da Silveira WA, Hazard ES, Chung D, Hardiman G (2019) Molecular Profiling of RNA Tumors Using High-Throughput RNA Sequencing: From Raw Data to Systems Level Analyses. Methods Mol Biol. 1908:185–204. https://doi.org/10.1007/978-1-4939-9004-7_13
18. Chen B, Butte AJ (2016) Leveraging big data to transform target selection and drug discovery. Clin Pharmacol Ther. 99(3):285–297. https://doi.org/10.1002/cpt.318

19. Kannan L, Ramos M, Re A, El-Hachem N, Safikhani Z, Gendoo DM et al (2016) Public data and open source tools for multi-assay genomic investigation of disease. Brief Bioinform. 17(4):603–615. https://doi.org/10.1093/bib/bbv080
20. Uhlen M, Zhang C, Lee S, Sjöstedt E, Fagerberg L, Bidkhori G et al (2017) A pathology atlas of the human cancer transcriptome. Science. 357(6352). https://doi.org/10.1126/science.aan2507
21. Gao W, Zhang ZW, Wang HY, Li XD, Peng WT, Guan HY et al (2022) TMED2/9/10 Serve as Biomarkers for Poor Prognosis in Head and Neck Squamous Carcinoma. Front Genet. 13:895281. https://doi.org/10.3389/fgene.2022.895281
22. Lánczky A, Győrffy B (2021) Web-Based Survival Analysis Tool Tailored for Medical Research (KMplot): Development and Implementation. J Med Internet Res. 23(7):e27633. https://doi.org/10.2196/27633
23. Tang Z, Kang B, Li C, Chen T, Zhang Z (2019) GEPIA2: an enhanced web server for large-scale expression profiling and interactive analysis. Nucleic Acids Res. 47(W1):W556–Ww60. https://doi.org/10.1093/nar/gkz430
24. Katona B, Lindskog C (2022) The Human Protein Atlas and Antibody-Based Tissue Profiling in Clinical Proteomics. Methods Mol Biol. 2420:191–206. https://doi.org/10.1007/978-1-0716-1936-0_15
25. Dey A, Balachandran I, Solis A, Sharp DJ (2022) Preclinical Research Strategy Development for RNAi-Based Therapies in Oncology Using Patient-Centered Information from Public Databases of Human Protein Atlas and R2Genomics Platform. Methods Mol Biol. 2423:181–185. https://doi.org/10.1007/978-1-0716-1952-0_17
26. Fan G, Tu Y, Wu N, Xiao H (2020) The expression profiles and prognostic values of HSPs family members in Head and neck cancer. Cancer Cell Int. 20:220. https://doi.org/10.1186/s12935-020-01296-7
27. Gao ZW, Yang L, Liu C, Wang X, Guo WT, Zhang HZ et al (2022) Distinct Roles of Adenosine Deaminase Isoenzymes ADA1 and ADA2: A Pan-Cancer Analysis. Front Immunol. 13:903461. https://doi.org/10.3389/fimmu.2022.903461

Chapter 23

Immunohistochemical, Flow Cytometric, and ELISA-Based Analyses of Intracellular, Membrane-Expressed, and Extracellular Hsp70 as Cancer Biomarkers

Ali Bashiri Dezfouli, Stefan Stangl, Gemma A. Foulds, Philipp Lennartz, Geoffrey J. Pilkington, A. Graham Pockley, and Gabriele Multhoff

Abstract

The major stress-inducible 70 kDa heat shock (stress) protein 70 (Hsp70) is frequently overexpressed in highly aggressive tumor cells and thus might serve as a tumor-specific biomarker of aggressive disease and/or therapeutic resistance. We have previously shown that, in contrast to normal cells, tumor cells present Hsp70 on their plasma membrane. In order to elucidate the role of intracellular, membrane-bound and extracellular Hsp70 as a potential tumor biomarker in cancer, herein we describe protocols for the staining of cytosolic Hsp70 in tumor formalin-fixed paraffin-embedded (FFPE) sections from patients with glioblastoma multiforme using immunohistochemistry, for detecting the expression of plasma membrane-bound Hsp70 by a range of cancer-derived cells using multi-parametric flow cytometry using the cmHsp70.1 monoclonal antibody (mAb) and for the measurement of free and vesicular-associated Hsp70 in the circulation of patients with cancer using a unique enzyme-linked immunosorbent assay (ELISA).

Key words Intracellular, Extracellular and membrane-bound heat shock protein 70 (Hsp70), Glioblastoma multiforme, Cancer, Formalin-fixed paraffin-embedded (FFPE) sections, Immunohistochemistry, Flow cytometry, Enzyme-linked immunosorbent assay (ELISA), Hsp70 antibody epitope

1 Introduction

The major stress-inducible 70 kDa heat shock (stress) protein Hsp70 is frequently overexpressed in the cytosol for a large variety of different tumor cells and elevated cytosolic Hsp70 levels are often associated with therapy resistance. We, therefore, propose Hsp70 as a potential biomarker of disease in cancer.

Using a unique monoclonal antibody (mAb, cmHsp70.1), the Multhoff laboratory made the seminal discovery that, in addition to its cytosolic localization, Hsp70 is also selectively expressed on the

Stuart K. Calderwood and Thomas L. Prince (eds.), *Chaperones: Methods and Protocols*, Methods in Molecular Biology, vol. 2693, https://doi.org/10.1007/978-1-0716-3342-7_23,

plasma membrane of tumor cells (but not normal tissue) [1–3]. The same laboratory also showed that the expression density of membrane Hsp70 on a broad profile of cancer cell lines is further increased in vitro by standard treatments such as radio(chemo)-therapy [4, 5] and that membrane Hsp70 expression reflects therapy resistance [6]. Membrane Hsp70 is more highly expressed in metastatic disease and relapsed tumors than its corresponding primary tumor [7], and its expression is associated with an unfavorable prognosis and a reduced overall survival for some tumors [8]. An ongoing screening program of over 1500 patients with various solid tumors in the Multhoff laboratory reveals that more than 50% of all patients have membrane Hsp70-positive tumors, as identified by flow cytometric analysis of isolated viable tumor cells using the cmHsp70.1 mAb.

These findings indicate that membrane-bound Hsp70 might act as a universal, selective tumor-specific marker of "aggressive" disease. Therefore, it is important that membrane Hsp70 expression can be determined to describe the status of a tumor, in addition to the more commonly analyzed cytosolic levels. This is highly relevant, as membrane-bound Hsp70 has also been shown to act as a tumor-specific therapeutic target which has underpinned the development of therapeutic approaches for treating aggressive and therapy-resistant cancers.

A number of studies have shown the capacity of ex vivo activated (low-dose IL-2 plus a 14-mer Hsp70-derived peptide, IL-2/TKD) NK cells to target a range of cancer cell types expressing the membrane form of Hsp70 in vitro, including leukemic blasts [9, 10], colorectal cancer cells [9, 11, 12], pancreatic cancer cells isolated from tumors grown in immunodeficient mice [13], and sarcoma cells [1, 2], as well as glioblastoma and lung cancer cells [14]. Preclinical animal studies have also demonstrated that allogeneic NK cells obtained from donor human peripheral blood mononuclear cells (PBMCs) that have been ex vivo activated with IL-2/TKD can recognize and control the growth of colorectal tumors [15, 16], metastatic pancreatic carcinoma [13], and glioblastoma and metastatic lung cancer [14] in preclinical mouse models.

The clinical safety and tolerability of IL-2/TKD-activated autologous (patient-derived) NK cells have been demonstrated in a Phase I clinical trial [17], and its safety and efficacy for the treatment of patients with advanced, inoperable NSCLC after standard-of-care radiochemotherapy has been demonstrated in a multi-center Phase II clinical trial [18, 19]. Of the seven patients treated with IL-2/TKD-activated autologous NK cells after radiochemotherapy, five experienced a clinical benefit (71%), including a complete remission/response (CR) and a partial response (PR, 28%). In contrast, of the seven patients who received standard-of-care radiochemotherapy, only two experienced a clinical benefit and there were no CRs [19].

Studies have demonstrated that the cytotoxic effect of IL-2/TKD-activated NK cells is manifested by the release and uptake of granzyme B into target cells via membrane Hsp70 [20]. We have also previously reported that recombinant human granzyme B binds to membrane Hsp70 on cancer cells and that its subsequent uptake, which results in the selective killing of membrane Hsp70 expressing cells and tumors, occurs via an Hsp70-mediated endocytosis process which is independent of perforin [7, 21]. The cmHsp70.1 membrane Hsp70 specific mAb itself can induce antibody-dependent cellular cytotoxicity (ADCC) [3].

Taken together, these findings make membrane Hsp70 an excellent and broadly applicable molecule on which to develop a portfolio of new immunotherapeutics based on ex vivo IL-2/TKD-activated NK cells [10, 12, 15, 17–19, 22], human recombinant granzyme B [7, 20, 21], or as a basis to the development of Chimeric Antigen Receptor (CAR) T cells and/or NK cells [23, 24].

The selective expression of membrane Hsp70 on cancer cells has also been used as a basis for the development of innovative imaging platforms for tumor detection and outcome monitoring which employ fluorescence or radionuclide-labeled cmHsp70.1 mAb or a 14-mer tumor penetrating peptide (TPP) [4, 25–29]. The 14-mer tumor penetrating peptide (TPP) is currently being developed for clinical use for PET/CT imaging [29].

Given the established association between membrane Hsp70 expression and disease aggressiveness, and the portfolio of "theranostic" approaches that are being developed to target tumors expressing membrane Hsp70, techniques for determining the expression of membrane Hsp70 expression by cancer cell lines and cells isolated from different tumor entities, by cancer tissues, and by tumors in patients have been established using the cmHsp70.1 mAb which specifically detects the inducible form of Hsp70 and does not cross-react with the highly homologous constitutive form Hsc70.

Another significant development in this field has been the discovery that tumor cells expressing membrane Hsp70 actively release Hsp70 in lipid vesicles with biophysical characteristics of exosomes [30]. The unique cmHsp70.1 mAb together with a second Hsp70 mAb (cmHsp70.2) form the basis for novel sandwich enzyme-linked immunoassays (ELISAs) for quantifying free and exosomal Hsp70 in the plasma/serum of patients with cancer – a liquid biopsy for determining the membrane Hsp70 expression status of a patient's tumor [31–34]. Using a format involving the cmHsp70.1 mAb as the detection antibody and the cmHsp70.2 mAb as the coating antibody, a comparison of the Hsp70 levels in patients with head and neck, colorectal, pancreatic cancer, glioblastoma, squamous cell or adeno non-small cell lung carcinoma (NSCLC) at diagnosis, or hematological malignancies and healthy human volunteers has revealed significantly higher levels in tumor patients [31, 34] and that serum levels of Hsp70 correlate with gross tumor volume in patients with squamous cell and adeno non-small cell lung cancer [32]. Furthermore, Hsp70 concentrations drop concomitantly with a decrease in viable tumor mass after irradiation [34].

This novel analytical platform, therefore, has the potential to be used as a liquid biopsy for identifying patients expressing the membrane Hsp70 tumor-specific biomarker, for risk assessment (i.e., differentiation of grade III vs. IV adeno NSCLC) [34], and for monitoring therapeutic outcomes and disease recurrence in a range of cancer settings.

Although great progress in the treatment of many cancers has been achieved, glioblastoma multiforme (GBM) remains a debilitating and life-threatening disease. Comprising nearly 75% of all cases, glioblastoma is the most frequent primary malignant tumor of the central nervous system (CNS) and continues to be associated with very poor long-term survival and the worst prognosis of all glial brain tumors (gliomas). Despite improvements in the treatment of glioblastoma using advanced neurosurgical techniques and radiation therapy, there remains no effective cure for this tumor [35]. The average life expectancy after diagnosis is less than 1 year, and fewer than 10% of patients survive longer than 5 years after diagnosis, even after total tumor resection [35–37]. Administration of the alkylating chemotherapeutic agent temozolomide (TMZ) only increases median survival rates to approximately 15 months [38–45]. Death from glioblastoma is due to a rapid, aggressive local infiltration of glioblastoma cells in the brain. Moreover, a high genetic tumor heterogeneity can mediate resistance to radio (chemo)therapy. An inherent or acquired resistance to radiochemotherapy drastically reduces life expectancy. It is therefore essential to have reliable tumor biomarkers which can identify patients who are likely to acquire resistance to TMZ early.

Immunohistochemical, flow cytometric, and ELISA-based approaches for determining tissue, cell membrane, and circulating levels of Hsp70 are described herein.

2 Materials

2.1 Patient-Derived Glioblastoma Multiforme Cell Lines

The human, low passage glioblastoma (UP007, UP029, and SEBTA027) and brain metastatic lung carcinoma (SEBTA028) cell lines, were established in culture at the University of Portsmouth from biopsy tissue obtained at King's College Hospital London under ethics permission, 11/SC/0048. All cell lines were cultured in a DMEM growth medium supplemented with 10% (v/v) fetal bovine serum (FBS).

- **UP-007** was derived from a 71-year-old treatment naïve male de novo GBM patient with wild-type IDH1.
- **UP-029** was derived from a 66-year-old treatment naïve female de novo GBM patient with wild-type IDH1 and IDH2.
- **SEBTA-027** was derived from a 59-year-old female with recurrent GBM in the right parieto-occipital region, with sarcomatous elements, grade IV, wild-type IDH1, unmethylated MGMT, and a 40% Ki67 labeling index and who had previously received radiotherapy and temozolomide chemotherapy.
- **SEBTA-028** was derived from a 52-year-old male with brain metastatic lung carcinoma in the right temporal–occipital–parietal region.

2.2 Patient- and Preclinical Model-Derived Paraffin-Embedded Tissue

The diagnosis of glioblastoma multiforme was based on hematoxylin/eosin staining of FFPE sections (2 μm) of two patients with glioblastoma multiforme WHO stage IV. Additionally, sections (2 μm) of orthotopically implanted GL261 mouse glioblastoma were used for the analysis. Ethical approval for access to patient tissue was obtained from the local ethical committee of the Technische Universität München, and informed consent was provided.

2.3 cmHsp70.1 Anti-Hsp70 Monoclonal Antibody (mAb)

The cmHsp70.1 mAb (multimmune GmbH, Munich, Germany), which selectively binds to membrane Hsp70 on tumor cells, was generated by immunization of mice with the 14-mer peptide TKDNNLLGRFELSG, termed "TKD," comprising amino acids 450–461 ($aa_{450-461}$) in the C terminus of the inducible Hsp70.

The "TKD" sequence which is exposed to the extracellular milieu of tumors resides in the C-terminally localized oligomerization domain which is part of the substrate binding domain of the Hsp70 molecule [46]. In contrast to other commercially available Hsp70 antibodies, the cmHsp70.1 mAb uniquely identifies the membrane form of Hsp70 on viable human and mouse tumor cells not only in vitro but also in vivo in tumor-bearing mice [25].

2.4 Immunohistochemistry (IHC) for Detecting Tissue Expression of Hsp70

2.4.1 Deparaffinization Reagents

1. Xylene
2. Ethanol absolute 100% (v/v)
3. Ethanol 96% (v/v)
4. Ethanol 70% (v/v)
5. Ethanol 50% (v/v)
6. H_2O_{dd}

2.4.2 Target Retrieval and Staining Reagents

1. Target retrieval solution: 1× H_2O_{dd}, 0.5× H_2O_{dd} (DAKO cat #: S1699)
2. Peroxidase blocking solution: 3% (v/v) H_2O_2/0.1% (w/v) sodium azide (NaN_3)/phosphate-buffered saline (PBS)

3. Protein blocking solution: 5% (v/v) rabbit serum/antibody diluent (DAKO REAL antibody diluent, cat #: S2022)
4. Phosphate-buffered saline (PBS)
5. Murine IgG1 cmHsp70.1 mAb (multimmune GmbH, Munich): working solution 1: 500 in antibody diluent
6. Horseradish peroxidase (HRP)-conjugated rabbit anti-mouse polyclonal antibody: DAKO EnVision+ system HRP-conjugated rabbit anti-mouse polyclonal antibody (cat #: DAKO K4001)
7. DAB (3,3′-diaminobenzidine): DAKO liquid DAB+ Substrate and Chromogen System (cat #: K3468)

2.5 Flow Cytometry for Detecting Cell Surface Expression of Membrane Hsp70

1. Cells of interest
2. Fc receptor (FcR)-blocking immunoglobulin (Ig; e.g., normal rat IgG) solution, 200 μg/ml or specific Fc blocking reagents that are suitable for use in the species of interest
3. Carboxyfluorescein or FITC-conjugated cmHsp70.1 mAb (multimmune GmbH)
4. Carboxyfluorescein or FITC-conjugated IgG1 murine mAb to Hsp70 (clone C92F3A, StressMarq Biosciences Inc. *See* **Notes**)
5. Appropriately conjugated murine immunoglobulin (Ig) isotype controls
6. Phosphate-buffered saline (PBS) supplemented with 0.1% (w/v) sodium azide (NaN_3) and bovine serum albumin (BSA) (PAB; see recipe below)
7. Nuclear/membrane integrity (viability) staining compound dissolved in PBS: 200 μg/ml PI [alternative options include 250 μg/ml 7-AAD, 250 μg/ml TO-PRO-3 (Molecular Probes), or 200 μg/ml PY(G)]—*see* **Notes**

2.5.1 PBS, pH 7.2, Supplemented with Sodium Azide (NaN_3) and Fetal Bovine Serum (FBS)

1. Dulbecco's PBS, Ca^{2+}- and Mg^{2+}-free (e.g., Life Technologies), containing:
   - 0.1% (w/v) sodium azide (NaN_3)
   - 10% (v/v) fetal bovine serum (FBS)
2. Adjust pH to 7.2 using 1 N NaOH or 1 N HCl.
3. Filter solution through 0.22 μm membrane.
4. Store for up to 1 month at 4 °C.
5. EDTA (1 mM) can be added to reduce cellular aggregation.

CAUTION: Always keep cells and reagents at 4 °C. Be extremely careful when handling sodium azide.

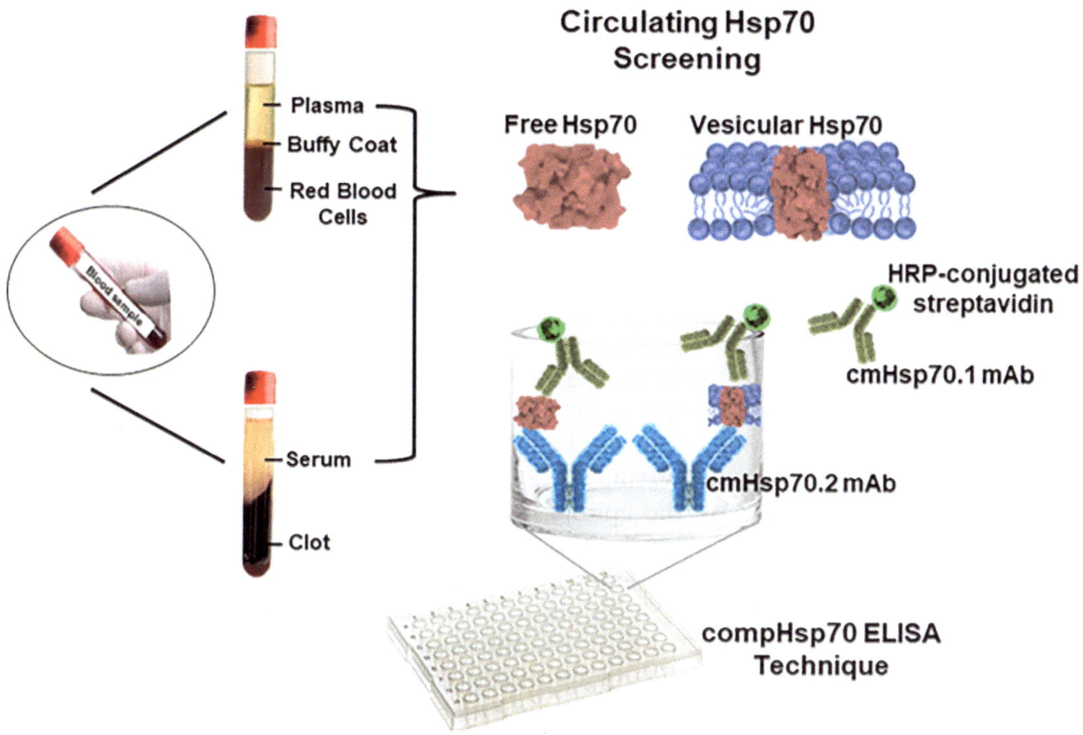

Fig. 1 Schematic representation of the compHsp70 sandwich ELISA for quantifying the circulating (free, exosomal) Hsp70 in the serum and plasma of mammalian blood samples

2.6 Enzyme-Linked Immunosorbent Assays (compHsp70 ELISA) for Detecting Extracellular Free and Exosomal Hsp70 (See Fig. 1)

1. Plasma/serum samples
2. Nunc™ MaxiSorp™ flat-bottom 96-well plates (Thermo Fisher Scientific)
3. Sodium carbonate "coating" buffer (0.1 M sodium carbonate, 0.1 M sodium hydrogen carbonate, pH 9.6; Sigma-Aldrich)
4. cmHsp70.2 mAb (multimmune GmbH, Munich, Germany)
5. cmHsp70.1-Biotin mAb (multimmune GmbH, Munich, Germany)
6. Phosphate-buffered saline (PBS; Life Technologies) supplemented with 0.05% (v/v) Tween-20 (Calbiochem, Merck, Darmstadt, Germany)
7. Blocking solution (Liquid Plate Sealer™, Candor Bioscience GmbH, Wangen i. Allgäu, Germany)
8. StabilZyme™ Select Stabilizer (Diarect GmbH, Freiburg i. Breisgau, Germany)
9. Recombinant Hsp70 protein (multimmune GmbH, Munich, Germany)
10. HRP-Protector™ (Candor Bioscience GmbH, Wangen i. Allgäu, Germany)

11. Horseradish peroxidase (HRP)-conjugated streptavidin (Senova GmbH, Weimar, Germany)
12. Substrate reagent (BioFX™ TMB Super Sensitive One Component HRP Microwell Substrate, Surmodics, Inc., Eden Prairie, MN, USA)
13. 2 N H_2SO_4

3 Methods

3.1 Immunohistochemistry (IHC) for Detecting Tissue Expression of Hsp70

IHC must be performed on formalin-fixed, paraffin-embedded (FFPE) tissue sections (2 μm) using a rabbit anti-mouse polyclonal antibody staining kit (DAKO). In brief, tissue sections were boiled by microwaving for 30 min in target retrieval buffer (DAKO) to unmask the epitope. Between each of the following incubation steps, sections were washed in PBS (Sigma-Aldrich). The nonspecific binding must be blocked in the protein blocking solution for 60 min. Slides must be incubated overnight at 4 °C with cmHsp70.1 mAb (multimmune GmbH) or with a corresponding IgG control (Sigma-Aldrich). After washing, sections are incubated with the HRP-conjugated anti-mouse polyclonal antibody, followed by a DAB chromogen reaction to detect the binding of the primary antibody. Nuclei are counterstained with hematoxylin, and sections are embedded in the embedding medium.

3.1.1 Rehydration

1. Xylene I & II 2 × 15 min.
2. Ethanol absolute I and II 2 × 10 min.
3. Ethanol 96% (v/v) 2 × 5 min.
4. Ethanol 70% (v/v) 1 × 5 min.
5. Ethanol 50% (v/v) 1 × 5 min.
6. H_2O_{dd} 1 × 5 min.

3.1.2 Target Retrieval and Staining

1. Target retrieval solution 1 × 30 min boiling (microwave).
2. Cooling down period 1 × 10 min in target retrieval solution.
3. H_2O_{dd} 1 × 2 min.
4. Marking section with a hydrophobic pen
5. Peroxidase blocking solution 1 × 30 min.
6. H_2O_{dd} 3 × 5 min.
7. Protein blocking solution 1 × 60 min.
8. PBS 2 × 5 min.
9. cmHsp70.1 mAb overnight (4 °C) (dilution 1:250–1: 1000).
10. PBS 2 × 5 min.

11. Envision+ System HRP-labeled anti-mouse polymer 1 × 30 min (RT).
12. PBS 1 × 5 min.
13. DAB 1 × 4 min.
14. H_2O_{dd} 1 × 5 min.
15. Counterstain with hematoxylin 1 × 30 s.
16. Rinse with warm tap water 1 × 5 min.

3.1.3 Dehydration and Embedding

1. Ethanol 50% (v/v) 1 × 5 min.
2. Ethanol 70% (v/v) 1 × 5 min.
3. Ethanol 96% (v/v) II & III 2 × 5 min.
4. Ethanol absolute III & IV 2 × 5 min.
5. Xylene III & IV 2 × 10 min.
6. Eukitt® mounting medium.

CAUTION: *Keep the time of the chromogen reaction exactly at 4 min, and use sections of the same thickness (2 μm) in order to be able to compare the staining intensities of different experiments. Always run reference sections with a defined staining intensity as an internal control.*

3.1.4 Scoring Criteria of IHC Using cmHsp70.1 mAb

The staining intensity must be analyzed semiquantitatively within the tumor area. The staining intensity of tumor cells is graded into different scores: normal (+), intermediate (++), and strong (+++). Normal tissues are graded as weak (+/−). The localization of the staining (cytosolic, nuclear) as well as the percentage of positively stained cells must be considered. The scoring must be performed by at least two independent researchers.

3.2 Flow Cytometry for Detecting Cell Surface Expression of Membrane Hsp70

1. Prepare a single-cell suspension of tumor cells in ice-cold PBS supplemented with 10% (v/v) FBS (PBS/FBS).
2. Dispense 1 × 10^5 cells into microcentrifuge tubes.
3. Pellet cells by centrifugation for 5 min at 500 × *g*, 4 °C.
4. Remove the supernatant by vacuum aspiration.
5. Add 4 μl ice-cold cmHsp70.1 mAb (1 μg/μl) to the pellet containing approximately 50 μl PBS/FBS, and resuspend and vortex tubes slightly.
6. Incubate tubes for 30 min on ice and wash in ice-cold PBS/FBS.
7. Resuspend cell pellet in 300 μl PBS/FBS.
8. Add 3 μl PI to the tube (*alternatively, add 4 μl 7-AAD, 4 μl TO-PRO-3, or 5 μl PY(G)*.
9. Incubate cells on ice for 5 min.

10. Analyze cells using a flow cytometer immediately thereafter.
11. Keep samples on ice during analysis.
12. *Prolonged periods before analyzing cells are generally not recommended.*

3.3 Enzyme-Linked Immunosorbent Assays (compHsp70 ELISA) for Detecting Extracellular Free and Exosomal Hsp70

1. Nunc™ MaxiSorp™ flat-bottom 96-well plates (Thermo Fisher Scientific) are coated by incubating overnight with 1 μg/mL cmHsp70.2 mAb (multimmune GmbH) in sodium carbonate buffer (0.1 M sodium carbonate, 0.1 M sodium hydrogen carbonate, pH 9.6; Sigma-Aldrich).
2. Plates are washed with PBS (Life Technologies) supplemented with 0.05% (v/v) Tween 20 (Calbiochem, Merck, Darmstadt, Germany).
3. Plates are blocked by incubation with blocking solution (Liquid Plate Sealer™, Candor Bioscience GmbH, Wangen i. Allgäu, Germany) for 30 min at room temperature.
4. After washing, serum or plasma samples diluted in StabilZyme™ Select Stabilizer (Diarect GmbH, Freiburg i. Breisgau, Germany) (25 μl sample; typical dilution 1:5), a total volume of 125 μl of the diluted samples is added to each well.
5. An eight-point concentration standard curve of Hsp70 protein (0–100 ng/ml) diluted in StabilZyme™ Select Stabilizer (Diarect GmbH, Freiburg i. Breisgau, Germany) is included in each assay.
6. After 30 min incubation at room temperature, plates are washed and incubated with 200 ng/ml of biotinylated cmHsp70.1 mAb (multimmune GmbH, Munich, Germany) diluted in HRP-Protector™ (Candor Bioscience GmbH, Wangen i. Allgäu, Germany) for 30 min at room temperature.
7. After another washing step, plates are incubated in the dark for 30 min at room temperature with 57 ng/ml horseradish peroxidase (HRP)-conjugated streptavidin (Senova GmbH, Weimar, Germany) diluted in HRP-Protector™ (Candor Bioscience GmbH, Wangen i. Allgäu, Germany).
8. After a final washing step, plates are incubated with substrate reagent (BioFX™ TMB Super Sensitive One Component HRP Microwell Substrate, Surmodics, Inc., Eden Prairie, MN, USA) for 15 min at room temperature.
9. The colorimetric reaction is stopped by adding 2 N H_2SO_4, and absorbance is read at 450 nm, corrected by absorbance at 570 nm, in a standard microplate spectrophotometer.

CAUTION: *Do not use serum and plasma aliquots that were frozen/thawed more than 3 times, and do not use serum and plasma samples with a hemoglobin concentration of more than 5 mg/dL since hemolysis disturbs the assay precision. Always run at least three reference serum or plasma samples with a defined Hsp70 value as an internal control.*

4 Notes

4.1 Immunohistochemistry (IHC) Scoring

Every nucleated human cell type expresses the stress-inducible Hsp70 at low levels in the cytosol, whereas the membrane-bound form of Hsp70 appears to be restricted to tumor cells. Given that the cmHsp70.1 mAb can stain both, the membrane-bound and cytosolic forms of Hsp70, and that approximately 70% to 90% of the total Hsp70 is residing intracellularly, it is nearly impossible to distinguish between the membrane-bound and cytosolic localizations of Hsp70 by IHC. Therefore, membrane-bound Hsp70 must be determined by flow cytometry of viable cells with intact plasma membranes, whereas cytosolic Hsp70 can be determined by IHC. It is well known that highly aggressive tumor cells show a higher expression density and a different subcellular localization of Hsp70 compared to normal cells. Therefore, the intensity and the subcellular distribution (cytosol, membrane, nucleus) of Hsp70 can be considered as tumor-specific characteristics. Since no human cell type is completely negative for cytosolic Hsp70, different Hsp70 staining intensities rather than Hsp70-negative and Hsp70-positive phenotypes can be distinguished. For a comparable scoring of different staining intensities in different IHC staining procedures, it is key to run control sections (tumor and normal tissue sections) as a reference. The thickness of the sections must be exactly 2 μm to be able to correctly score different staining intensities. Apart from the staining intensity, many glioblastomas show a heterogenous staining pattern within the tumor tissue. Therefore, it is also necessary to determine the percentage of the field which shows a certain staining intensity. Finally, the subcellular distribution of the Hsp70 staining also needs to be considered. Figure 2 provides a typical example of strong and weak staining intensity in the nucleus and the cytosol in the tumor of a single patient with GBM.

4.2 Flow Cytometry for Detecting Cell Surface Expression of Membrane Hsp70

The inclusion of the murine IgG1 Hsp70 mAb (clone C92F3A, StressMarq Biosciences Inc.) in the staining protocol is essential to determine the amount of Hsp70 which is bound to receptors such as TLRs or scavenger receptors on the cell surface of tumor cells. This antibody does not stain the confirmation of lipid-bound Hsp70 in the plasma membrane of tumor cells.

Although cell viability should routinely be assessed in all flow cytometric analyses, as nonviable cells and debris can nonspecifically

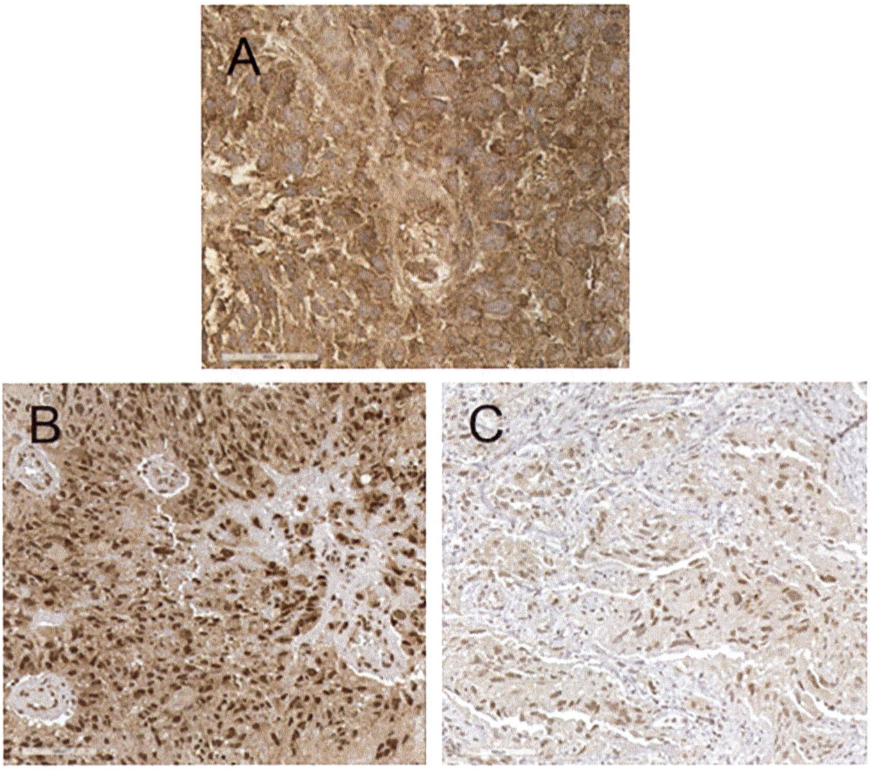

Fig. 2 Representative immunohistochemical (IHC) staining patterns of sections (2 μm) of GL261 mouse glioblastoma (**a**), human glioblastoma multiforme WHO grade IV (b, c) with cmHsp70.1 mAb. Staining intensity scores are as follows: (**a**) +++, (**b**) +++, (**c**) +. Panel a shows a typical cytosolic staining pattern; panels b and c show representative examples of a cytosolic and nuclear staining pattern

take up and bind probes and antibodies and so must be excluded from phenotypic analysis, it is particularly essential for the specific analysis of membrane Hsp70 expression by flow cytometry, as cells having a compromised cell membrane will be highly positive for intracellular Hsp70 expression. Non viable cells might also exhibit a higher level of autofluorescence and/or release DNA which can promote cell clumping and aggregation. Cell aggregation can be inhibited by including a small amount of DNAse in the staining solution, and nonviable cells can be excluded from the analysis using viability stains.

The viability of cultured cells can be quite low; however, cell death during sample processing can be minimized by performing all cell preparations and staining procedures (including wash steps) at 4 °C or on ice and in the presence of bovine serum albumin (BSA) or heat-inactivated fetal bovine serum (FBS). Cells should be analyzed in a flow cytometer as soon as possible. *Cells should be viable and NOT be fixed when analyzing the expression of membrane Hsp70.*

The best method for excluding dead cells from data analysis is to use a vital DNA dye in all samples. Although some of the more common vital dyes used in multicolor analyses are propidium iodide (PI), 7-aminoactinomycin D (7-AAD), TO-PRO-3, and pyronin Y(G) [PY(G)], a range of options are now available from commercial suppliers. A note of caution is that dyes such as PI and 7-AAD can be taken up into viable cells over time, and so these stains should be added immediately prior to analysis.

The relatively broad emission spectrum of 7-AAD can result in spectral overlap with other detectors and require considerable compensation in large multiparameter panels. It is also quite a "dim" fluorescent molecule.

The far-red viability dye DRAQ7™ (BioStatus Ltd., UK) can be used in similar settings to PI and 7-AAD. A particularly useful feature of DRAQ7™ is that its dual excitation characteristics allow multibeam excitation and the exclusion of dead (DRAQ7™+) cells without "consuming" what could be a vital additional fluorescent channel.

Figure 3 provides representative flow cytometric histograms illustrating membrane Hsp70 expression by the murine GL261 GBM-like cell line and a panel of cell lines derived from patients with GBM, as described above.

4.3 Enzyme-Linked Immunosorbent Assays (compHsp70 ELISA) for Detecting Extracellular Free and Exosomal Hsp70

Since the human and murine "TKD" sequences only differ in one amino acid (TKDNNLLGRFELSG; mouse T**R**DNNLLGRFELSG) [47] and the minimal antibody epitope (NNLLGRF) is identical among different species [3], the cmHsp70.1 and cmHsp70.2 mAbs show cross-reactivity for human, mouse, canine, feline, horse, and porcine tumors.

In contrast to the cmHsp70.1 and cmHsp70.2 mAbs, other commercially available mAbs directed against Hsp70 fail to detect the membrane form of Hsp70, which is expressed on the surface of viable tumor cells and on exosomes [34]. Therefore, the measurement of the compHsp70 ELISA must be interpreted to include both free and exosomal Hsp70 found on circulating vesicles.

The assay is suitable for the measurement of both plasma and serum samples, both of which provide similar results with no significant differences [34]. The assay is suitable for the measurement of plasma and serum samples of different mammalian species.

Figure 4 shows levels of free and exosomal Hsp70 in the serum of healthy individuals ($N = 108$) and patients with glioblastoma multiforme ($N = 34$) and lung tumors ($N = 166$), as determined with the ELISA.

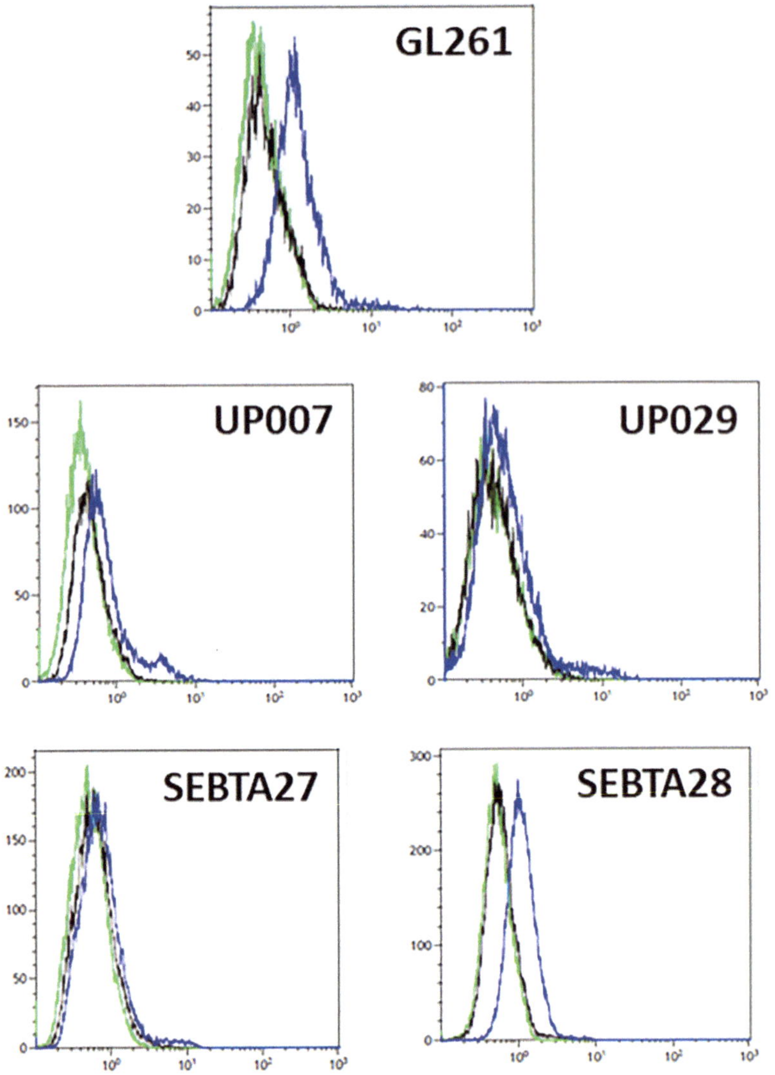

Fig. 3 Representative flow cytometric histograms illustrating membrane Hsp70 expression by a panel of patient-derived cell lines, as described in the main body of the text. GL261 tumors were induced via intracranial injection of methylcholanthrene followed by serial intracranial and subcutaneous transplantations of tumor fragments into syngeneic C57BL/6 mice. Green, unstained control; black, stained using the C92F3A clone mAb; blue, stained using the cmHsp70.1 clone mAb

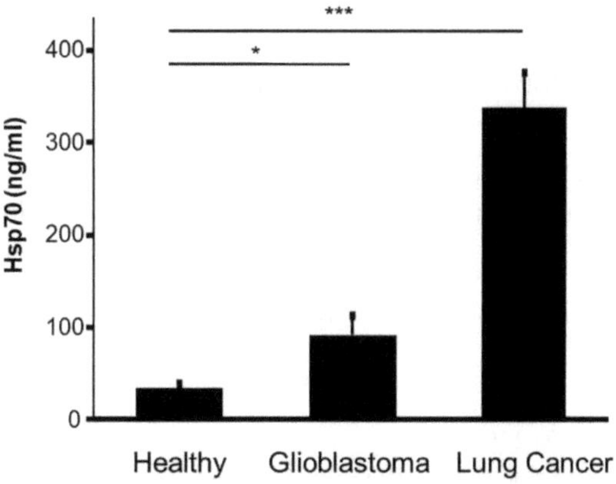

Fig. 4 Levels of free and exosomal Hsp70 levels in the serum of healthy individuals (N = 108) and patients with glioblastoma multiforme (N = 34) and lung tumors (N = 166) as determined with the compHsp70 ELISA. The circulating Hsp70 levels in healthy human volunteers (35.06 ± 3.99 ng/mL) were significantly lower than that in glioblastoma patients (91.82 ± 21.28 ng/mL) and lung cancer patients (332.19 ± 37.90 ng/ml). * $p < 0.05$, *** $p < 0.001$. Patient details are provided by Werner et al. [34]

Acknowledgments

Gabriele Multhoff has received funding from the Deutsche Forschungsgemeinschaft (SFB824/2; DFG INST95/980-1 FUGG, INST411/37-1 FUGG, DFG KU3500/2-1, DFG STA1520/1-1), the DFG Cluster of Excellence: Munich-Centre for Advanced Photonics (MAP), the Bundesministerium für Forschung und Technologie (BMBF Innovative Therapies, 01GU0823; Kompetenzverbund Strahlenforschung 02NUK038A), the Bundesministerium für Wirtschaft und Energie BMWi (ZF4320104AJ8, ZF4320102CS7) EU-CELLEUROPE (315963), and the German Cancer Consortium Radiation Oncology Group, Munich (DKTK-ROG). The John van Geest Cancer Research Centre is supported by funding received from the John and Lucille van Geest Foundation and the Headcase Cancer Trust. GJP was awarded grant funding from the Ollie Young Foundation and Brain Tumour Research.

Conflict of Interest Statement AGP is CEO of multimmune GmbH; GM is the founder, former CEO, and current CSO of multimmune GmbH.

References

1. Multhoff G, Botzler C, Wiesnet M, Eissner G, Issels R (1995) $CD3^{-}$ large granular lymphocytes recognize a heat-inducible immunogenic determinant associated with the 72-kD heat shock protein on human sarcoma cells. Blood 86:1374–1382
2. Multhoff G, Botzler C, Wiesnet M, Muller E, Meier T, Wilmanns W, Issels RD (1995) A stress-inducible 72-kDa heat-shock protein (HSP72) is expressed on the surface of human tumor cells, but not on normal cells. Int J Cancer 61:272–279
3. Stangl S, Gehrmann M, Riegger J, Kuhs K, Riederer I, Sievert W, Hube K, Mocikat R, Dressel R, Kremmer E, Pockley AG, Friedrich L, Vigh L, Skerra A, Multhoff G (2011) Targeting membrane heat-shock protein 70 (Hsp70) on tumors by cmHsp70.1 antibody. Proc Natl Acad Sci USA 108: 2 733-738
4. Gehrmann M, Stangl S, Foulds GA, Oellinger R, Breuninger S, Rad R, Pockley AG, Multhoff G (2014) Tumor imaging and targeting potential of an hsp70-derived 14-mer peptide. PLoS One 9(8):e105344. https://doi.org/10.1371/journal.pone.0105344
5. Gehrmann M, Pfister K, Hutzler P, Gastpar R, Margulis B, Multhoff G (2002) Effects of antineoplastic agents on cytoplasmic and membrane-bound heat shock protein 70 (Hsp70) levels. Biol Chem 383:1715–1725
6. Murakami N, Kuhnel A, Schmid TE, Ilicic K, Stangl S, Braun IS, Gehrmann M, Molls M, Itami J, Multhoff G (2015) Role of membrane Hsp70 in radiation sensitivity of tumor cells. Radiat Oncol 10:149. https://doi.org/10.1186/s13014-015-0461-1
7. Gehrmann M, Stangl S, Kirschner A, Foulds GA, Sievert W, Doss BT, Walch A, Pockley AG, Multhoff G (2012) Immunotherapeutic targeting of membrane Hsp70-expressing tumors using recombinant human granzyme B. PLoS One 7(7):e41341. https://doi.org/10.1371/journal.pone.0041341
8. Pfister K, Radons J, Busch R, Tidball JG, Pfeifer M, Freitag L, Feldmann HJ, Milani V, Issels R, Multhoff G (2007) Patient survival by Hsp70 membrane phenotype: association with different routes of metastasis. Cancer 110: 926–935
9. Stangl S, Gross C, Pockley AG, Asea AA, Multhoff G (2008) Influence of Hsp70 and HLA-E on the killing of leukemic blasts by cytokine/Hsp70 peptide-activated human natural killer (NK) cells. Cell Stress Chaperones 13(2):221–230
10. Gross C, Holler E, Stangl S, Dickinson A, Pockley AG, Asea AA, Mallappa N, Multhoff G (2008) An Hsp70 peptide initiates NK cell killing of leukemic blasts after stem cell transplantation. Leuk Res 32(4):527–534
11. Multhoff G, Botzler C, Jennen L, Schmidt J, Ellwart J, Issels R (1997) Heat shock protein 72 on tumor cells: a recognition structure for natural killer cells. J Immunol 158:4341–4350
12. Multhoff G, Pfister K, Gehrmann M, Hantschel M, Gross C, Hafner M, Hiddemann W (2001) A 14-mer Hsp70 peptide stimulates natural killer (NK) cell activity. Cell Stress Chaperones 6(4):337–344
13. Stangl S, Wortmann A, Guertler U, Multhoff G (2006) Control of metastasized pancreatic carcinomas in SCID/beige mice with human IL-2/TKD-activated NK cells. J Immunol 176:6270–6276
14. Shevtsov M, Pitkin E, Ischenko A, Stangl S, Khachatryan W, Galibin O, Edmond S, Lobinger D, Multhoff G (2019) *Ex vivo* Hsp70-activated NK cells in combination with PD-1 inhibition significantly increase overall survival in preclinical models of glioblastoma and lung cancer. Front Immunol 10:454. https://doi.org/10.3389/fimmu.2019.00454
15. Multhoff G, Pfister K, Botzler C, Jordan A, Scholz R, Schmetzer H, Burgstahler R, Hiddemann W (2000) Adoptive transfer of human natural killer cells in mice with severe combined immunodeficiency inhibits growth of Hsp70-expressing tumors. Int J Cancer 88:791–797
16. Moser C, Schmidbauer C, Gurtler U, Gross C, Gehrmann M, Thonigs G, Pfister K, Multhoff G (2002) Inhibition of tumor growth in mice with severe combined immunodeficiency is mediated by heat shock protein 70 (Hsp70)-peptide-activated, CD94 positive natural killer cells. Cell Stress Chaperones 7(4):365–373
17. Krause SW, Gastpar R, Andreesen R, Gross C, Ullrich H, Thonigs G, Pfister K, Multhoff G (2004) Treatment of colon and lung cancer patients with *ex vivo* heat shock protein 70-peptide-activated, autologous natural killer cells: a clinical phase I trial. Clin Cancer Res 10: 3699–3707
18. Specht HM, Ahrens N, Blankenstein C, Duell T, Fietkau R, Gaipl US, Gunther C, Gunther S, Habl G, Hautmann H, Hautmann M, Huber RM, Molls M, Offner R, Rodel C, Rodel F, Schutz M, Combs SE, Multhoff G (2015) Heat shock protein 70 (Hsp70) peptide activated natural killer (NK) cells for the treatment of patients

with non-small cell lung cancer (NSCLC) after radiochemotherapy (RCTx) - from preclinical studies to a clinical phase II trial. Front Immunol 6:162. https://doi.org/10.3389/fimmu.2015.00162

19. Multhoff G, Seier S, Stangl S, Sievert W, Shevtsov M, Werner C, Pockley AG, Blankenstein C, Hildebrandt M, Offner R, Ahrens N, Kokowski K, Hautmann M, Rödel C, Fietkau R, Lubgan D, Huber R, Hautmann H, Duell T, Molls M, Specht H, Haller B, Devecka M, Sauter A, Combs SE (2020) Targeted natural killer cell based adoptive immunotherapy for the treatment of patients with NSCLC after radiochemotherapy – a randomized phase II clinical study. Clin Cancer Res 26:5368–5379
20. Gross C, Koelch W, Demaio A, Arispe N, Multhoff G (2003) Cell surface-bound heat shock protein 70 (Hsp70) mediates perforin-independent apoptosis by specific binding and uptake of granzyme B. J Biol Chem 278(42): 41173–41181
21. Gehrmann M, Doss BT, Wagner M, Zettlitz KA, Kontermann RE, Foulds G, Pockley AG, Multhoff G (2011) A novel expression and purification system for the production of enzymatic and biologically active human granzyme B. J Immunol Methods 371(1–2):8–17. https://doi.org/10.1016/j.jim.2011.06.007
22. Milani V, Stangl S, Issels R, Gehrmann M, Wagner B, Hube K, Mayr D, Hiddemann W, Molls M, Multhoff G (2009) Anti-tumor activity of patient-derived NK cells after cell-based immunotherapy - a case report. J Transl Med 7: 50. https://doi.org/10.1186/1479-5876-7-50
23. Bashiri Dezfouli A, Yazdi M, Pockley AG, Khosravi M, Kobold S, Wagner E, Multhoff G. (2021) NK cells armed with Chimeric Antigen Receptors (CAR): Roadblocks to successful development. Cells 10:12. https://doi.org/10.3390/cells10123390
24. Bashiri Dezfouli A, Yazdi M, Benmebarek MR, Schwab M, Michaelides S, Micciche A, Geerts D, Stangl S, Klapproth S, Wagner E, Kobold S, Multhoff G (2022) CAR T cells targeting membrane-bound Hsp70 on tumor cells mimic Hsp70-primed NK cells. Front Immunol 13:883694. https://doi.org/10.3389/fimmu.2022.883694
25. Stangl S, Gehrmann M, Dressel R, Alves F, Dullin C, Themelis G, Ntziachristos V, Staeblein E, Walch A, Winkelmann I, Multhoff G (2011) *In vivo* imaging of CT26 mouse tumours by using cmHsp70.1 monoclonal antibody. J Cell Mol Med 15(4):874–887
26. Stangl S, Themelis G, Friedrich L, Ntziachristos V, Sarantopoulos A, Molls M, Skerra A, Multhoff G (2011) Detection of irradiation-induced, membrane heat shock protein 70 (Hsp70) in mouse tumors using Hsp70 Fab fragment. Radiother Oncol 99(3): 313–316
27. Stangl S, Varga J, Freysoldt B, Trajkovic-Arsic-M, Siveke JT, Greten FR, Ntziachristos V, Multhoff G (2014) Selective *in vivo* imaging of syngeneic, spontaneous, and xenograft tumors using a novel tumor cell-specific hsp70 peptide-based probe. Cancer Res 74(23):6903–6912
28. Gehrmann MK, Kimm MA, Stangl S, Schmid TE, Noel PB, Rummeny EJ, Multhoff G (2015) Imaging of Hsp70-positive tumors with cmHsp70.1 antibody-conjugated gold nanoparticles. Int J Nanomedicine 10:5687–5700
29. Stangl S, Tei L, De Rose F, Reder S, Martinelli J, Sievert W, Shevtsov M, Ollinger R, Rad R, Schwaiger M, D'alessandria C, Multhoff G (2018) Preclinical evaluation of the Hsp70 peptide tracer TPP-PEG24-DFO[(89)Zr] for tumor-specific PET/CT imaging. Cancer Res 78(21): 6268–6281
30. Gastpar R, Gehrmann M, Bausero MA, Asea A, Gross C, Schroeder JA, Multhoff G (2005) Heat shock protein 70 surface-positive tumor exosomes stimulate migratory and cytolytic activity of natural killer cells. Cancer Res 65(12):5238–5247
31. Breuninger S, Erl J, Knape C, Gunther S, Regel I, Rödel F, Gaipl US, Thorsteinsdottir J, Giannitrapani L, Dickinson AM, Multhoff G (2014) Quantitative analysis of liposomal heat shock protein 70 (Hsp70) in the blood of tumor patients using a novel lipHsp70 ELISA. J Clin Cell Immunol 5:264. https://doi.org/10.4172/2155-9899.1000264
32. Gunther S, Ostheimer C, Stangl S, Specht HM, Mozes P, Jesinghaus M, Vordermark D, Combs SE, Peltz F, Jung MP, Multhoff G (2015) Correlation of Hsp70 serum levels with gross tumor volume and composition of lymphocyte subpopulations in patients with squamous cell and adeno non-small cell lung cancer. Front Immunol 6:556. https://doi.org/10.3389/fimmu.2015.00556
33. Rothammer A, Sage EK, Werner C, Combs SE, Multhoff G (2019) Increased heat shock protein 70 (Hsp70) serum levels and low NK cell counts after radiotherapy – potential markers for predicting breast cancer recurrence? Radiat

Oncol 14(1):78. https://doi.org/10.1186/s13014-019-1286-0

34. Werner C, Stangl S, Salvermoser L, Schwab M, Shevtsov M, Xanthopoulos A, Wang F, Dezfouli AB, Thölke D, Ostheimer C, Medenwald D, Windberg M, Bache M, Schlapschy M, Skerra A, Multhoff G (2021) Hsp70 in liquid biopsies - A tumor-specific biomarker for detection and response monitoring in cancer. Cancers (Basel) 13:3706. https://doi.org/10.3390/cancers13153706
35. Burton EC, Prados MD (2000) Malignant gliomas. Curr Treat Options in Oncol 1(5): 459–468
36. Nieder C, Grosu AL, Molls M (2000) A comparison of treatment results for recurrent malignant gliomas. Cancer Treat Rev 26(6): 397–409
37. Dazzi C, Cariello A, Giannini M, Del Duca M, Giovanis P, Fiorentini G, Leoni M, Rosti G, Turci D, Tienghi A, Vertogen B, Zumaglini F, De Giorgi U, Marangolo M (2000) A sequential chemo-radiotherapeutic treatment for patients with malignant gliomas: a phase II pilot study. Anticancer Res 20(1B):515–518
38. Barrié M, Couprie C, Dufour H, Figarella-Branger D, Muracciole X, Hoang-Xuan K, Braguer D, Martin PM, Peragut JC, Grisoli F, Chinot O (2005) Temozolomide in combination with BCNU before and after radiotherapy in patients with inoperable newly diagnosed glioblastoma multiforme. Ann Oncol 16(7): 1177–1184
39. Combs SE, Gutwein S, Thilmann C, Debus J, Schulz-Ertner D (2005) Reirradiation of recurrent WHO grade III astrocytomas using fractionated stereotactic radiotherapy (FSRT). Strahlenther Onkol 181(12):768–773
40. Combs SE, Thilmann C, Edler L, Debus J, Schulz-Ertner D (2005) Efficacy of fractionated stereotactic reirradiation in recurrent gliomas: long-term results in 172 patients treated in a single institution. J Clin Oncol 23(34): 8863–8869
41. Combs SE, Wagner J, Bischof M, Welzel T, Wagner F, Debus J, Schulz-Ertner D (2008) Postoperative treatment of primary glioblastoma multiforme with radiation and concomitant temozolomide in elderly patients. Int J Radiat Oncol Biol Phys 70(4):987–992
42. Combs SE, Wagner J, Bischof M, Welzel T, Edler L, Rausch R, Wagner F, Zabel-Du Bois A, Debus J, Schulz-Ertner D (2008) Radiochemotherapy in patients with primary glioblastoma comparing two temozolomide dose regimens. Int J Radiat Oncol Biol Phys 71(4):999–1005
43. Frenay M, Lebrun C, Lonjon M, Bondiau PY, Chatel M (2000) Up-front chemotherapy with fotemustine (F)/cisplatin (CDDP)/etoposide (VP16) regimen in the treatment of 33 non-removable glioblastomas. Eur J Cancer 36(8):1026–1031
44. Stupp R, Mason WP, Van Den Bent MJ, Weller M, Fisher B, Taphoorn MJ, Belanger K, Brandes AA, Marosi C, Bogdahn U, Curschmann J, Janzer RC, Ludwin SK, Gorlia T, Allgeier A, Lacombe D, Cairncross JG, Eisenhauer E, Mirimanoff RO (2005) Radiotherapy plus concomitant and adjuvant temozolomide for glioblastoma. N Engl J Med 352(10):987–996
45. Stupp R, Hegi ME, Mason WP, Van Den Bent MJ, Taphoorn MJ, Janzer RC, Ludwin SK, Allgeier A, Fisher B, Belanger K, Hau P, Brandes AA, Gijtenbeek J, Marosi C, Vecht CJ, Mokhtari K, Wesseling P, Villa S, Eisenhauer E, Gorlia T, Weller M, Lacombe D, Cairncross JG, Mirimanoff RO (2009) Effects of radiotherapy with concomitant and adjuvant temozolomide versus radiotherapy alone on survival in glioblastoma in a randomised phase III study: 5-year analysis of the EORTC-NCIC trial. Lancet Oncol 10(5): 459–466
46. Fouchaq B, Benaroudj N, Ebel C, Ladjimi MM (1999) Oligomerization of the 17-kDa peptide-binding domain of the molecular chaperone HSC70. Eur J Biochem 259(1–2): 379–384
47. Zhang H, Liu R, Huang W (2007) A 14-mer peptide from HSP70 protein is the critical epitope which enhances NK activity against tumor cells *in vivo*. Immunol Investig 36(3):233–246

INDEX

Stuart K. Calderwood and Thomas L. Prince (eds.), *Chaperones: Methods and Protocols*, Methods in Molecular Biology, vol. 2693, https://doi.org/10.1007/978-1-0716-3342-7,

R

S

T

U

Y